Quan Bai and Naoki Fukuta (Eds.)

Advances in Practical Multi-Agent Systems

Studies in Computational Intelligence, Volume 325

Editor-in-Chief
Prof. Janusz Kacprzyk
Systems Research Institute
Polish Academy of Sciences
ul. Newelska 6
01-447 Warsaw
Poland
E-mail: kacprzyk@ibspan.waw.pl

Further volumes of this series can be found on our homepage: springer.com

Vol. 302. Bijaya Ketan Panigrahi, Ajith Abraham, and Swagatam Das (Eds.)
Computational Intelligence in Power Engineering, 2010
ISBN 978-3-642-14012-9

Vol. 303. Joachim Diederich, Cengiz Gunay, and James M. Hogan
Recruitment Learning, 2010
ISBN 978-3-642-14027-3

Vol. 304. Anthony Finn and Lakhmi C. Jain (Eds.)
Innovations in Defence Support Systems, 2010
ISBN 978-3-642-14083-9

Vol. 305. Stefania Montani and Lakhmi C. Jain (Eds.)
Successful Case-Based Reasoning Applications-1, 2010
ISBN 978-3-642-14077-8

Vol. 306. Tru Hoang Cao
Conceptual Graphs and Fuzzy Logic, 2010
ISBN 978-3-642-14086-0

Vol. 307. Anupam Shukla, Ritu Tiwari, and Rahul Kala
Towards Hybrid and Adaptive Computing, 2010
ISBN 978-3-642-14343-4

Vol. 308. Roger Nkambou, Jacqueline Bourdeau, and Riichiro Mizoguchi (Eds.)
Advances in Intelligent Tutoring Systems, 2010
ISBN 978-3-642-14362-5

Vol. 309. Isabelle Bichindaritz, Lakhmi C. Jain, Sachin Vaidya, and Ashlesha Jain (Eds.)
Computational Intelligence in Healthcare 4, 2010
ISBN 978-3-642-14463-9

Vol. 310. Dipti Srinivasan and Lakhmi C. Jain (Eds.)
Innovations in Multi-Agent Systems and Applications – 1, 2010
ISBN 978-3-642-14434-9

Vol. 311. Juan D. Velásquez and Lakhmi C. Jain (Eds.)
Advanced Techniques in Web Intelligence, 2010
ISBN 978-3-642-14460-8

Vol. 312. Patricia Melin, Janusz Kacprzyk, and Witold Pedrycz (Eds.)
Soft Computing for Recognition based on Biometrics, 2010
ISBN 978-3-642-15110-1

Vol. 313. Imre J. Rudas, János Fodor, and Janusz Kacprzyk (Eds.)
Computational Intelligence in Engineering, 2010
ISBN 978-3-642-15219-1

Vol. 314. Lorenzo Magnani, Walter Carnielli, and Claudio Pizzi (Eds.)
Model-Based Reasoning in Science and Technology, 2010
ISBN 978-3-642-15222-1

Vol. 315. Mohammad Essaaidi, Michele Malgeri, and Costin Badica (Eds.)
Intelligent Distributed Computing IV, 2010
ISBN 978-3-642-15210-8

Vol. 316. Philipp Wolfrum
Information Routing, Correspondence Finding, and Object Recognition in the Brain, 2010
ISBN 978-3-642-15253-5

Vol. 317. Roger Lee (Ed.)
Computer and Information Science 2010
ISBN 978-3-642-15404-1

Vol. 318. Oscar Castillo, Janusz Kacprzyk, and Witold Pedrycz (Eds.)
Soft Computing for Intelligent Control and Mobile Robotics, 2010
ISBN 978-3-642-15533-8

Vol. 319. Takayuki Ito, Minjie Zhang, Valentin Robu, Shaheen Fatima, Tokuro Matsuo, and Hirofumi Yamaki (Eds.)
Innovations in Agent-Based Complex Automated Negotiations, 2010
ISBN 978-3-642-15611-3

Vol. 320. xxx

Vol. 321. Dimitri Plemenos and Georgios Miaoulis (Eds.)
Intelligent Computer Graphics 2010
ISBN 978-3-642-15689-2

Vol. 322. Bruno Baruque and Emilio Corchado (Eds.)
Fusion Methods for Unsupervised Learning Ensembles, 2010
ISBN 978-3-642-16204-6

Vol. 323. Yingxu Wang, Du Zhang, Witold Kinsner (Eds.)
Advances in Cognitive Informatics, 2010
ISBN 978-3-642-16082-0

Vol. 324. Alessandro Soro, Vargiu Eloisa, Giuliano Armano, and Gavino Paddeu (Eds.)
Information Retrieval and Mining in Distributed Environments, 2010
ISBN 978-3-642-16088-2

Vol. 325. Quan Bai and Naoki Fukuta (Eds.)
Advances in Practical Multi-Agent Systems, 2010
ISBN 978-3-642-16097-4

Quan Bai and Naoki Fukuta (Eds.)

Advances in Practical Multi-Agent Systems

Springer

Quan Bai
Tasmanian ICT Centre, CSIRO
GPO Box 1538, Hobart
TAS 7001, Australia
Email: quan.bai@csiro.au

Naoki Fukuta
Faculty of Informatics
Shizuoka University
351 Johoku
Hamamatsu Shizuoka 432-8011
Japan
E-mail: fukuta@cs.inf.shizuoka.ac.jp

ISBN 978-3-642-16097-4 e-ISBN 978-3-642-16098-1

DOI 10.1007/978-3-642-16098-1

Studies in Computational Intelligence ISSN 1860-949X

Library of Congress Control Number: 2010936511

Typeset & Cover Design: Scientific Publishing Services Pvt. Ltd., Chennai, India.

Printed on acid-free paper

9 8 7 6 5 4 3 2 1

springer.com

Preface

This book constitutes the thoroughly refereed post-proceedings of workshops under the 12th International Conference on Principles of Practice in Multi-Agent Systems (PRIMA2009). Twenty-eight revised full papers were included in this book.

The International Conference on Principles of Practice in Multi-Agent Systems (PRIMA) is a conference that has developed from a series of leading international workshops in the Pacific Rim region for the presentation of research and the latest developments on principles, practices, and theories in multi-agent system (MAS) technologies, including the application of MAS to problems of social, industrial, and economic importance. PRIMA 2009 was the 12th in the series of conferences/workshops and was held in Nagoya, Japan, December 13-16, 2009.

The PRIMA2009 workshops were designed to provide a forum for researchers and practitioners to present and exchange the latest developments at the MAS frontier. Workshops are more specialized and on a smaller scale than conferences to facilitate active interaction among the attendees. They encourage MAS theories to meet reality, MAS researchers to work with practitioners, and vice versa. Through workshops, both sides get exposure to evolving research and tools, and to the practical problems to which MAS can be applied. As an excellent indicator of the current level of active research and development activities, PRIMA2009 included a total of five workshops that were grouped as two joint workshops and one stand alone workshop:

- International Workshop on Agent-based Collaboration, Coordination, and Decision Support;
- Joint workshop on Multi-Agent Simulation in Finance and Economics & Applied Agent based simulator Engineering for Complex System study; and
- Joint Workshop on Agent Technology for Disaster Management & Intelligent Agents in Sensor Networks and Sensor Web.

The output from those workshops has created this unique collection that is presented in three parts. Part I reports the use of agent technologies in assisting the collaboration among humans and software mechanisms, coordination among software agents and systems, and decision support on knowledge discovery; Part II covers agent-based simulation in finance, economics and social science; and Part III discusses agent-based applications for environmental monitoring and disaster management.

Each workshop paper was accepted after review by at least two experts. Further improvements were included in many papers in preparation for this collection. Readers will be able to explore a diverse range of topics and detailed discussions related to the five important themes in our ever changing world. This collection plays an important role in bridging the gap between MAS theory and practice. It emphasizes the importance of MAS in the research and development of smart power grid systems, decision support systems, optimization and analysis systems for road traffic and markets, environmental monitoring and simulation, and in many other real-world applications and publicizes and extends MAS technology to many domains in this fast moving information age.

The organizers of the workshops did an excellent job in bringing together many MAS researchers and practitioners from the Pacific-Asia region and from all over the world. The well-received workshops at PRIMA2009 and the publications included in this collection convincingly demonstrate the significance and practical impact of the latest work presented in MAS.

We are certain that this collection will stimulate increased MAS research and applications, influence many graduate students to conduct research and development in MAS, and have a positive impact towards making our future better by creating an increasingly smarter world.

December 2009

Quan Bai
CSIRO, Australia
Naoki Fukuta
Shizuoka University, Japan

Acknowledgements

The editors would like to thank the organizers and program committee members of the five PRIMA2009 workshops and all the workshop participants for their support and contribution. We would also like to thank Dr Stephen Giugni and Dr Greg Timms from the CSIRO Tasmanian ICT Centre for their help in improving this book.

PRIMA2009 Workshop Organizers

ACCDS 2009:
Fenghui Ren, University of Wollongong, Australia
Katsuhide Fujita, Nagoya Institute of Technology, Japan

MASFE 2009:
Takao Terano, Tokyo Institute of Technology, Japan
Tohgoroh Matsui, Tohgoroh Machine Learning Research Institute, Japan
Kiyoshi Izumi, National Institute of Advanced Industrial Science and Technology, Japan

AAECS 2009:
Alexis Drogoul UMI 209 UMMISCO, IRD, UPMC, MSI-IFI, Vietnam
Benoit Gaudou UMI 209 UMMISCO, IRD, MSI-IFI, Vietnam
Nicolas Marilleau UMI 209 UMMISCO, IRD, UPMC, France

ATDM 2009:
Serge Stinckwich, UMI UMMISCO IRD/UPMC/MSI-IFI, Vietnam
Sarvapali D. Ramchurn, University of Southampton, UK

IASNW 2009:
Andrew Terhorst, CSIRO ICT Centre, Australia
Siddeswara Guru, CSIRO ICT Centre, Australia

Contents

Part II: Agent-Based Simulation for Complex Systems: Application to Economics, Finance and Social Sciences

Part I
Agent-Based Collaboration, Coordination and Decision Support

The autonomy and intelligence of software agents have greatly enhanced automation in many operational domains. Benefits of using software agents are the ability to assist in the collaboration among humans and software mechanisms, coordination among software agents and systems, and decision support on knowledge discovery. In general, agents may perform different behaviors for different kinds of expectations. In cooperative systems, agents will perform tasks more cooperatively in order to achieve their common goals, while in competitive systems, agents may compete with others in order to reach individual aims. However, in most real world situations, both cooperation and competition exist between agents. Agents need to be carefully organized and managed in order to maximize the efficiency of the whole system, and to balance benefits between others.

We solicited all topics related to the interdisciplinary aspects of agent-based collaboration, coordination, and decision support as much as these are in some way relevant for or applicable to software agents and multi-agent systems. The following incomprehensive listing gives examples of such potential topics:

- Agent-based collaboration
- Agent-based coordination
- Methodologies, languages and tools to support agent collaboration
- Protocols and approaches for agent-mediated collaboration and decision support
- Teamwork, coalition formation and coordination in multi-agent systems
- Distributed problem solving
- Electronic markets and institutions
- Game theory (cooperative and non-cooperative)
- Auction and mechanism design
- Argumentation, negotiation and bargaining
- Trust and reputation
- Agent-based decision support
- Agent-based knowledge discovery
- Agent-based collective intelligence
- Agent-based situation awareness

These issues are being explored by researchers from different communities in Autonomous Agents and Multi-Agent systems, distributed problem solving, complex systems as well as agent-based applications. This part collects nine high quality papers, all involving some important aspect of agent-based collaboration, coordination and decision support. The idea is to try to motivate the researchers to explore cutting-edge challenges in the area and accelerate progress towards scaling up to larger and more realistic applications.

Katsuhide Fujita
Fenghui Ren
ACCDS'09 Workshop Chairs

DGF: Decentralized Group Formation for Task Allocation in Complex Adaptive Systems

Dayong Ye, Minjie Zhang, and Danny Sutanto

Abstract. In this paper, a decentralized group formation algorithm for task allocation in complex adaptive systems is proposed. Compared with current related works, this decentralized algorithm takes system architectures into account and allows not only neighboring agents but also indirect linked agents in the system to help with a task. A system adaptation strategy is also developed for discovering effective system structures for task allocation. Moreover, a set of experiments was conducted to demonstrate the efficiency of our methods.

1 Introduction

Allocating tasks to agents in complex adaptive systems is a significant research issue. Task allocation problem can be simply described as that when an agent has a task which it cannot complete by itself, the agent then attempts to discover other agents which contain the appropriate resources and assigns the task, or part of the task, to those agents. Recently, many approaches have been proposed for task allocation. Some of them [6] [9] [14] [16] are based on the centralized fashion which assumes that there is a central controller to assign tasks to agents. An efficient agent group formation algorithm [9] was proposed by Manisterski et al. for completing complex tasks. Their work took both cooperative and non-cooperative settings into account and pointed out

Dayong Ye · Minjie Zhang
School of Computer Science and Software Engineering,
University of Wollongong, Australia
e-mail: {dy721,minjie}@uow.edu.au

Danny Sutanto
School of Electrical, Computer and Telecommunications Engineering,
University of Wollongong, Australia
e-mail: danny@elec.uow.edu.au

Q. Bai and N. Fukuta (Eds.): Advances in Practical Multi-Agent Systems, SCI 325, pp. 3–19.
springerlink.com

what can and cannot be achieved. However, their algorithm assumed there is a central controller, which uses global information, to manage the group formation. Zheng and Koenig [16] provided reaction functions for task allocation to cooperative agents. The objective of their approach is to find a solution with a small team cost and assign each task to the exact number of different agents. Their work also assumed that there is a central planner to allocate tasks to agents. Centralized fashion can make the allocation process effective in a small scale network, since the central planner has a global view of the system and is, therefore, aware of the capabilities of each agent. In that case, communication overhead during allocation processes could be reduced. However, the centralized fashion has also several notable disadvantages. Firstly, in some complex adaptive systems, it is difficult to have such a central controller, such as social networks or sensor networks. In these networks, it is hardly for a single agent to have the global view with regard to the network but only the local prospect about directly linked neighbors. Secondly, when the central planner is out of order or cracked by some attackers, task allocation might fail accordingly. The last drawback is that the scalability of a system with centralized control is limited because when too many agents exist, the central controller has to maintain much information to hold the global view and respond huge amount of messages from other agents. These situations might drastically raise the CPU and memory usage of the central controller and also consume much network bandwidth.

In order to overcome the shortcomings of centralized fashion, some distributed task allocation methods were developed based on decentralized style. Compared with centralized fashion, decentralized style is more scalable and robust. However, there are still a few limitations in current decentralized task allocation methods. Some of these methods do not assume the existence of any agent networks, but emphasize only on the behaviors of resource suppliers and demanders, e.g. [13] [11] [8] [1] [2] and [3]. On the other hand, several distributed task allocation mechanisms have to employ some global information to achieve their goals, although the execution procedure of these mechanisms are distributed. For example, in [10], the proposed resource allocation method supposed that the resource load of resource suppliers is global information which are known by all resource demanders. Furthermore, some other approaches, e.g. Greedy Distributed Allocation Protocol (GDAP) presented in [15], only allow agents to request help for tasks to their directly linked neighbors. In this way, the possibility of task allocation failure would be increased due to the limited resources available from agents' neighbors.

Some other group or coalition formation schemes have also been proposed, although they were not devised for task allocation. Nonetheless, there are more or less limitations of these schemes, and, therefore, they cannot be directly used for task allocation in complex adaptive systems. A coalition formation based self-organization technique was presented in [12]. The self-organization process is through negotiation strategies, whose aim is maximizing the global utility of a sensor network. The disadvantage of this technique is that agents in the sensor network were supposed to be homogeneous, while agents in our work are considered to be heterogeneous. Kraus et al. [7] developed a coalition formation protocol and several revenue distribution strategies with incomplete information. However, they supposed that the capabilities of agents are common knowledge in the environment, which is employed as global information. In addition, their approach focused on self-interested agents.

In this paper, we propose a decentralized group formation algorithm, called DGF, for task allocation. Compared with centralized control approach, like the one proposed in [16], our method does not need a central planner. In contrast to the approach, developed in [3], which overlooked the system architecture, our method takes the system architecture into account during task allocation process. Opposite to the protocol introduced in [7], our method does not need global information, and our research stresses on cooperative agents rather than self-interested agents. Unlike GDAP presented in [15] which allows only neighboring agents to help with a task, our method enables agents to allocate tasks not only to their neighbors but also other agents in the system based on the proposed group formation algorithm. In this way, the agents would have more opportunities to achieve solution for their tasks. Moreover, in order to discover effective system structures for task allocation, a system adaptation strategy will be performed after a predefined number of task allocation rounds. In our model, tasks are generated periodically and agents attempt to form groups to accomplish these tasks by using only local information. A set of tasks is introduced in the system at a fixed time interval, where the length of the interval between the introduction of two sets of tasks is predefined. Thus, a task allocation round is a procedure of allocating a set of tasks to appropriate agents.

The rest of this paper is organized as follows. Section 2 formalizes the problem and describes the decentralized group formation algorithm. Section 3, then, depicts the system adaptation strategy. Section 4 demonstrates the experimental results and analyzes the quality and performance of our method. Finally, we discuss and conclude our work in Section 5.

2 Decentralized Group Formation Algorithm

To cope with the issue of allocating tasks in complex adaptive systems, a Decentralized Group Formation (DGF) algorithm is elaborated in this section.

2.1 Problem Description

We formalize the description of task allocation problem in this subsection. The definition of *complex adaptive system* is first introduced.

Definition 1. Complex Adaptive System. *A complex adaptive system consists of a set of interdependent cooperative agents, namely* $A = \{a_1, ..., a_n\}$, *and a* Compatible Relation R $(R \subseteq A \times A)$. *The meaning of* R *is* "a neighbor of", *so we denote an ordered couple* $\langle a_i, a_j \rangle \in R$ *if and only if* a_j *is a neighbor of* a_i. *Since* R *is a* Compatible Relation, *namely that* R *is* reflexive *and* symmetric, *it can be achieved that* $\forall a_i : a_i \in A \Rightarrow \langle a_i, a_i \rangle \in R$ *and* $\forall a_i, a_j \in A : \langle a_i, a_j \rangle \in R \Rightarrow \langle a_j, a_i \rangle \in R$. *Examples of such systems include social networks and sensor networks.*

Each agent $a \in A$ is composed of five tuples $< AgentID(a), Neig(a), Resource(a), State(a), Pop(a) >$, where $AgentID(a)$ is the identity of agent a, $Neig(a)$ is a set which indicates the neighbors of agent a, $Resource(a)$ is the resource which agent a contains, $State(a)$ demonstrates the state of agent a, and $Pop(a)$ exhibits the popularity of agent a. $Pop(a)$, presented in [5], is described as the ratio between the number of successful group joined and the number of attempted group joined during a predefined number of task allocation rounds, and can be calculated via Equation 1.

$$Pop(a) = \frac{\#\ of\ successful\ group\ joined}{\#\ of\ attempted\ group\ joined} \tag{1}$$

In this paper, each agent $a \in A$ is assumed to have a single fixed resource, $r_a \in [1, \varepsilon]$, where ε is the number of different resources that are present in the system. Before introducing the states of an agent, we provide the definition of three terms used throughout this paper, i.e. *Initiator*, *Participant* and *Mediator*.

Definition 2. Initiator, Participant *and* **Mediator.** *In a complex adaptive system, the agent which requests help for its task is called* **Initiator**, *the agent which accepts and performs the announced task is called* **Participant**, *and the agent that receives another agent's commitments for assistance to find* **Participant***s is called* **Mediator**.

Suppose there is a set of tasks $T = \{t_1, ..., t_m\}$ arrives at a complex adaptive system. Each task $t \in T$ consists of three tuples, namely $< TaskID(t), TTL(t), Rsc(t) >$. $TaskID(t)$ is the identity of the task, $TTL(t)$ demonstrates the number of hops with which the task t could be committed, and $Rsc(t) = \{r_t^1, ..., r_t^j\}$ is a set that indicates the resources which are needed for successfully completing the task. The number of resources required for a given task, $t \in T$, are chosen uniformly from $[1, \varepsilon]$, as a result $|Rsc(t)| \leq \varepsilon$. Then, for each $r_t^l \in Rsc(t)$, it is selected randomly from $[1, \varepsilon]$. Therefore, for each task, it is necessary for agents to form a group to handle the task. The term *group* is defined as follows.

Definition 3. Group. *A* group g *is a set of agents, i.e.* $g \subseteq A$, *which cooperate together in order to complete a complex task* $t \in T$. *Each group is associated with a task, and then there is a set of groups* $G = \{g_1, ..., g_h\}$ *which are* *associated with tasks,* $t_1, ..., t_h$*, respectively. In addition, a valid group should satisfy the situation that the resources of agents in the group should cover the required resources of the associated task, i.e.* $\bigcup_{a_i \in g_j} Resource(a_i) \supseteq Rsc(t_j)$.

It is now ready to describe the states of an agent.

Definition 4. States. *There are three states in a complex adaptive system, i.e.* $States = \{BUSY, COMMITTED, IDLE\}$*, and an agent can be only in one of the three states at any time step. When an agent is an* Initiator *or* Participant*, the state of that agent is* BUSY*. When an agent is a* Mediator*, the agent is in* COMMITTED *state. An agent in* IDLE *state is available and not assigned or committed to any task.*

For efficient task allocation, it is supposed that only an *IDLE* agent can be assigned to a new task as an *Initiator* or a partial fulfilled task as a *Participant*, or committed to a partial fulfilled task as a *Mediator*. A partial fulfilled task is a task, for which a full group is in formation procedure and has not yet formed. After formalizing the task allocation problem, in the next subsection, the principle of DGF will be depicted.

2.2 The Principle of Decentralized Group Formation (DGF) Algorithm

In a complex adaptive system, agents can make decisions based only on local information about the system, and the decision making process of agents is autonomous without external control. Hence, we define another set $P = \{P_1, ..., P_n\}$. P is defined as a *partition* of the *Compatible Relation* R which has been described in Definition 1. Accordingly, it can be obtained that $\bigcup_{1 \leq i \leq n} P_i = R$ and $\forall P_i, P_j \in P : i \neq j \Rightarrow P_i \cap P_j = \emptyset$. The set P can be generated by using **Algorithm 1**.

Algorithm 1: Create a partition, P, on the relation, R
input:
  $a_i \in A$: a set of agents
  R: a *compatible relation* defined on A
output:
  $P = \{P_1, P_2, ..., P_n\}$: a *partition* of R
begin:
(1) **for** $\forall a_i : a_i \in A$ in sequential order
(2)   **if** $\exists a_j \in A : \langle a_i, a_j \rangle \in R$ **then**
(3) $P_i \leftarrow P_i \cup \{\langle a_i, a_j \rangle\}$;
end

From **Algorithm 1**, it can be found that P_i is composed of ordered couples, and each ordered couple dictates an agent which is a neighbor of a_i. It seems that P_i and $Neig(a_i)$ demonstrate the same meaning, but, actually, they are used for different purposes. P_i represents not only neighbors of agent a_i but also other agents which have indirect connections with a_i that are established during future task allocation processes, although P_i initially denotes only neighbors of agent a_i. On the other hand, $Neig(a_i)$ stores only directly linked neighboring agents of agent a_i. The idea of DGF is illustrated as follows.

The *Initiator* agent, denoted as a_I, $a_I \in A$, first checks whether its resource $Resource(a_I)$ could satisfy any one of the resource requirements of its task, denoted as t_I and $t_I \in T$. Thereafter, the *Initiator* agent a_I attempts to find its neighboring agents to help with the rest resource requirements of t_I which cannot be satisfied by $Resource(a_I)$. *Initiator* a_I then sends resource query messages, $ResQueryMess = \langle AgentID(a_I), TaskID(t_I), Resource(t_I)\rangle$, to its neighbors one by one. These neighboring agents, which are requested by the *Initiator* a_I, will respond with information about the types of resources they contain, the state of them and the identities of them, namely $RespMess = \langle AgentID(a), Resource(a), State(a)\rangle$. If *Initiator* a_I is satisfied with any neighbor's resource and the state of that neighboring agent is *IDLE*, *Initiator* a_I will make a contract with that satisfying agent, denoted as a_P, and the state of a_P will correspondingly be changed to *BUSY*. A contract is composed of four tuples, $Contract =< AgentID(a_I), AgentID(a_P), TaskID(t_I), Resource(a_P) >$.

After obtaining requested information, the *Initiator* agent a_I then compares the available resources from its neighbors, i.e. $Reso_{neig}(a_I)$, with the resources required for its task t_I, namely $Rsc(t_I)$. (Here, $Reso_{neig}(a_I) = \bigcup_{a_i \in Neig(a_I)} Resource(a_i)$). This comparison would result in one of the following two cases.

Case One: ($Reso_{neig}(a) \supseteq Rsc(t_I)$) In this situation, *Initiator* a_I can form a full group g_I for task t_I directly with its neighboring agents.

Case Two: ($Reso_{neig}(a) \subset Rsc(t_I)$) In this condition, *Initiator* a_I can only form a partial group for task t_I. It then commits the task t_I to one of its neighbors. The commitment selection is based on the number of neighbors each neighbor of a_I maintaining. The more neighbors an agent has, the higher probability that agent could be selected as a *Mediator* agent to commit the task t_I. For example, *Initiator* a_I has two neighbors, say a_1 and a_2, and a_1 and a_2 have 2 and 4 neighboring agents, separately. Thereby, the selection probability distributions on a_1 and a_2 are $\frac{2}{2+4} = \frac{1}{3}$ and $\frac{4}{2+4} = \frac{2}{3}$, respectively. After selection, the *Initiator* agent a_I commits its partial fulfilled task t_I to the *Mediator* agent, denoted as a_M. A commitment consists of four tuples, $Commitment = \left\langle AgentID(a_I), AgentID(a_M), TaskID(t_I), Rsc(t_I)' \right\rangle$, where $Rsc(t_I)'$ is a subset of $Rsc(t_I)$, which contains the unfulfilled required resources. Afterward, the *Mediator* a_M subtracts 1 from $TTL(t_I)$ and attempts to discover the agents with available resources from its neighbors. If any agents satisfy resource requirement, the *Mediator* a_M will send a response message, *RespMess*, back to the *Initiator* a_I. The *Initiator* a_I then directly makes contract with the agents which satisfy the resource requirement. If the neighboring agents of the *Mediator* a_M cannot satisfy the resource requirement either, the *Mediator* a_M will commit the partial fulfilled task t_I to one of its neighbors again. This process will continue until all of the resource requirements of task t_I are satisfied, or the $TTL(t_I)$ reaches 0, or there is no more *IDLE* agent among the neighbors. Both of the last two conditions, i.e. $TTL(t_I) = 0$ and no more *IDLE* agent, demonstrate the failure of task allocation. In these two conditions, the *Initiator* a_I disables the assigned contracts with the *Participants*, and the states of these *Participants* are reset to *IDLE*.

When finishing an allocation for one task, the system is restored to its original status and each agent's state is reset to *IDLE*.

From the above description, it can be found that the proposed decentralized group formation algorithm, DGF, enables *Initiators* to request help not only from neighbors but also other agents through commitment if needed. **Algorithm 2** gives the pseudocode of the decentralized group formation algorithm employed by each *Initiator* during a task allocation process.

Algorithm 2: Decentralized Group Formation for Task Allocation
input:
  $a_i \in A$: a set of agents
  R: a *compatible relation* defined on A
  $T = \{T_1, ..., T_m\}$: a set of tasks
  $P' = \{P'_1, ..., P'_n\}$ and $\forall P'_i \in P' : P'_i = \emptyset$
output:
  $G = \{g_1, ..., g_h\}$: a set of groups and each is associated with a task
begin:
Call Algorithm 1 to generate P;
  for $\forall t_i : t_i \in T$ in sequential order
    randomly select an agent, $a_i \in A$, as *Initiator*, **and** $State(a_i) = IDLE$
    $State(a_i) \leftarrow BUSY;$
    for $TTL(t_i)$ **from** a predefined integer **to** 0
      if $\forall \langle a_i, a_j \rangle \in P_i : State(a_j) \neq IDLE$ **then**
        break;
      else
        for $\exists a_j \in A : \langle a_i, a_j \rangle \in P_i$
          if $State(a_j) = IDLE$ **and** $\exists r^l_{t_i} \in Rsc(t_i) : r^l_{t_i} = Resource(a_j)$
           and $r^l_{t_i}$ is unsatisfied **then**
            $g_i \leftarrow g_i \cup \{a_j\};$
            $State(a_j) \leftarrow BUSY;$
        if $\forall r^l_{t_i} \in Rsc(t_i) : r^l_{t_i}$ is satisfied
         or $\forall a_k \in Neig(a_i) : State(a_k) \neq IDLE$ **then**

(14) **break**;
(15) **else**
(16) **select** a_k as *Mediator* based on the number of a_k's neighbours
: $a_k \in Neig(a_i)$ **and** $State(a_k) = IDLE$
(17) $P_i \leftarrow P_i \circ P_k$;
(18) $State(a_k) \leftarrow Mediator$;
(19) **for** $\forall P_i, P_i^{'} : P_i \in P \wedge P_i^{'} \in P^{'}$ in sequential order
(20) $P_i^{'} \leftarrow P_i^{'} \cup P_i$
end

In **Algorithm 2**, line (17), the notation "$\circ$" is *Relational Composition* (line 17). The meaning of this notation is that $\forall \langle x_i, y_i \rangle \in X, \langle y_j, z_j \rangle \in Y : y_i = y_j \Rightarrow \langle x_i, z_j \rangle \in Z$. In this paper, *Relational Composition* is utilized to model establishing indirect connections between *Initiator*s and *Participants* via *Mediators*. Through this way, an *Initiator* can request not only its directly linked neighbors but also other indirectly connected agents for help. Furthermore, in **Algorithm 2**, $P_i^{'}$ records a set of ordered couples, and each couple displays an agent that has cooperated with or, at least, been requested by agent a_i during one task allocation round (lines 19 and 20). In other words, $P_i^{'}$, which is used for adapting the system structure, records the interaction history of agent a_i during one task allocation round.

3 System Adaptation Strategy

In this section, a Novel System Adaptation Strategy (NSAS) is described which is for discovering effective system architecture. The strategy is based on the popularity of each agent in the system. The idea is motivated by and derived from the previous findings on multi-agent systems [6] [5].

As depicted in Subsection 2.2, popularity of each agent can be calculated by using Equation 1. Each agent maintains its *popularity* as a local information. After each task allocation round, each agent, $a_i \in A$, considers to adapt its system structure. An agent a_i is only valid to adapt its system structure, if its popularity is lower than the average popularity of its neighbors. If agent a_i is valid, it will remove a connection from its immediate neighbors, i.e. $Neig(a_i)$, which has the lowest popularity. Thereafter, the agent a_i chooses an agent from $P_i^{'}$ but not in $Neig(a_i)$ as a new neighbor, which has the highest popularity, and creates a direct connection with it. The purpose of this manner is to keep the average degree of agents stable in the system. Degree of an agent refers to the number of neighbors the agent has, and, thus, average degree is the average number of neighbors each agent has in a system. **Algorithm 3** formally describes the system adaption strategy.

Algorithm 3: System Adaption Strategy
input:
  $a_i \in A$: a set of agents
  $P' = \{P'_1, ..., P'_n\}$: records interaction history of each agent
output:
  R: a new *Compatible Relation*
  $Neig(a)$: a new neighbor list of each agent
begin:
(1)**for** $\forall a_i : a_i \in A$ in sequential order
(2) **if** $Pop(a_i) < \frac{1}{|Neig(a_i)|} \sum_{a_l \in Neig(a_i)} Pop(a_l)$ **then**
(3) **select** $a_j : a_j \in Neig(a_i)$ **and** $Pop(a_j)$ is the lowest
(4) $R \leftarrow R - \{\langle a_i, a_j \rangle\}$;
(5) $Neig(a_i) \leftarrow Neig(a_i) - \{a_j\}$;
(7) **select** $a_k : \langle a_i, a_k \rangle \in P'_i$ **and** $a_k \notin Neig(a_i)$ **and** $Pop(a_k)$ is the highest
(8) $R \leftarrow R \cup \{\langle a_i, a_k \rangle\}$;
(9) $Neig(a_i) \leftarrow Neig(a_i) \cup \{a_k\}$;
end

After each agent adapts its neighbors linkage, it resets its counters of both group joining attempts and group joining successes to 0, in an effort to establish an estimate of popularity with the new system structure.

4 Experiment and Analysis

To evaluate the performance of DGF, we compare DGF with the Greedy Distributed Allocation Protocol (GDAP) presented in [15]. GDAP is utilized for allocating tasks in a distributed environment, but it only allows neighboring agents to help with a task. In this section, we first depict GDAP briefly. Then, the settings of an experiment environment and the evaluation criteria are introduced. Finally, the experiment results and the relevant analysis are illustrated.

4.1 Greedy Distributed Allocation Protocol

Greedy Distributed Allocation Protocol (GDAP) [15] is employed to handle task allocation problem in agent social networks. The task allocation process of GDAP is described briefly as follows. All manager agents try to find neighboring contractors for help with their tasks. Manager agents start with offering the most efficient task. Out of all tasks offered, contractors select the task with the highest efficiency, and send a bid to the related manager. A bid includes all the resources the agent is able to supply for this task. If sufficient resources have been offered, the manager selects the required resources and informs all contractors of its choice. When a task is allocated, or when a manager has received offers from all neighbors but still cannot satisfy its task, the task is removed from its task list.

4.2 Experiment Setting

In order to compare the two approaches, DGF and GDAP, we set an experiment environment for testing them. Scale-free network [4] is used for the evaluation of DGF through comparison with GDAP. The feature of scale-free network is that although the average degree of each agent is constant, there are few agents with very high connectivity and most other agents with few neighbors.

There are four different setups used in the experiment.

Setup 1: The number of agents and tasks introduced in the experiment environment are 50 and 30, respectively. The number of different resources is 10, i.e. $\varepsilon = 10$, and each agent is randomly assigned one of them. As mentioned in Subsection 2.1, the number of resources required for a given task are chosen uniformly from $[1, \varepsilon]$, and each resource of a task is selected randomly from $[1, \varepsilon]$. The *TTL* value of each task is set to 4. The tasks are distributed uniformly on each *IDLE* agent one by one. The average number of neighbors of each agent, i.e. the average degree of each agent, is fluctuated from 6 to 12. This setup is designed to show how different average degree influences the performance of the two approaches.

Setup 2: In this setting, the average degree of each agent is fixed at 10. The number of agents fluctuates from 100 to 400 and the ratio between the number of agents and tasks is confirmed at 5/3, namely that the number of tasks varies from 60 to 240 which depends on the number of agents. The proportion of the number of agents and resources is set to 5/1, namely that the number of resources fluctuates from 20 to 80. In order to match the fluctuation of the number of agents, the *TTL* value for each task transforms from 4 to 10. Table 1 lists the details of this setting. This setup is used for demonstrating the scalability of the two approaches in different scale networks with a fixed average degree.

Setup 3: *Setups 1* and *2* are used to test various aspects of DGF and GDAP in a static environment. This setting is devised for testing DGF in a dynamic environment. The number of agents in the system is 50. During each task allocation round, the number of tasks is 30. The number of types

Table 1 The Details of Setup 2

# of agents	# of tasks	# of resources	*TTL*
100	60	20	4
200	120	40	6
300	180	60	8
400	240	80	10

of different resources is 10, i.e. $\varepsilon = 10$. The TTL value of each task is set to 4. The average degree is fixed at 6. In order to achieve the popularity of each agent, the task allocation process will be run in 2000 rounds as an initial period with no system adaption. Afterward, the system is allowed to be evolved with the adaption strategy described in Section 3. For comparison, we utilize another adaption strategy as a standard, i.e. Agent Organized Network (AON) strategy, which was presented in [5]. The AON strategy is briefly described as follows. If an agent, a_i, decides to adapt its neighboring linkage, the agent will remove the connection from the neighboring agent, a_j, which has the lowest popularity. The agent a_i then requests a referral from its neighbor with the highest popularity, say a_l. Finally, the agent a_i will establish a new direct link with the agent a_k that is the most popular neighbor of a_l. It is apparent that the AON strategy depends on neighboring agents in most extent. An agent with the adaptation strategy proposed in this paper, i.e. Novel System Adaptation Strategy (NSAS), can choose a new neighbor from those candidate agents which cooperated with or were requested by the agent (mentioned in Section 3). Thus, compared with AON strategy, NSAS let an agent have more candidates when the agent decides to adapt its neighbors linkage.

In this experiment, three criteria are used to evaluate the performance of the two approaches.

CpR (Completion Rate of tasks): The proportion of the number of successfully allocated tasks to the total number of tasks in the experiment environment, namely:

$$CpR = \frac{\#\ of\ successfully\ completed\ tasks}{|T|} \tag{2}$$

where T is a set of tasks in the experiment environment and $|T|$ represents the number of tasks as described in Subsection 2.1. Higher CpR demonstrates that more tasks can be allocated, so the performance is better.

CmC (Communication Cost): The entire number of the communication messages transferring in the system during a task allocation round. It should be recalled that a task allocation round is the process of allocating a set of tasks to appropriate agents. Lower CMC indicates that less communication messages are generated and transferred in the system. Therefore, the burden of a system can be remitted more.

Proposition 5. *For a complex adaptive system with m tasks, n agents, ε resources and TTL for each task, the complexity of CmC is $O(TTL{\cdot}m(n{+}\varepsilon))$.*

Proof. In each iteration in the worst case (i.e. a fully connected system), for each of the $O(m)$ *Initiators*, $O(n)$ resource query messages are sent. Then, the $O(n)$ available resource response messages are generated and sent back to the *Initiators*. The messages generated during this phase are $O(2mn)$. Next, each of the $O(m)$ *Initiators* forms $O(m)$ groups with their $O(n)$ neighbors.

Thus, the messages created at this stage are $O(m\varepsilon)$. After that, each of the $O(m)$ *Initiators* commits $O(m)$ tasks to one of their neighboring agents, and therefore sends $O(m)$ messages. This process will be iterated in $O(TTL)$ hops. Hence, the messages generated during commitment process are $O(TTL \cdot (2mn + m\varepsilon + m))$. Totally, during a task allocation round, the number of communication messages created are $O(2mn+m\varepsilon+TTL\cdot(2mn+m\varepsilon+m)) = O(TTL \cdot m(n + \varepsilon))$.

CoP (Coefficient of Performance): The ratio between the number of communication messages (CmC) and the number of successfully completed tasks during a task allocation round which can be calculated by using Equation 3.

$$CoP = \frac{CmC}{\#\ of\ successfully\ completed\ tasks} \tag{3}$$

Generally, the performance of the approach is more desirable if higher CpR could be achieved and lower number of communication messages (CmC) are created. Hence, observing only either CpR or CmC is not enough to determine the quality of the approach. We, therefore, employ the CoP as the third metric to evaluate the two approaches. Lower CoP indicates better performance since lower CoP means completing each task needs fewer communication messages.

4.3 Experiment Results

The experiment is performed on the four aforementioned setups for DGF and GDAP. In order to achieve precise results, each evaluation step was executed 30 times and the average data were obtained.

4.3.1 Experiment Results from Setup 1

This experiment is done on *Setup 1* as described in Subsection 4.2. The number of agents and tasks are 50 and 30, respectively. The number of different resources is 10. The TTL value of each task is set to 4. The average degree of each agent is fluctuated from 6 to 12. The purpose of this setting is to test the performances of both the two approaches which are influenced by the fluctuation of average degree.

In Fig. 1(a), it is demonstrated that the *Completion Rate of tasks* (CpR) of DGF in different conditions is much higher and more stable than that of GDAP. This is because task allocation with GDAP is only depending on neighbors of *Initiator*. Therefore, more neighbors can provide more opportunities for tasks to be solved. Comparatively, DGF relies on not only neighbors but also other agents if needed. This feature results in steady performance of DGF. Fig. 1(a) also showed that with higher average degree, the performance of GDAP is improved continuously. The reason of this situation is that when

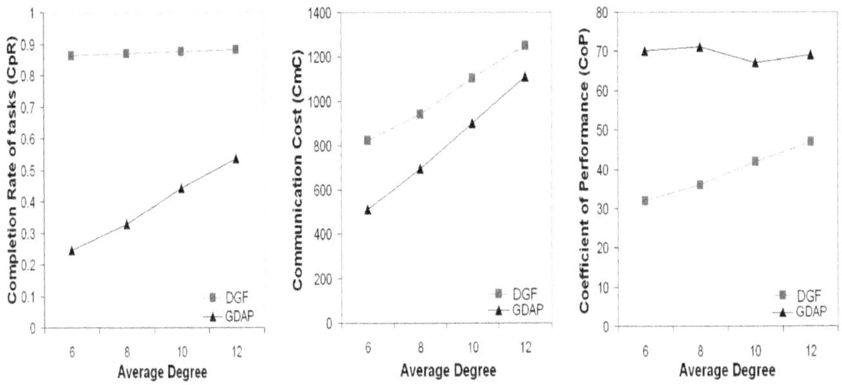

(a) Completion Rate of tasks (CpR)

(b) Communication Cost (CmC)

(c) Coefficient of Performance (CoP)

Fig. 1 Performance of two approaches on different average degree

there are more neighbors, *Initiator* has higher probability to derive sufficient resources to allocate its tasks successfully.

Fig. 1(b) displays the number of messages (CmC) of the two approaches in different situations generated in a task allocation round. As DGF would request other agents for help when resources from neighbors are insufficient, the communication cost of DGF is higher than that of GDAP. Thus the presentation of GDAP in this test is relatively good due to its considering only neighbors which could decrease the number of messages created during task allocation processes. It should also be noticed that with the increase of average degree, the distinction between DGF and GDAP is reducing. This situation could be explained that with more average number of neighbors, *Initiator*s have more opportunity to solve their tasks based on only their neighbors in both approaches. Therefore, for DGF, some *Initiator*s do not need to commit their tasks to other agents, and consequently save communication cost.

Fig. 1(c) shows the *Coefficient of Performance* (CoP) of the two approachs in different cases. With the average degree ascending, the CoP of GDAP declines, while the CoP of DGF increases gradually. It should be recalled that lower CoP means better performance. For GDAP, although communication cost rises with the increase of average degree, the number of allocated tasks also increases. Therefore, for each allocated task, the communication cost remains constant in principle. On the other hand, with the increase of average degree for DGF , the completion rate of tasks keeps steady displayed in Fig. 1(a), while the number of messages, i.e. communication cost, mounts up persistently demonstrated in Fig 1(b). Hence, for each successfully completed task, the communication cost increases. However, DGF still has lower CoP,

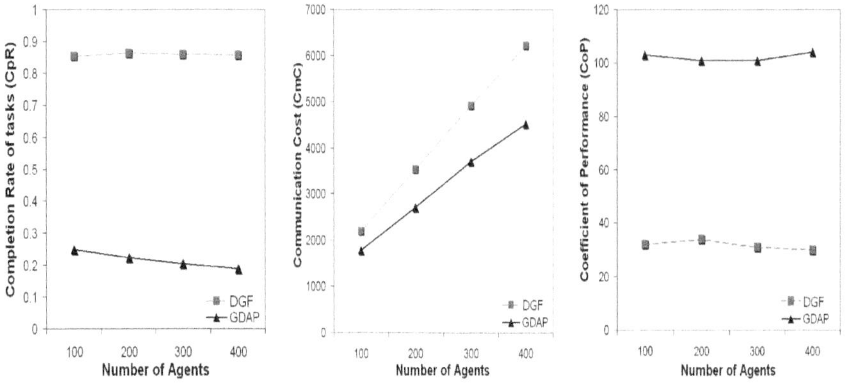

(a) Completion Rate of tasks (CpR)

(b) Communication Cost (CmC)

(c) Coefficient of Performance (CoP)

Fig. 2 Performance of two approaches on different network scales

i.e. better performance, in contrast to GDAP throughout the test process. This can be explained that the number of successfully completed tasks based on GDAP is few, so GDAP generates many useless communication messages during task allocation processes which thus leads to low *CoP*.

4.3.2 Experiment Results from Setup 2

This experiment is based on *Setup 2*. The aim of this setup is to test the scalability of the two approaches in different network scales. The average degree of each agent is fixed at 10. The number of agents fluctuates from 100 to 400. The number of tasks varies from 60 to 240, the number of resources fluctuates from 20 to 80, and the *TTL* value for each task transforms from 4 to 10, all of which depend on the number of agents. Details of *Setup 2* can be found in Table 1.

According to Fig. 2(a), we can see that with the spread of network scale, the Completion Rate of tasks (*CpR*) of GDAP is continually descending while *CpR* of DGF keeps stable and high. This case can be argued that with the network scale expanding, the numbers of tasks and resources also rise proportionally, while the average degree of each agent is fixed. Therefore, the more resources each task requires, the more possible the task allocation would be failure if *Initiators* request only neighboring agents, e.g. GDAP. Compared with GDAP, benefited from requesting other agents, DGF can preserve decent performance.

Fig. 2(b) shows the Communication Cost (*CmC*) of the two approaches in different network scales. With the expansion of network scale, both of the two approaches generate more communication messages, since there are

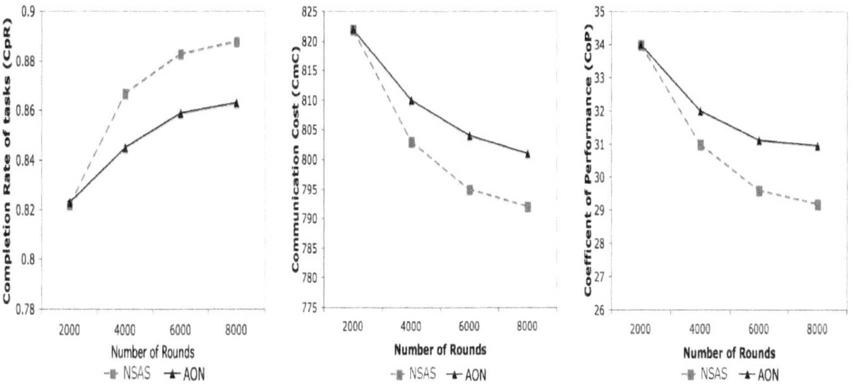

(a) Completion Rate of tasks (CpR)

(b) Communication Cost (CmC)

(c) Coefficient of Performance (CoP)

Fig. 3 The performance of DGF with different adaptation strategies

more agents and tasks in the network, and in order to allocate these tasks, more communication steps cannot be avoided which results in communication messages rising.

Fig. 2(c) demonstrates the Coefficient of Performance (CoP) of the two approaches. It can be found that the CoP of both DGF and GDAP keeps almost steady with the increase of network scale. The reason of this result is that even though the number of communication messages ascends, the absolute number of successfully completed tasks also increases, and, thus, the CoP can keep stable.

The conclusion can be achieved that DGF has better scalability than GDAP, since both CpR and CoP of DGF can be preserved while GDAP can maintain only CoP but CpR of GDAP decreases gradually.

4.3.3 Experiment Results from Setup 3

As described in Subsection 4.2, this setup is designed for testing DGF in a dynamic environment. The number of agents in the system is 50. During each task allocation round, the number of tasks is 30. The number of types of different resources is 10, i.e. $\varepsilon = 10$. The TTL value of each task is set to 4. The average degree is fixed at 6. The task allocation process will be run in 2000 rounds as an initial period with no system adaption to establish each agent's popularity. Then, two system adaptation strategies, i.e. NSAS and AON strategy, are applied.

According to Fig. 3, with the number of rounds increasing, both adaptation strategies achieve better performance. In contrast to AON strategy, the outcome accomplished by NSAS is better on all the three metrics, i.e. CpR,

CmC and *CoP*. This can be explained that an agent with AON strategy only chooses one of its neighbors' neighbors as its new neighboring agent, while an agent with NSAS has more candidates when selecting a new neighbor, and, hence, the agent is very likely to create a better neighboring linkage. It should also be observed that, for both strategies, the ratio with which performance is improving slows down over time. This implies that the system structure converges to stability with the increase of task allocation rounds.

4.4 Experiment Analysis

From the above description, it is obvious that the performance of DGF is better than that of GDAP since more tasks can be successfully completed by using DGF, i.e. higher *CpR*. Although, in all the cases, the absolute number of communication messages generated by using DGF is more than that of GDAP, namely higher *CmC*, DGF has less average number of communication messages for each completed task, i.e. lower *CoP*, compared with that of GDAP. Furthermore, the adaptation strategy developed in this paper is more efficient than AON strategy, as our strategy can achieve higher *CpR* and *CoP*, and, meanwhile, lower *CmC*. Therefore, DGF is more effective and scalable compared with GDAP, and NSAS is more efficient than AON strategy.

5 Conclusion and Future Work

In this paper, we proposed a Decentralized Group Formation (DGF) algorithm and a Novel System Adaptation Strategy (NSAS) for complex adaptive systems. Compared with current related works, DGF has several advantages. First, DGF does not have a central controller so as to avoid single point failure and network congestion. Second, DGF takes agent network architectures into account during task allocation processes. Third, DGF allows *Initiators* to request help for tasks not only from neighboring agents but also other indirect linked agents if needed. In the future, we plan to improve DGF by considering that each agent in the system has multiple resources instead of only one. Another stream is to enhance the system adaptation strategy, NSAS, by utilizing some multi-agent learning methods.

Acknowledgements. This research is supported by Australia Research Council Linkage Projects (LP0991428) and a URC Research Partnerships Grants Scheme, from the University of Wollongong with the collaboration of TransGrid, Australia.

References

1. Abdallah, S., Lesser, V.: Modeling task allocation using a decision theoretic model. In: Proceedings of AAMAS 2005, Utrecht, Netherlands, pp. 719–726 (2005)

2. Abdallah, S., Lesser, V.: Learning the task allocation game. In: Proceedings of AAMAS 2006, Hakodate, Hokkaido, Japan, pp. 850–857 (2006)
3. Bachrach, Y., Rosenschein, J.S.: Distributed multiagent resource allocation in diminishing marginal return domains. In: Proceedings of AAMAS, Estoril, Portugal, pp. 1103–1110 (2008)
4. Barabasi, A.L., Albert, R.: Emergence of scaling in random networks. Science 286(5439), 509–512 (1999)
5. Gaston, M.E., desJardins, M.: Agent-organized networks for dynamic team formation. In: Proceedings of AAMAS 2005, Utrecht, Netherlands, pp. 230–237 (2005)
6. Kraus, S., Shehory, O., Taase, G.: Coalition formation with uncertain heterogeneous information. In: Proceedings of AAMAS 2003, Melbourne, Australia, pp. 1–8 (2003)
7. Kraus, S., Shehory, O., Taase, G.: The advantages of compromising in coalition formation with incomplete information. In: Proceedings of AAMAS 2004, New York, USA, pp. 588–595 (2004)
8. Lerman, K., Shehory, O.: Coalition formation for large-scale electronic markets. In: Proceedings of ICMAS 2000, Boston, Massachusetts, USA, pp. 167–174 (2000)
9. Manisterski, E., David, E., Kraus, S., Jennings, N.R.: Forming efficient agent groups for completing complex tasks. In: Proceedings of AAMAS 2006, Hakodate, Hokkaido, Japan, pp. 834–841 (2006)
10. Schlegel, T., Kowalczyk, R.: Towards self-organising agent-based resource allocation in a multi-server environment. In: Proceedings of AAMAS 2007, Honolulu, Hawai'i, USA, pp. 74–81 (2007)
11. Shehory, O., Kraus, S.: Methods for task allocation via agent coalition formation. Artificial Intelligence 101(1-2), 165–200 (1998)
12. Sims, M., Goldman, C.V., Lesser, V.: Self-organization through bottom-up coalition formation. In: Proceedings of AAMAS 2003, Melbourne, Australia, pp. 867–874 (2003)
13. Smith, R.G.: The contract net protocol: High-level communication and control in a distributed problem solver. IEEE Trans. Computers C-29(12), 1104–1113 (1980)
14. Theocharopoulou, C., Partsakoulakis, I., Vouros, G.A., Stergiou, K.: Overlay network for task allocation and coordination in dynamic large-scale networks of cooperative agents. In: Proceedings of AAMAS 2007, pp. 295–302 (2007)
15. Weerdt, M.D., Zhang, Y., Klos, T.: Distributed task allocation in social networks. In: Proceedings of AAMAS 2007, Honolulu, Hawaii, USA, pp. 500–507 (2007)
16. Zheng, X., Koenig, S.: Reaction functions for task allocation to cooperative agents. In: Proceedings of AAMAS 2008, Estoril, Portugal, pp. 559–566 (2008)

Cellular Automata and Immunity Amplified Stochastic Diffusion Search

Duncan Coulter and Elizabeth Ehlers

Abstract. Nature has often provided the inspiration needed for new computational paradigms and metaphors [1,16]. However natural systems do not exist in isolation and so it is only natural that hybrid approaches be explored. The article examines the interplay between three biologically inspired techniques derived from a plethora of natural phenomena. Cellular automata with their origins in crystalline lattice formation are coupled with the immune system derived clonal selection principle in order to regulate the convergence of the stochastic diffusion search algorithm. Stochastic diffusion search is itself biologically inspired in that it is an inherently multi-agent oriented search algorithm derived from the non-stigmergic tandem calling / running recruitment behaviour of ant species such as *Temnothorax albipennis*. The paper presents an invesitigation into the role cellular automata of differing complexity classes can play in order to establish a balancing mechanism between exploitation and exploration in the emergent behaviour of the system...

1 Background

Respect for the resourcefulness of the ant runs deep in humankind flowing through a gambit of cultures crossing East and West. The Korean flood myth tells how one of the last human survivors was assisted by ants in sifting between grains of rice and of sand [18]. In the Middle East the Book of Proverbs, Chapter 6, admonishes the reader to *go to the ant* who without *commander*, *overseer* or *ruler* manages to store provisions in summer. While in the West one of the the greek fables of Æsop tells of the industrious ant and the lazy grasshopper.

Duncan Coulter · Elizabeth Ehlers
University of Johannesburg, Johannesburg, South Africa
e-mail: {dcoulter, emehlers}@uj.ac.za

Q. Bai and N. Fukuta (Eds.): Advances in Practical Multi-Agent Systems, SCI 325, pp. 21–32.
springerlink.com

1.1 *Stochastic Diffusion Search*

Stochastic Diffusion Search (SDS) is an inherently multi-agent oriented algorithm for exploring a solution space. The biological basis for SDS has its roots in the swarming behaviour of certain social insects. SDS differs from many conventional swarm intelligence algorithms in that it is non-stigmergic. Stigmergy refers to communication by means of environmental modification.

1.1.1 Biological Basis

Algorithms such as Ant Colony Optimization are said to be stigmergic [15] in that they use the metaphor of pheremonic trails laid down by exploratory agents in search of high utility values across the solution space. These trails then bias the line of inquiry followed by subsequent agents to instead follow the path laid out for them as opposed their initial semi-random exploratory walk. The intensity of these trials fades over time unless reinforced by subsequent agents. Eventually, as a result of a combination of random exploration guided by reinforced exploitation, the agents converge on desirable optima.

Environmental pheromones are not however the only communication mechanism employed across all ant species. Certain species such as *Temnothorax albipennis* use direct one-to-one communication, by means of interlocked antennae, as part of a a nest mate recruitment process known as tandem calling [6]. This tandem calling mechanism forms the basis for the metaphor employed by SDS. It is also worth noting that explicit message passing such as this makes SDS far more amenable to implentation according to modern message-centric agent standards such as FIPA [5, 2] instead of requiring the creation of an explicit stigmergic abstraction [15].

1.1.2 SDS Agent Algorithm with Incremental Improvements

One primary requirement for SDS is that the overall problem be decomposable into a set of independently executable partial evaluation functions [12]. It is for this reason that not all solution spaces are trivially amenable to search by way of SDS.

Once the solution space has been defined or transformed in such a way as to allow the above mentioned functional decomposition the SDS algorithm will then proceed as described [9, 12, 13]. A pool of agents is maintained such that each agent:

1. ... maintains its own hypothesis about the solution to the overall problem under consideration (i.e. has a position within the current solution landscape).
2. ... is able to *test* its current hypothesis by means of a partial evaluation function which is well defined at its current position within the solution landscape.

3. ... is in one of two possible states *active* or *passive* based on the result of its most recent hypothesis test.

The system then proceeds to enter into a series of states, as an aggregation of the states of its individual agents, as described by the pseuduocode in Figure 1. It is also worth noting that the bulk of these phases may occur in parallel as illustrated by Figure 2

1. *Initialisation:* During the initialisation phase each agent adopts a random initial hypothesis i.e. assumes its position in the search space.
2. *Testing:* Every agent applies the partial evaluation function defined at its location within the solution space to its current hypothesis. In the event of this function's result falling within a given utility tolerance the agent's state will shift to being that of an active agent. It is possible that active agents may receive a reduced utility during the testing phase as a result of operating within a dynamic environment. While the system's response in such a case is not prescribed by the meta-heuristic, and in fact some implementations do not consider it [9].The first incremental enhancement to the SDS algorithm which deals with this situation will now be described.
   a. Evaluate the historical utility received from previous tests of related hypotheses over a predefined interval.
   b. Should the historical utility fall below a threshold value then slightly modify the current hypothesis
   c. Should repeated modifications of the current hypothesis fail to achieve a satisfactory utility then select a new hypothesis at random.
3. *Diffusion:* This is the phase which reflects the tandem calling behaviour in the algorithm's natural analogue. Each active agent then seeks to recruit a passive agent to explore its neighbourhood. This is accomplished by the passive agent adopting the hypothesis of the active agent. The second small enhancement to the basic SDS algorithm would be to probabilistically introduce some minor variation to the transmitted hypothesis to avoid the agents becoming stuck on local maxima. Lastly a third opportunity to optimise the algorithm presents itself in the fact that the standard SDS algorithm does not make provision for active agents to benefit from recruitment to higher utility areas by other active agents.

```
AgentPool.initialise()
while(AgentPool.activeAgents < Threshold)
        AgentPool.foreach((x) => x.test())
        AgentPool.foreach((x) => x.diffuse())
end while
```

Fig. 1 Pseudocode for SDS

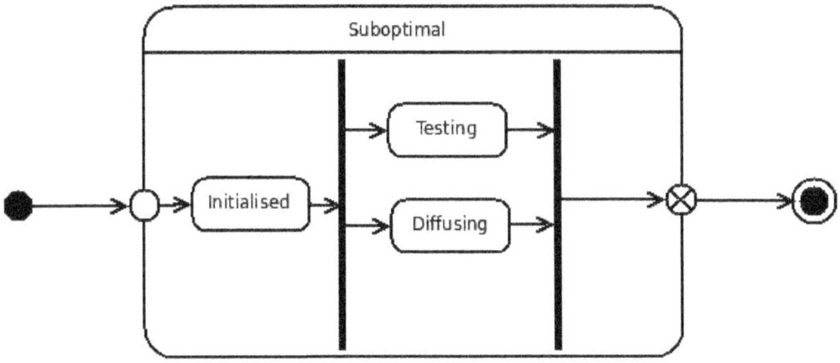

Fig. 2 UML 2.0 state diagram for SDS

1.2 Cellular Automata

Cellular Automata (CA) are a special class of finite state automata with their origins in models of crystalline and bacterial growth [17]. Each automaton is composed of a set of *cells* which at each iteration may each be in one of a finite set of *states*. These cells are related to each other by way of a topological function defining for each cell a *neighbourhood* of other cells in the CA. At each iteration of the CA, the state of each cell is simultaneously updated by a *local update rule* based on the current state of its neighbouring cells.

Due to the rapidly emergent complex behaviour of certain instances of these systems the future state of such a CA can often not be calculated from its initial state without expending an equivalent amount of computational effort as would have been spent running the automaton.

1.2.1 Elementary Cellular Automata

The emergent complexity characteristics of CA with a single dimensional von Neumann neighbourhood topology and a binary state set have been exhaustively enumerated [17]. As a result of this work elementary CA have been placed along a four point system according to the complexity class of their emergent behaviour.

- *Class I* initial patterns rapidly devolve into static homogenous states
- *Class II* initial patterns rapidly devolve into repetitive oscillating structures
- *Class III* initial patterns rapidly devolve into pseudo-random states
- *Class IV* initial patterns rapidly evolve into complex interacting structures. Instances of class IV cellular automata have been shown to be *turing complete*

1.2.2 Elementary CA and Multi-agent Evolution

In [3] the interaction between elementary CA and multi-agent evolutionary computation was explored for the problem of collaborative prior art detection by a community of gene expression programming driven agents. Cellular automata were used for the purpose of defining the accessibility relationship between candidate solution sub-pools for evolving improved prior art detection functions by agents operating on a suitably prepared patent corpus.

1.3 Clonal Selection

Artificial Immune Systems (AIS) is a new computational intelligence approach which draws its inspiration from the immune system [4]. The mammalian adaptive immune system is formed from a complex set of interacting subsystems which together allow for mammals to prove resilient against assault from an ever adapting array of potential pathogens. The system is remarkable in that in order to fulfill its role in ensuring the destruction of potentially harmful foreign elements it is not aware of what constitutes a foreign object *a-priori*. The system employs a sophisticated and robust pattern recognition system in order to discriminate between *self* and *non-self*. In contrast to this the system must also selective amplify the recognition of pathogenic patterns. The failure of self / non-self discrimination in the system causes various degenerative autoimmune conditions. This is not to say that AIS are limited to security / intruder detection tasks. They have successfully been applied to large scale optimisation problems as well [7].

The immune system is inherently agent oriented immunity information is embedded within the states of the various lymphocytes. Each lymphocyte is keyed to bind to one single antigenic protein pattern known as its affinity. In order to reduce interference between the two pattern recognition goals the immune system employs two opposing mechanisms: *negative selection* triggers cell death in those lymphocytes cells which fail in self / non-self discrimination while *clonal selection* selectively multiplies the number of those lymphocytes which bind to a non-self antigen. This selective duplication of those cells and their retention within the blood and lymph is what produces the phenomenon of tolerance where subsequent exposures to a given pathogen are less severe.

2 The CAIA–SDS Model

The section introduces the *Cellular Automata / Immunity Amplified Stochastic Diffusion Search* (CAIA–SDS) model. The model is first introduced followed by a brief discussion of its implementation details. The following section discusses some of the results obtained from the prototype system.

2.1 Model Development

Apart from the three incremental improvements introduced in section 1.1.2 the two major improvements proposed each depend on a different biologically inspired paradigm.

The first opportunity for significant improvement stems from the nature of inter-agent communication in SDS and conventional swarm intelligence. Stigmergic communication is one-to-many but inherently undirected relying as it does on an exploratory agent picking up the pheremonic trail. SDS on the other hand employs one-to-one inter-agent communication but this communication is directed from high utility agents to lower utility agents. Since any single recruitment message will result in an increase in the utility of a passive agent a naive optimisation of SDS would attempt to maximise the number of recipients of each message.

Such a modification would however be unwise in the medium to long term for the system. The first agent to test a hypothesis with a utility exceeding the threshold value would immediately end the inquiry into other lines of reasoning by agents in other sections of the solution space. In effect the system would move to adopt a highly exploitative behaviour rapidly ascending the first positive utility gradient encountered during exploration and then rapidly becoming stuck on the current local maximum. It is clear then that simply maximising the number of recipients is not sufficient. The question then remains what mechanism should be used in order to select a subset of the agent population in such a way as would allow for explicit control over the degree of exploitation versus exploration.

Selecting some fixed subset of the agent population would fragment the search space by causing clustering on a slightly larger set of local maxima than would have been achieved by one-to-one communication alone. While it may be tempting to simply select a random subset of the agent population during the diffusion phase of the algorithm problems of agent inquiry bias would then emerge based on the underlying distribution employed by the pseudo random number generator. Such bias reduces the explorative nature of the algorithm potentially excluding high utility hypotheses from consideration. A further defect of this approach lies in the ephemeral nature of the relationships between agents. In utility landscapes with a high degree of noise communication between proximal agents may facilitate the overcoming of very localised maxima.

The new approach proposed makes use of the emergent complexity of CA to overcome the aforementioned approaches' shortcomings. This is effected by creating an indirection communication layer through which all recruitment messages must pass. The indirection layer maintains an elementary cellular automaton with each agent in the system being represented by a single cell. The agent's communication neighbourhood is then defined as being those contiguous cells with the same state as the current agent as illustrated in

Figure 3. The advantages of using elementary CA to define the neighbourhood are as follows...

1. They are simple to describe: Using only a single identifying integer number (the Wolfram Code [17]) the entire set of production rules which constitute the local update rule can be inferred.
2. They are flexible: By specifying a local update rule from any of the four complexity classes the emergent properties of the system may be guided. From static preselection through pseudo-random selection through to persistent complex patterns all may be accommodated via a single mechanism.
3. Their properties are well established: Through exhaustive enumeration the runtime properties of every single possible elementary CA has been exhaustively enumerated [17]
4. Since it is possible for an active agent to communicate with a different active agent there is now an opportunity for the active agent with the lower utility hypothesis to adopt the hypothesis of the other agent.

The second substantial amplification of SDS draws from the Clonal Selection algorithm described in the previous section. In AIS the intelligence of the system is embedded in the state of the lymphocyte analogues. As each lymphocyte has an affinity for only a single antigen as determined by the bone marrow algorithm a strong parallel exists between AIS and SDS. To clarify the point, every lymphocyte in the AIS can be mapped onto an individual agent within the SDS solution space. The hypothesis maintained by each SDS agent is analogous to the antigen affinity by the lymphocyte.

The primary difference lies in the agent population itself. In conventional SDS the agent pool is fixed in that a predetermined number of agents are exploring the utility landscape for the entire duration of the system's lifespan. In clonal selection agents with a high affinity for a given candidate solution (i.e. those which bonded to an antigen in the blood) are selectively multiplied.

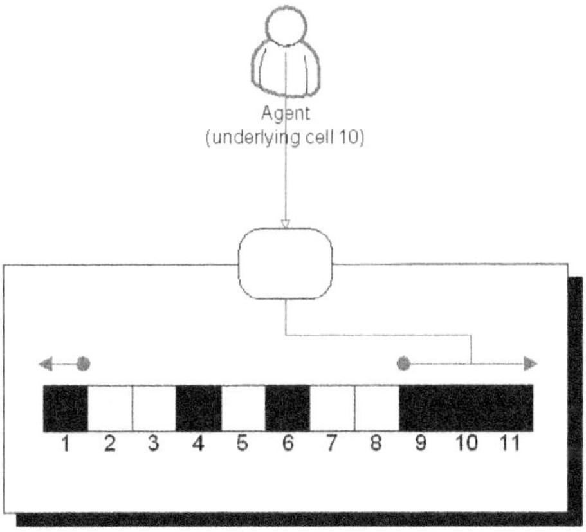

Fig. 3 Agent neighbourhood definition via a cellular automaton.

SDS may be improved by allowing for the dynamic introduction of new agents during the diffusion phase. When an agent applies the partial evaluation function to its current hypothesis and receives a utility value higher than the threshold then its recruitment behaviour may proceed in one of two different ways based on the current available system resources.

Should sufficient system resources be available an active agent is allowed to spawn a new copy of itself (with a slightly modified hypothesis) in lieu of utilising the one-to-many communication pattern described earlier. This allows the amplified SDS algorithm to embed a memory of the utility landscape which may prove useful in accommodating temporary perturbations in dynamic environments. In living organisms cells have limited lifespans the failure of this mechanism leading to cancer through unwarranted cell immortality and a similar effect would be true in amplified SDS should the lifespan of newly spawned agents not be strictly controlled. Three approaches may be adopted in amplified SDS...

- Firstly, the lifespan of each agent in the system may be limited to a certain number of diffusion phases. In this case it is very important that the probability of new agent instantiation be high enough to prevent the system from losing all agents and hence entering into an irrecoverable state.
- Secondly, the total size of the agent pool may be expressed as a function of the available system resources with agents being spawned and culled as needed on a phase by phase basis.
- Lastly, an agent may be permitted the ability to perform only a limited number of new agent instantiations before being removed.

2.2 Model Implementation

The nature of the model had a direct bearing on the selection of implementation language. CA, SDS and clonal selection are concepts which center very heavily on the concept of state and state modification. CA are explicitly a form of state machine while each lymphocyte agent in clonal selection and SDS must maintain its own individual state. In fact agent orientation may itself be viewed as an extension to object orientation and hence is rooted in the notion of private mutable state. All of these features lean heavily towards the use of an imperative language for the creation of a prototype system.

However the need for the functional decomposition of the solution into partial evaluation functions and the need for the dynamic modification and transmission of hypotheses would benefit from the support of functions as first class members offered by Functional Programming languages. Unfortunately purely functional programming languages completely forbid the side effect inducing modification of state required by agent orientation and CA.

The Scala [14] programming language offered a good compromise between allowing mutable state, discouraging unnecessary side effects and supporting functions as first class objects at a programming language level. In addition the languages standard library supports a robust support for actors as a unit of concurrency. Actors also provided a convenient mechanism with which to implement the message passing required by SDS. Initially the JADE library [2] was considered in order to provide FIPA compatible communication as the Scala language compiles directly to Java bytecode which runs on Java's Virtual Machine and hence supports a high degree of interoperability with existing Java classes. However JADE imposes a layered behavioural decomposition on the agent design. While such a design definitely possible for both SDS and clonal expansion was considered not to be the most suitable choice considering that interoperability with existing FIPA systems was not a requirement for the prototype.

This does not suggest that actors equate with agents by any means. Actors and software agents, while sharing superficial similarities such as a use of message passing are not equivalent. Actors are simply a concurrency control mechanism and do not equate with agents in the same way that threads do not equate with agents. An agent may well be implemented using one or more actors or indeed one or more threads.

Similar to the technique employed in [3] the source of the neighbourhood topology is obscured from the participating agents according to the Facade design pattern. As an improvement to the previously utilised technique the topology facade was this time realised using a trait rather than a conventional Java interface. This allows the trait to maintain some state of its own instead of simply defining a contact of interaction. The state maintained in this case is the mapping between agents and cells in the underlying CA. Two approaches to the maintenance of the CA in light of the dynamic addition and removal of agents necessitated by clonal selection were considered.

1. Additional agents may be simply chained to the same cell as the agent which instantiated them. This approach while converging some memory and avoiding the question of how to initialise the new cells in an expanded CA does suffer from a serious drawback. Since each newly instantiated agent occupies the same cell as its originator they are eternally neighbours. Over time in high churn scenarios (configurations where there is large degree of agent addition and removal) the causal relationship between originator agents and their offspring will come to dominate neighbourhood determination to a degree which completely marginalises the underlying complexity class of the automaton.
2. Alternatively, the CA can be expanded so that the one-to-one mapping of agents to cells is maintained. The question then needs to be addressed regarding how to initialise the newly created cells. The danger of not initialising them is that the newly added agents are immediately part of the same contiguous cell block and hence neighbourhood. This means that in class I or class II automata the danger of new agents immediately getting stuck on the local maximum of the highest utility agent remains high. It is the case that often a completely random configuration would disrupt any persistent structures created by a class IV automaton. In the end a

compromise was selected in line with the original configuration pattern imposed on the automaton in the initialisation phase of SDS. A compromise between disruption and stagnation can be achieved by setting the elements at `l + (nl % ol)` to true where `l` represents the length of the old CA, `nl` represents the number of new agents and `ol` represents the original length of the automaton.

The solution space is defined by a functionoid (functional object) which mixes in the appropriate optimisation function to be used via a trait. For the case of the prototype system the function was defined across two dimensions. The agents hypothesis was simply the tuple formed by its x and y coordinate. The hypothesis testing function simply combined the two values of the tuple together into a single value and then repeatedly sampled a probability function brought in by the trait. The utility of the hypothesis was evaluated as being the sum of the positive samples from the distribution.

3 Results

The results presented here show the aggregate behaviour of the system in terms of rate of convergence across several fitness landscapes defined by instances of each of the four main classes of CA for solution space defined on a power-law distribution as illustrated in Figure 4. For the purposes of the simulation the threshold value for terminating the system was kept unattainably high in order to ensure an indefinitely long run.

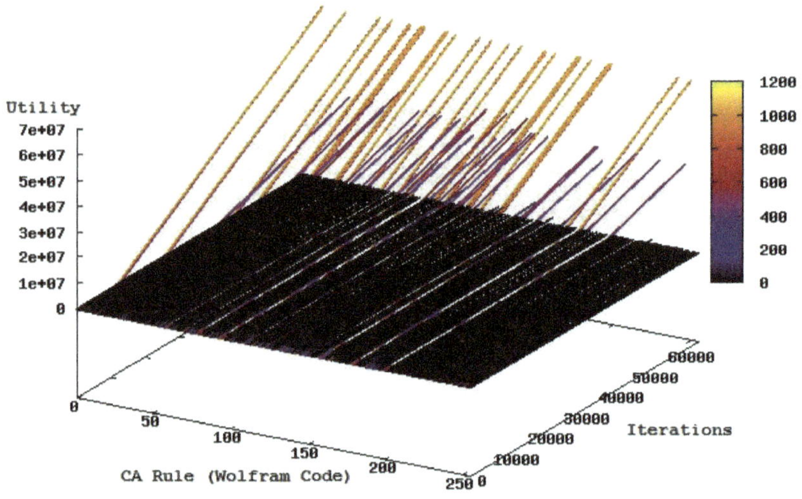

Fig. 4 The CA derived agent topology's effect on exploration and exploitation. Brighter colours denote a stronger bias towards exploration.

4 Conclusion

The paper explored several incremental improvements to the SDS algorithm. These improvements took the form of clarifying the potential shortfalls of "fixing" the hypotheses of active agents in dynamic environments along the steps to be taken when such changes occur. In this case the option of relaxing the allowable recipients of recruitment messages to include other active agents was broached. The paper also discussed the benefit of introducing minor variations into the propagated hypotheses during the diffusion phase of the algorithm in the interests of avoiding local maxima.

Clonal selection has an unambiguous amplification effect on the exploitative nature of the system by way of increasing the population of active agents. In biological immune systems the amplifying effect of clonal selection is tempered by the lifetime of individual lymphocytes [4] . The agent topology's effect on convergence is directly tied to the underlying complexity class of its defining CA.

1. Class I CA had a negative effect on the exploitative nature of the system without a comparative increase in the explorative nature of the search.
2. Class II CA had neither a positive effect of exploration nor much effect on exploitation
3. Class III CA had a positive effect on exploration and no major effect on exploitation
4. Class IV CA appear to provide the best compromise between exploration and exploitation although more research is needed in this regard.

To summarize the paper presents the use of several techniques to regulate the emergent exploration versus exploitation properties of the stochastic diffusion search algorithm. In keeping with biologically inspired metaphor of the original algorithm the major approaches described in the paper draw their inspiration from the natural world. Clonal expansion provides for a mechanism to enhance exploitation while cellular automata defined recruitment communication channels provide for a mechanism to regulate exploration versus exploitation, at run time, simply by specifying a single integer parameter.

References

1. Michalek, R., Tarantello, G.: Design Patterns from Biology for Distributed Computing. ACM Transactions on Autonomous and Adaptive Systems 1, 26–66 (2006)
2. Bellifemine, F., Caire, G., Greenwood, D.: Developing Multi-Agent Systems with JADE. John Wiley & Sons Ltd., Chichester (2007)
3. Coulter, D., Ehlers, E.: Federated Patent Harmonisation: An evolvable agent-wise approach. In: Proceedings of TMCE Symposium, vol. 2, pp. 1391–1393 (2008)
4. de Castro, L., Timmis, J.: Artificial Immune Systems A New Computational Intelligence Approach. Springer, Heidelberg (2002)

5. FIPA contributors FIPA ACL Message Structure Specificatio (2002), http://www.fipa.org/specs/fipa00061/CitedAugust2009
6. Franks, N.R., Richardson, T.: Teaching in tandem-running ants. Nature 493, 153 (2006)
7. Gong, M., Jiao, L., Ma, W.: Large-scale Optimization Using Immune Algorithm. In: GEC 2009, vol. 2, pp. 1391–1393 (2009)
8. Hilaire, V., Koukam, A., Rodriguez, S.: An Adaptative Agent Architecture for Holonic Multi-Agent Systems. ACM Transactions on Autonomous and Adaptive System (2008) doi: 10.1145/1342171.1342173
9. Hurley, S., Whitaker, R.M.: An Agent Based Approach to Site Selection for Wireless Networks. In: Proceedings of the 2002 ACM symposium on Applied computing (2002) doi: 10.1145/508791.508902
10. Hong, L.: On the Convergence Rates of Clonal Selection Algorithm. Information Science and Engieering (2008) doi: 10.1109/ISISE.2008.63
11. Karakasis, V.K., Stafylopatis, A.: Efficient Evolution of Accurate Classification Rules Using a Combination of Gene Expression Programming and Clonal Selection. IEEE Transactions on Evolutionary Computation (2008) doi: 10.1109/TEVC.2008.920673
12. De Meyer, K., Bishop, J.M., Nasuto, S.J.: Small-World Effects in Lattice Stochastic Diffusion Search. In: Dorronsoro, J.R. (ed.) ICANN 2002. LNCS, vol. 2415, pp. 147–152. Springer, Heidelberg (2002)
13. De Meyer, K., Nasuto, S.J., Bishop, M.: Stochastic Diffusion Search: partial function evaluation in swarm intelligence dynamic optimisation. In: Abraham, A., Grosam, C., Ramos, V. (eds.) Swarm intelligence and data mining, ch. 2, vol. 2 (2006)
14. Odersky, M., Spoon, L., Venners, B.: Programming in Scala. Artima Press (2008)
15. OReilly, G.B., Ehlers, E.M.: The Artificial Collective Engine Utilising Stigmergy (ACEUS) a Framework for Building Adaptive Software Systems. IJCSNS International Journal of Computer Science and Network Security (2008) doi: 10.1.1.104.3513
16. Shen, H., Zhu, Y., Zhou, X., Guo, H., Chang, C.: Bacterial Foraging Optimization Algorithm with Particle Swarm Optimization Strategy for Global Numerical Optimization. In: Proceedings of the first ACM/SIGEVO Summit on Genetic and Evolutionary Computation (2009) doi: 10.1145/1543834.1543901
17. Wolfram, S.: A New Kind of Science. Wolfram Media, Inc. (2002)
18. In-Sob, Z.: Folk Tales from. Routledge & Kegan Paul Ltd., London (1952)

Related Word Extraction Algorithm for Query Expansion – An Evaluation

Tetsuya Oishi, Tsunenori Mine, Ryuzo Hasegawa,
Hiroshi Fujita, and Miyuki Koshimura

Abstract. When searching for information a user wants, search engines often return lots of results unintended by the user. Query expansion is a promising approach to solve this problem. In the query expansion research, one of the biggest issues is to generate appropriate keywords representing the user's intention. The Related Word Extraction Algorithm (RWEA) we proposed extracts such keywords for the query expansion. In this paper, we evaluate the RWEA through several experiments considering the types of queries given by the users. We compare the RWEA, Robertson's Selection Value (RSV) which is one of the famous relevance feedback methods, and the combination of RWEA and RSV. The results show that as queries become more ambiguous, the advantage of the RWEA becomes higher. From the points of view of query types, the RWEA is appropriate for informational queries and the combined method is for navigational queries. For both query types, RWEA helps to find relevant information.

1 Introduction

The World Wide Web, a treasure house of knowledge, is widely used not only by professionals for supporting their research or business but also by non professionals in their daily life. In fact, it is very common for people these days to use a search engine before looking into dictionaries or textbooks when they want to know something.

However, search engines often fail to give results that users want. One of the reasons is that the keyword (or query) submitted by the user is so ambiguous for the search engine to grasp his/her intention. Accordingly the

Tetsuya Oishi · Tsunenori Mine · Ryuzo Hasegawa · Hiroshi Fujita ·
Miyuki Koshimura
Faculty of Information Science and Electrical Engineering, Kyushu University,
Motooka 744, Nishi-ku, Fukuoka-shi, Fukuoka-ken 819-0395 Japan
e-mail: {oishi,hasegawa,fujita,koshi}@ar.is.kyushu-u.ac.jp,
mine@al.is.kyushu-u.ac.jp

Q. Bai and N. Fukuta (Eds.): Advances in Practical Multi-Agent Systems, SCI 325, pp. 33–48.
springerlink.com

search engine returns enormous number of pages concerning the ambiguity that the user can hardly browse all of them. As a result, the user cannot identify his/her desired pages that may be scattered among a number of irrelevant pages.

One way to solve the above problems is to add more keywords to make the meaning of the query clear so that the number of retrieved pages can significantly be reduced. The more suitable keywords were added to the initial query, the more desired pages would be obtained. However, users may not always be able to give such keywords by themselves.

In this paper, we present an algorithm which extracts words from some relevant texts that are supposed to be strongly related to the initial query just given by the user without any extension. We call this algorithm the "Related Word Extraction Algorithm" (RWEA). Moreover, Robertson's Selection Value (RSV) [4], a well known method for relevance feedback is used, where each word is weighed by using our algorithm in place of those used in the original method, and query expansion is performed based on the results of each method. RWEA evaluates a word in a document and RSV evaluates a word among some documents. When RWEA and RSV is combined, a useful result may be obtained. In a specific condition, we may get another result. Then we compare RWEA, RSV, and the method combining RWEA and RSV.

The combined method works effectively for all queries on the average. Especially, when the user inputs such initial query whose Average Precision (AP) is under 0.6, this method obtains the highest Mean Average Precision (MAP) [5]. Focusing ambiguous queries [2] [4], RWEA obtains the highest MAP. For informational queries [1], RWEA obtains the highest MAP. However, for navigational queries [1], the combined method obtains the highest MAP. The experimental results show that the useful methods for the query expansion vary depending on the types of queries, though the RWEA always worked effectively.

Section 2 discusses some related work of the topics touched on in this paper. We explain our query expansion system in Section 3. In section 4, we present RWEA and RSV, and describe a method to calculate the importance value of a word to be added to a new query based on the algorithm. We discuss the experimental results in Section 5.

2 Related Work

There have been a lot of studies on the query expansion. Wang and Davison [13] built an auto-tagging system to assign tags to Web pages. The system used a support vector machine (SVM) for classifying tags. After successfully having constructed the auto-tagging system with good performance, they took advantage of it for suggesting query expansion to the user. The result pages of expanded queries created by the system were significantly better than those of the Google.

Oyama et al [11] made a domain-specific Web search engines that uses keyword spices for locating recipes, restaurants, and used cars. Here, the spices are the words to be added to the initial query. They accomplished Web retrieval effectively by specifying domain. Against their method, we expand a query without specifying domain. Nabeshima et al [8] made Web search engines which hardly depend on users. Our method depends on users, but we will create an automatic system such as Nabeshima's system in future.

Zighelnic and Kurland [14] addressed the query drift problem of query expansion methods caused by the pseudo relevance feedback. To remedy this, they fused the results retrieved in response to the original query and to its expanded form. Experimental results showed the validity of their approach.

Chirita et al [3] proposed the system that expands a query by using the user's desktop information in order to automatically extract additional keywords related to the query. They expanded queries using the Local Desktop Analysis, which is related to the Pseudo Relevance Feedback, and the Global Desktop Analysis, which relies on information from across the entire personal Desktop.

Okabe and Yamada proposed a system [6] that tries to improve the search efficiency by expanding the queries using the minimum feedback from the user. This system is realized with a machine learning method called the transductive learning. This method improves recall ratio by 0.07 compared to the conventional method without using transductive learning. Masada et al [7] sorts the original retrieved results by query expansion using RSV in which top-R pages of retrieved results are used as the relevant documents and the rest as non-relevant documents. They found a suitable pair of the parameters used by RSV that is useful to improve precision of the retrieved results.

Oishi et al [8] proposed the user-schedule-based Web page recommendation system. This system searches for a Web page that matches the user's situation by analyzing the user's schedule. It employs an algorithm similar to the related word extraction algorithm we discuss in this paper. When a user inputs a schedule with its title and details, the related words are extracted from them using the algorithm. The algorithm needs a set of keywords and a text, the former is extracted from the title of the schedule and the latter is just the details of the schedule. This system succeeded in presenting documents which fit the user's demand at higher rank in the retrieved results. The difference between the algorithm and the one we discuss here is that the former pays attention to the distance between words, whereas the latter to the distance between sentences. Here the distance means the textual distance.

3 Experimental Condition

We evaluate the result of the query expansion in our experiment. In this section, we explain the query expansion system. Additionally, we show the types of the queries used for the query expansion.

3.1 Query Expansion System

The outline of our system is as follows:

1. Input Query
   The user inputs a query into the system.
2. Decide Relevant Documents and Non-Relevant Documents
   The system performs Web search, and the user decides whether or not retrieved documents are relevant to the query.
3. Extract the Candidates for the Related Words
   The system extracts the candidates of the related words from the documents obtained in 2 by using a Morphological Analyzer MeCab[1]. Some written languages like Chinese, Japanese does not have single-word boundaries, so any significant text parsing usually requires the identification of word boundaries, which is often a non-trivial task. Mecab identifies word boundaries.
4. Calculate the Importance Value of the Candidates
   The system calculates the importance value of candidates as the related words extracted in 3 by using Related Word Extraction Algorithm (RWEA) and Robertson's Selection Value (RSV). RWEA pays attention to the distance between sentences. We explain it in detail in Section 4.1. RSV pays attention to the frequency of relevant documents and non-relevant documents. We explain it in Section 4.2.
5. Generate the Expanded Query
   The system generates an expanded query based on the importance value calculated with two methods in 4. The expanded query is the one expanded based on the original query the user gave. For example, if the original query is "A" and "B", then the expanded query "A", "B", and "C" is obtained by adding "C" to the original one.
6. Show the Retrieved Results
   The system shows the retrieved result by using the expanded query.
   The system submits the expanded query, which is obtained in 5, to the search engine and shows the retrieved results to the user.

3.2 Types of Queries

Three evaluators conducted Web retrieval with the same 150 queries in Japanese. Thus 450 queries were totally used. These are original queries we created. These queries are evenly divided into 50 one-word queries, 50 two-word queries, and 50 three-word queries.

[1] "Yet Another Part-of-Speech and Morphological Analyzer: MeCab", http://mecab.sourceforge.jp/

Chirita et al. [2] [4] classified the queries into three types: the clear queries ($P@10 \geq 70$)[2], the semi-ambiguous queries ($20 \leq P@10 < 70$) and the ambiguous queries ($P@10 < 20$). From the points of view of ambiguity of the queries, these are divided into 27 clear queries, 48 semi-ambiguous queries, and 75 ambiguous queries.

Broader [1] classified the queries into three types (informational, navigational, and transactional query) based on the user's needs. The intent of the informational query is to acquire some information assumed to be present on one or more web pages. The intent of the navigational query is to reach a particular site that the user has in mind. The intent of the transactional query is to perform some web-mediated activity, such as shopping, downloading various type of file, accessing certain database, finding services and so on. From the user intent behind queries, these are divided into 99 informational queries and 51 navigational queries.

4 Score of Similarity between Related Words and User's Intent

In this section, we describe RWEA and RSV methods to calculate the score of the related words.

4.1 RWEA

The algorithm extracts, from the text T, a group of words that are related to a given set of keywords K. The extracted words are thus expected to be relevant to K. Based on the textual distance from a keyword in K, we evaluate each word in T, and output the words ranked according to the evaluation.

Importance of the distance between sentences
We regard the distance between sentences as important when considering relevance of words. Moreover we pay attention to the order of the sentences that contain some specific word (for example, the query word used for Web retrieval, the word in the title of schedule, and so on) appearing in a document. Our algorithm is based on the following idea: "If a certain word A in a certain sentence is an important word for the user, the word nearby this sentence is also important."
Preparations
We define some symbols as follows:

- K: a set of keywords that give a basis for extracting related words.
- T: a text where some keywords related to K are extracted.

[2] This formula means that the precision at 10 retrieved documents is larger than 70%.

Table 1 (Top) An example of scoring sentences using distance; (Bottom) An example of calculating $EBV(t_j)$ using the expected value $EBV(j)$ at occurrence position j

keyword K	A B				
t_j in text T	A G B	E G	A F C	F H	D E
$BV_A(t_j)$	5	4	3	2	1
$BV_B(t_j)$	5	4	3	2	1
$BV_A(t_j)$	3	4	5	4	3
$BV(t_j)$	13	12	11	8	5
$EBV(j)$	3	3.6	3.8	3.6	3
$EBV(t_j)$	4.33	3.33	2.89	2.22	1.67

- $k_i(i = 1, \cdots, m)$: m keywords appearing in K, where the keywords $k_1, k_2, \cdots, k_m$ appear in this order.
- $t_j(j = 1, \cdots, n)$: n sentences appearing in T, where the sentences $t_1, t_2, \cdots, t_n$ appear in this order.
- $w_l(l = 1, \cdots, o)$: o words appearing in T, where the words $w_1, w_2, \cdots, w_o$ appear in this order.
- $c_p(p = 1, \cdots, q)$: q words appearing in T, where the words $c_1, c_2, \cdots, c_q$ are different from each other.

If a word appears more than once, each occurrence of the word is considered to be distinct. w_l is actually a noun which is extracted from a text T by MeCab. t_j is a sequence of nouns that appear in T.

4.1.1 Score of Word

Score of words in the text T is calculated as follows:

1. Calculate basic value $BV(t_j)$ of $t_j(j = 1, \cdots, n)$ with $k_i(i = 1, \cdots, m)$ as a criterion.
2. Flatten $BV(t_j)$.
3. Calculate a final value $S(c_p)$ using word frequeincies.

Calculating $BV(t_j)$(see the top of Table 1)
We define $BV_{k_i}(t_j)$ as the score of t_j with respect to k_i.
After having scored every $BV_{k_i}(t_j)$ $(i = 1, \cdots, m$ and $j = 1, \cdots, n)$, the algorithm calculates $BV(t_j)$.
Flattening $BV(t_j)$(see the bottom of Table 1)
The above way of determining $BV(t_j)$ is unfair when considering the position of sentence t_j in T. The expected values of $BV(t_j)$ differ depending on the position j at which t_j occurs.
To remedy this problem, we flatten the obtained
$BV(t_j)$ using $EBV(j)$, the expected evaluation value of t_j at position j.

Table 2 (Top)A process of calculating $S(c_p)$; (Bottom) $S(c_p)$ of each words

keyword K	A B											
t_j in text T	A	G	B	E	G	A	F	C	F	H	D	E
$WEBV(w_l)$	4.33	4.33	4.33	3.33	3.33	2.89	2.89	2.89	2.22	2.22	1.67	1.67
$AveEBV(c_p)$	3.61	3.83	4.33	2.50	3.83	3.61	2.56	2.89	2.56	2.22	1.67	2.50
$tf(c_p)$	2	2	1	2	2	2	2	1	2	1	1	2
$V(c_p)$	1.28	1.28	1	1.28	1.28	1.28	1.28	1	1.28	1	1	1.28
$S(c_p)$	4.62	4.90	4.33	3.19	4.90	4.62	3.27	2.89	3.27	2.22	1.67	3.19

c_p	A	B	C	D	E	F	G	H
$S(c_p)$	4.62	4.33	2.89	1.67	3.19	3.27	4.90	2.22

Let $EBV(t_j)$ be the value evaluated after flattening. It is calculated using the following formula;
$EBV(t_j) = BV(t_j)/EBV(j)$.
Calculating $S(c_p)$(see Table 2)
In addition to the evaluation based on the distance between sentences, we take into consideration the concept of *Term Frequency* that is often used in the TF/IDF method. First, we compute the average $AveEBV(c_p)$ of the expected evaluation values $WEBV(w_l)$, for word w_l that appears several times. $WEBV(w_l) = EBV(t_j)$ when w_l in t_j.
Moreover, we compute the weight $V(c_p)$ using the tf value of word c_p as $V(c_p) = 1 + \{tf(c_p)/o\} \log tf(c_p)$.
The parameters used in the above formula are as follows. o: the total number of occurrences of words in T (mentioned above), and $tf(c_p)$: the number of occurrences of c_p in T.
Using $AveEBV(c_p)$ and $V(c_p)$, we calculate the evaluation value $S(c_p)$ of c_p in T as $S(c_p) = AveEBV(c_p) \times V(c_p)$. We assume that $S(c_p)$ is the evaluation value of word c_p in text T in this algorithm.

4.2 RSV

RSV gives higher weight to the word w which only appears in the relevant documents and does not appear in non-relevant documents than others.

RSV_w is calculated as follows:

$$RSV_w = \left(\frac{df_w^+}{R^+} - \frac{df_w^+ + df_w^-}{R^+ + R^-}\right)\{\alpha \log \frac{R^+ + R^-}{df_w^+ + df_w^-} + (1-\alpha) \log \frac{(df_w^+ + 0.5)/(R^+ - df_w^+ + 0.5)}{(df_w^- + 0.5)/(R^- - df_w^- + 0.5)}\} \quad (1)$$

The parameters used in formula (1) are as follows.

- R^+: the number of relevant documents
- df_w^+: the number of relevant documents with word w

- R^-: the number of non-relevant documents
- df_w^-: the number of non-relevant documents with w
- α: the control parameter ($0 \leq \alpha \leq 1$)

The constant 0.5 in the formula (1) is to get a reasonable value of natural logarithm. In addition, used log is a natural logarithm.

4.3 The Method Combining RWEA and RSV

Let RSV_w be the score of the word w obtained by RSV. RSV requires several documents (we use 10 documents in our work) to calculate RSV_w and since RSV_w is independent to the position of w in the document, the words weighted by RSV_w sometimes do not appropriate in a new query. To remedy these problems, we use the RWEA. Let $S(w)$ be the evaluation value of the word w obtained by RWEA. The score of w obtained by the method combining RWEA and RSV is calculated as $Score(w) = S(w) \times RSV_w$.

We assume that as a word appears at a shorter distance with query words in retrieved documents, the word becomes more relevant to the query and obtains higher level of importance to the query, even if there are only a few documents to calculate RSV_w.

5 Experiment

In this section, we explain our experiments. The setting of our experiments is described in Section 5.1. In Section 3, we show the results of our experiments.

5.1 Experimental Methods

First three evaluators evaluate the results retrieved by the search engine "goo"[3]. Secondly, they evaluate the results retrieved by the expanded queries using RSV, RWEA, and the method that combines RSV and RWEA. The details are as follows:

5.1.1 Web Retrieval Using "goo"

1. Evaluate the Web pages
   The top-20 Web pages are obtained by submitting a query to the search engine "goo". The evaluators mark each page as follows.

   Very Good : when the pages include the information that is strongly relevant to the query and easily found by the evaluators.
   Good : when the pages include the information relevant to the query.

[3] http://www.goo.ne.jp/ goo partly uses the engine of Google.

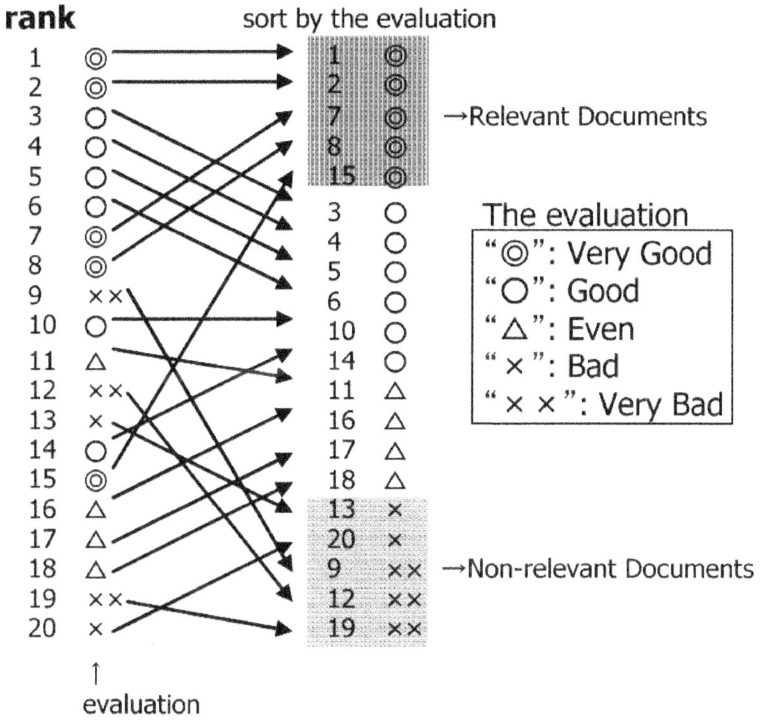

Fig. 1 Sorting the Web pages according to the order of the user's evaluation

Even : when the pages include the information partially relevant to the query.
Bad : when the pages do not include the information relevant to the query.
Very Bad : when the pages include the information completely irrelevant to the query.

2. Sort the Web pages
   The system sorts the Web pages according to the order of the user's evaluation by using the explicit feedback. The explicit feedback is the method performed based on the user's evaluation. For pages which have the same evaluation, the order is preserved as in the original rank. Figure 1 shows how the system sorts the evaluated Web pages according to the order of evaluation value.
3. Decide the relevant documents and non-relevant documents
   Here we decide the relevant and non-relevant documents by using the pseudo or explicit feedback. The pseudo feedback means that we take top-5 Web pages as relevant documents and bottom-5 Web pages as non-relevant from the unsorted Web pages. The explicit feedback means that

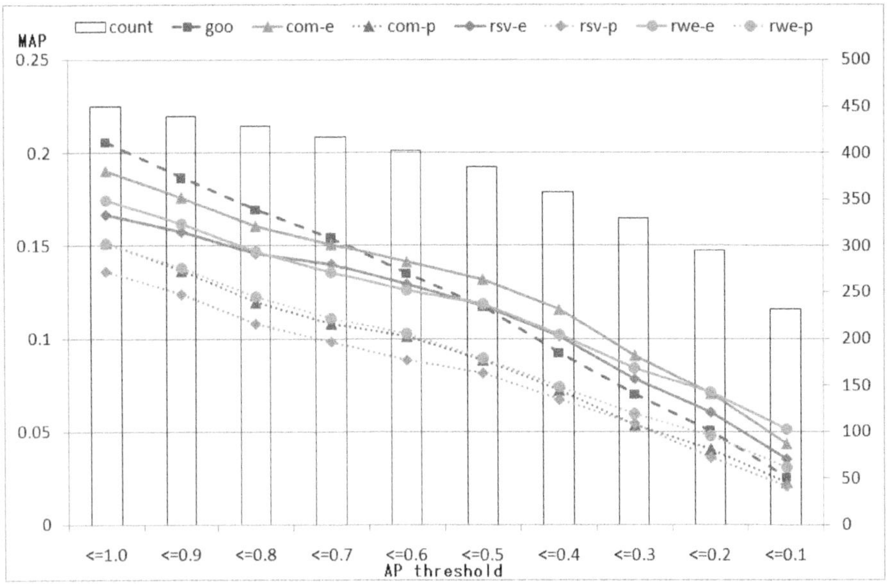

Fig. 2 The Results by the Original Rank.

we use the Web pages sorted based on the user's evaluation to decide the relevant and non-relevant Web pages.

5.1.2 Query Expansion

We apply each of three methods; RSV, RWEA and Combined RSV and RWEA to the relevant documents and non-relevant documents, and then we expand the queries used in Section 5.1.1. The expanded query is made by adding to original query the word which has the highest importance value by using each method as follows.

1. RSV
   We apply the RSV to the relevant documents and non-relevant documents.
2. RWEA
   We apply RWEA to the relevant documents. Although RWEA is basically applied to a single document, we use it for more than one document as follows. Let $d_k (k = 1, 2, \cdots, n)$ be the relevant documents and $S(w_{d_k})(k = 1, 2, \cdots, n)$ be importance values of word w in each document, where $S(w_{d_k})$ is given by RWEA. Then the importance value $S(w)$ of word w is given as follows:

$$S(w) = \frac{\sum_{k=1}^{n} S(w_{d_k})}{n} \tag{2}$$

3. The Method Combining RSV and RWEA
   We first apply RWEA only to the relevant documents, and then apply RSV to both the relevant documents and non-relevant documents.

5.2 Result

We compare the three methods for the query expansion by calculating MAP (Mean Average Precision). We assume that the number of all correct answers for calculating MAP is 20 because the number of all retrieved results for each query is 20. In other words, the number of all correct answers is not bigger than 20.

First, we describe a general trend of the methods for all the queries (in 5.2.1). Next, we evaluate the methods based on the types of the queries (in 5.2.2).

5.2.1 All Queries

In figure 2, we show MAP of each method for all queries. Here, for calculating MAP, we adopted only the documents marked "Very Good" as the relevant ones. In this Figure, the left vertical axis is MAP, the right vertical axis is the number of queries, and the horizontal axis is the threshold of AP. For example, when the threshold of AP is "<=0.8", we plot MAP for each method with all queries for which AP of "goo" is smaller than or equal to 0.8 in the figures. The explanatory notes "goo", "com", "rsv", and "rwe" present the MAP of the retrieved results with the query expanded by using the search engine "goo", the combined method, RSV, and RWEA respectively. Then indices after the notes; "e" and "p" present the explicit and pseudo feedback respectively.

In all the methods, the results of query expansion with the explicit feedback are better than those with the pseudo feedback. Moreover, the query expansion with the pseudo feedback shows worse results than that with the initial query ("goo"). However, it should be noted that the explicit feedback imposes unpleasant chore to the user. We will reduce the chore for the feedback in the future. Furthermore, MAP of the combined method with the explicit feedback ("com-e") is always higher than MAP of "goo" with the queries whose AP is lower than 0.6. This means that the query expansion by "com-e" works better than "goo" when a good search is not obtained by the initial query.

We re-ranked each retrieved result using Rank Freezing (RF) [6]. RF re-ranks the results retrieved with the expanded query by fixing the rank of Web pages obtained by the initial query. RF is the method that the results retrieved by the initial query are important. Figure 3 shows MAP with the results re-ranked by RF as well as Figure 2. We explain a difference between Figure 2 and Figure 3 by using Table 3 because these figures look very similar. Table 3 shows the MAP values we plotted in Figure 2 and those in Figure 3,

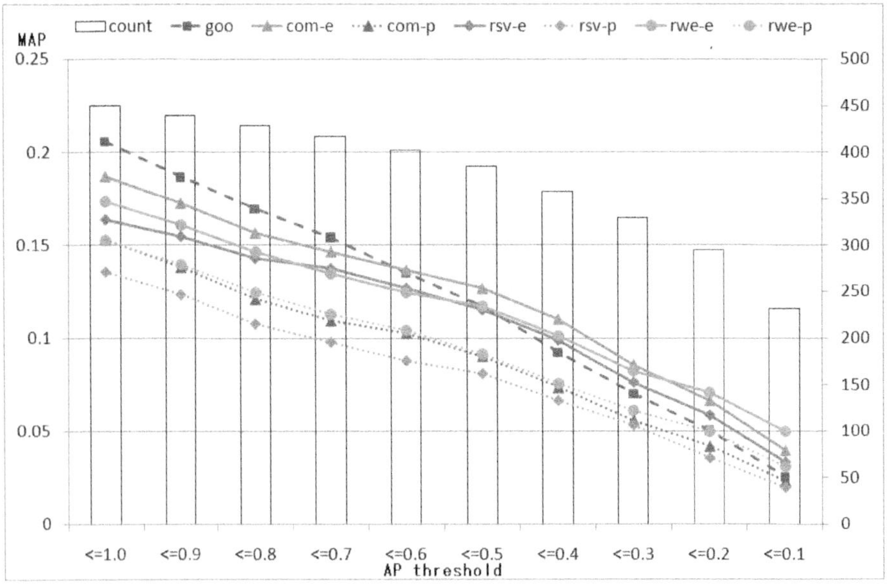

Fig. 3 The Results by the Rank Freezing.

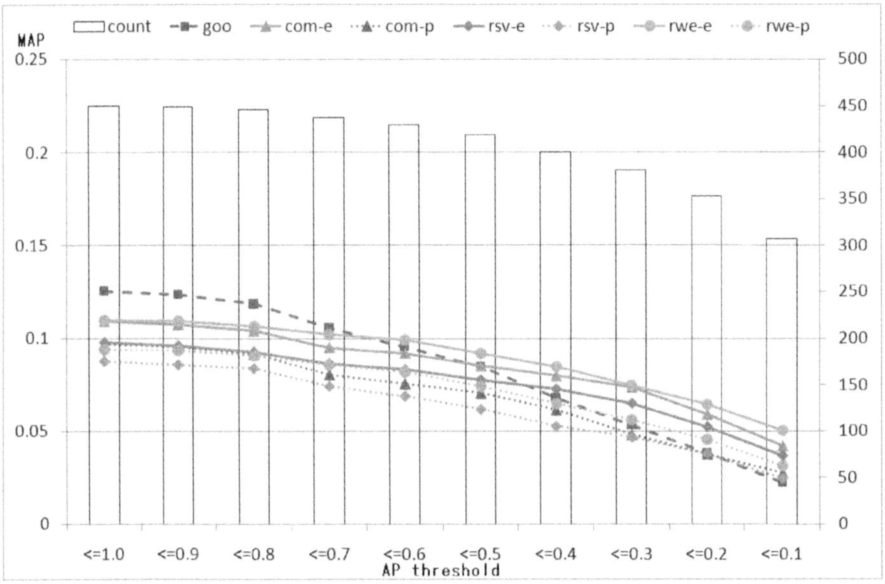

Fig. 4 The Results by the Residual Collection.

Table 3 MAP in Figure 2 and Figure 3

AP threshold (# of queries)		Initial Query	Expanded Query											
			com-e		com-p		rsv-e		rsv-p		rwe-e		rwe-p	
		goo	ori	rf	ori	rf	ori	rf	ori	rf	ori	rf	ori	rf
$AP \leq 1.0$	MAP	0.206	0.190	0.187	0.151	0.153	0.167	0.164	0.136	0.136	0.174	0.174	0.151	0.153
(450)	Ratio	-	1.017		0.989		1.018		1.002		1.005		0.990	
$AP \leq 0.9$	MAP	0.187	0.176	0.172	0.136	0.138	0.157	0.154	0.124	0.124	0.162	0.161	0.138	0.140
(439)	Ratio	-	1.020		0.987		1.019		1.001		1.004		0.988	
$AP \leq 0.8$	MAP	0.169	0.160	0.156	0.119	0.121	0.146	0.143	0.108	0.108	0.147	0.146	0.123	0.124
(428)	Ratio	-	1.025		0.986		1.020		1.000		1.004		0.985	
$AP \leq 0.7$	MAP	0.154	0.151	0.146	0.108	0.110	0.140	0.137	0.098	0.098	0.135	0.135	0.111	0.113
(417)	Ratio	-	1.029		0.986		1.020		1.003		1.005		0.985	
$AP \leq 0.6$	MAP	0.135	0.141	0.137	0.101	0.103	0.129	0.127	0.088	0.088	0.126	0.125	0.103	0.104
(402)	Ratio	-	1.033		0.981		1.019		1.004		1.010		0.984	
$AP \leq 0.5$	MAP	0.117	0.132	0.127	0.089	0.090	0.118	0.115	0.081	0.081	0.119	0.117	0.090	0.092
(385)	Ratio	-	1.040		0.979		1.023		1.003		1.011		0.980	
$AP \leq 0.4$	MAP	0.092	0.115	0.110	0.072	0.074	0.101	0.099	0.067	0.067	0.102	0.101	0.074	0.076
(358)	Ratio	-	1.049		0.980		1.024		1.005		1.010		0.979	
$AP \leq 0.3$	MAP	0.070	0.091	0.086	0.054	0.056	0.078	0.076	0.054	0.054	0.084	0.083	0.060	0.061
(329)	Ratio	-	1.059		0.956		1.028		1.013		1.021		0.980	
$AP \leq 0.2$	MAP	0.050	0.070	0.066	0.041	0.042	0.061	0.059	0.036	0.036	0.071	0.071	0.048	0.050
(295)	Ratio	-	1.063		0.966		1.030		1.011		1.004		0.963	
$AP \leq 0.1$	MAP	0.026	0.044	0.040	0.023	0.023	0.036	0.034	0.021	0.020	0.051	0.050	0.031	0.031
(232)	Ratio	-	1.099		0.967		1.058		1.033		1.026		1.001	

In this table, "Ratio" is defined as Ratio = (MAP of "ori") / (MAP of "rf")
"ori" is MAP for the original rank and "rf" is MAP for the rank by the Rank Freezing

and the ratio between the MAP values. The ratio for the method using the explicit feedback ("com-e", "rsv-e", and "rwe-e") is larger than 1.0. In other words, applying RF to the original rank makes MAP worse.

We re-ranked each retrieved result using Residual Collection (RC) [6]. RC re-ranks the results retrieved with the expanded query except for the Web pages retrieved with the query before the expansion. RC is the method that the newly retrieved results are important. In other words, we calculate MAP only for the Web pages newly retrieved with the expanded query. Figure 4 shows MAP with the results re-ranked by RC as well as Figure 2. We can see that "rwe-e" always shows the highest MAP among the query expansion methods. This shows that the expanded query created by RWEA can get new useful Web pages.

5.2.2 Types of Queries

We compare the results according to the classification of Chirita et al. Table 4(1) shows that the query expansion effectively works for the ambiguous queries. In particular, RWEA is the most effective. In the rest of this section, we argue about the ambiguous queries.

RWEA effectively works for one-word and three-word query, and the combined method works well only for two-word query. This is because the number of results retrieved by one word query was too large and that by three word query was too small to get appropriate documents for using the combined method.

In this paper, all queries are divided into informational (Table 4(3-1)) and navigational (Table 4(3-2)) queries. RWEA works effectively for informational queries and the combined method works well for navigational queries. In other words, when the intended page is not defined clearly, RWEA works effectively. Otherwise, the combined method works well.

Table 4 Retrieved Results classified by Query Type

(1)ALL Query

Queries (# of queries)		Initial Query goo	Expanded Query com-e	com-p	rsv-e	rsv-p	rwe-e	rwe-p
clear query (81)	MAP	0.664	0.504	0.474	0.435	0.428	0.465	0.472
	Improvement	-	0.759	0.713	0.656	0.644	0.700	0.711
semi-ambiguous query (144)	MAP	0.229	0.236	0.169	0.218	0.150	0.199	0.158
	Improvement	-	**1.031**	0.737	0.951	0.654	0.867	0.691
ambiguous query (225)	MAP	0.026	0.048	0.024	0.037	0.022	0.054	0.031
	Improvement	-	**1.851**	0.947	**1.425**	0.860	**2.111**	**1.210**

(2-1)One-Word Query

Queries (# of queries)		Initial Query goo	Expanded Query com-e	com-p	rsv-e	rsv-p	rwe-e	rwe-p
clear query (10)	MAP	0.712	0.444	0.494	0.416	0.466	0.352	0.337
	Improvement	-	0.623	0.694	0.584	0.654	0.495	0.474
semi-ambiguous query (16)	MAP	0.197	0.091	0.130	0.094	0.094	0.154	0.049
	Improvement	-	0.464	0.660	0.478	0.476	0.781	0.248
ambiguous query (124)	MAP	0.012	0.030	0.011	0.020	0.008	0.032	0.016
	Improvement	-	**2.452**	0.895	**1.600**	0.679	**2.592**	**1.284**

(2-2)Two-Word Query

Queries (# of queries)		Initial Query goo	Expanded Query com-e	com-p	rsv-e	rsv-p	rwe-e	rwe-p
clear query (34)	MAP	0.649	0.483	0.488	0.414	0.438	0.453	0.471
	Improvement	-	0.744	0.752	0.638	0.675	0.698	0.725
semi-ambiguous query (65)	MAP	0.232	0.251	0.163	0.225	0.135	0.217	0.197
	Improvement	-	**1.082**	0.702	0.970	0.583	0.935	0.848
ambiguous query (51)	MAP	0.054	0.099	0.052	0.082	0.048	0.092	0.048
	Improvement	-	**1.828**	0.954	**1.514**	0.885	**1.696**	0.878

(2-3)Three-Word Query

Queries (# of queries)		Initial Query goo	Expanded Query com-e	com-p	rsv-e	rsv-p	rwe-e	rwe-p
clear query (37)	MAP	0.664	0.540	0.455	0.460	0.408	0.505	0.510
	Improvement	-	0.812	0.684	0.692	0.615	0.761	0.768
semi-ambiguous query (63)	MAP	0.234	0.257	0.184	0.242	0.179	0.191	0.146
	Improvement	-	**1.099**	0.788	**1.034**	0.765	0.815	0.625
ambiguous query (50)	MAP	0.031	0.040	0.030	0.033	0.030	0.073	0.053
	Improvement	-	**1.297**	0.986	**1.089**	0.996	**2.385**	**1.738**

(3-1)Informational Query

Queries (# of queries)		Initial Query goo	Expanded Query com-e	com-p	rsv-e	rsv-p	rwe-e	rwe-p
clear query (56)	MAP	0.700	0.543	0.500	0.459	0.461	0.494	0.505
	Improvement	-	0.776	0.715	0.655	0.658	0.705	0.721
semi-ambiguous query (96)	MAP	0.220	0.216	0.143	0.210	0.151	0.181	0.150
	Improvement	-	0.982	0.649	0.954	0.687	0.822	0.681
ambiguous query (145)	MAP	0.026	0.045	0.025	0.036	0.026	0.060	0.033
	Improvement	-	**1.754**	0.979	**1.387**	0.991	**2.333**	**1.265**

(3-2)Navigational Query

Queries (# of queries)		Initial Query goo	Expanded Query com-e	com-p	rsv-e	rsv-p	rwe-e	rwe-p
clear query (25)	MAP	0.584	0.416	0.413	0.383	0.354	0.400	0.400
	Improvement	-	0.713	0.708	0.656	0.607	0.686	0.686
semi-ambiguous query (48)	MAP	0.247	0.276	0.220	0.234	0.147	0.233	0.174
	Improvement	-	**1.118**	0.892	0.947	0.596	0.945	0.707
ambiguous query (80)	MAP	0.026	0.052	0.023	0.039	0.016	0.044	0.029
	Improvement	-	**2.026**	0.889	**1.493**	0.624	**1.711**	**1.111**

In this table, "Improvement" is defined as follows:
Improvement = (MAP of Expanded Query) / (MAP of Initial Query)

6 Conclusion

In this paper, we have discussed three types of query expansion methods; the Related Word Extraction Algorithm (RWEA), Robertson's Selection Value (RSV) and the method combining RWEA and RSV considering several types of queries.

The explicit feedback worked better than the pseudo feedback. For all queries, the combined method works best on the average.

According to Chirita's classification, the MAP of RWEA for the ambiguous queries was the highest. Based on the classification by the number of words in a query, RWEA effectively worked for one-word and three-word queries, and the combined method worked well for two-word queries. Finally, considering the Broder's classification, RWEA got the best score for the informational queries, while the combined method worked best for the navigational queries.

The results show that the best query expansion method depends on the type of queries. Therefore, we (or system) should select an appropriate expansion method according to the type of query.

Although we obtained better results for ambiguous or semi-ambiguous queries with explicit feedback than those returned by search engine "goo" whose main engine is "Google", the explicit feedback imposes unpleasant chore to the user. Therefore the future work is to reduce such chore for the feedback.

Acknowledgements. We would like to thank Koji Kurakado, Yuichi Tashiro, and Toru Nakamura of the Graduate School of Information Science and Electrical Engineering, Kyushu University for their cooperation in the experiments.

References

1. Broder, A.: A taxonomy of web search. SIGIR Forum 36(2), 3–10 (2002), http://doi.acm.org/10.1145/792550.792552
2. Chirita, P.A., Firan, C.S., Nejdl, W.: Summarizing local context to personalize global web search. In: CIKM 2006: Proceedings of the 15th ACM international conference on Information and knowledge management, pp. 287–296. ACM, New York (2006), http://doi.acm.org/10.1145/1183614.1183658
3. Chirita, P.A., Firan, C.S., Nejdl, W.: Personalized query expansion for the web. In: SIGIR 2007: Proceedings of the 30th annual international ACM SIGIR conference on Research and development in information retrieval, pp. 7–14. ACM, New York (2007), http://doi.acm.org/10.1145/1277741.1277746
4. Chirita, P.A., Nejdl, W., Paiu, R., Kohlschütter, C.: Using odp metadata to personalize search. In: SIGIR 2005: Proceedings of the 28th annual international ACM SIGIR conference on Research and development in information retrieval, pp. 178–185. ACM, New York (2005), http://doi.acm.org/10.1145/1076034.1076067

5. Cormack, G.V., Lynam, T.R.: Statistical precision of information retrieval evaluation. In: SIGIR 2006: Proceedings of the 29th annual international ACM SIGIR conference on Research and development in information retrieval, pp. 533–540. ACM, New York (2006), http://doi.acm.org/10.1145/1148170.1148262
6. Hull, D.: Using statistical testing in the evaluation of retrieval experiments. In: SIGIR 1993: Proceedings of the 16th annual international ACM SIGIR conference on Research and development in information retrieval, pp. 329–338. ACM, New York (1993), http://doi.acm.org/10.1145/160688.160758
7. Masada, T.: Improving web search by query expansion with a small number of terms. In: NTCIR-5 Workshop Meeting 2005 (2005), http://ci.nii.ac.jp/naid/10021846115/
8. Nabeshima, H., Miyagawa, R., Suzuki, Y., Iwanuma, K.: Rapid synthesis of domain-specific web search engines based on semi-automatic training-example generation. In: Proceedings of IEEE/WIC/ACM International Conference on Web Intelligece (WI 2006), pp. 769–772 (2006)
9. Oishi, T., Kuramoto, S., Nagata, H., Mine, T., Hasegawa, R., Fujita, H., Koshimura, M.: User-schedule-based web page recommendation. In: The 2007 IEEE/WIC/ACM International Conference on Web Intelligence, Silicon Valley, USA, pp. 776–779 (2007)
10. Okabe, M., Yamada, S.: Query expansion with the minimum user feedback by transductive learning. Transactions of the Japanese Society for Artificial Intelligence 21(4), 398–405 (2006), http://ci.nii.ac.jp/naid/10022006623/, doi:10.1527/tjsai.21.398
11. Oyama, S., Kokubo, T., Ishida, T.: Domain specific search with keyword spices. IEEE Transactions on Knowledge and Data Engineering 16(1), 17–27 (2004)
12. Robertson, S.E.: On term selection for query expansion. Journal of Documentation 46(4), 359–364 (1990), http://ci.nii.ac.jp/naid/10020789082/
13. Wang, J., Davison, B.D.: Explorations in tag suggestion and query expansion. In: SSM 2008: Proceeding of the 2008 ACM workshop on Search in social media, pp. 43–50. ACM, New York (2008), http://doi.acm.org/10.1145/1458583.1458592
14. Zighelnic, L., Kurland, O.: Query-drift prevention for robust query expansion. In: SIGIR 2008: Proceedings of the 31st annual international ACM SIGIR conference on Research and development in information retrieval, pp. 825–826. ACM, New York (2008), http://doi.acm.org/10.1145/1390334.1390524

Verification of the Effect of Introducing an Agent in a Prediction Market

Takuya Yamamoto and Takayuki Ito

Abstract. In recent years, attention to "prediction markets", which make predictions of the future using market mechanisms, has been increasing. A prediction market applies techniques of experimental markets that have been used in the field of experimental economics to make predictions. A prediction market is a market in which participants make trades of securities predicting the result of a certain event that will be decided in the future. A security provides a dividend based on the result of the event, and the price of the security serves as predictor of the event's realization probability. A participant predicts the event's result from various sources of information related to the target phenomenon, and he trades to gain profits from his predictions. From the market price of the result we can suppose think that all the members predictions have been unified. Some of the prediction markets in the U.S. are the Iowa Electronic Market (IEM) and the Hollywood Stock Exchange (HSX). Some of the prediction markets in Japan are at sites such as Shuugi.in and Kounaru. The IEM prediction market in the United States has been effective in predicting election outcomes. It has correctly predicted 75% of the results of elections traded on its exchange, a success rate that compares favorably with that of opinion polls. Thus, in this research, we studied what kind of influence the use of an agent had on a prediction market. For example, we studied how much influence an agent would have on predictive accuracy through an increase in trading volume.

1 Introduction

Attention to "prediction markets", which make predictions of the future using market mechanisms has been increasing. A prediction market applies

Takuya Yamamoto · Takayuki Ito
Nagoya Institute of Technology, Gokiso, Showa-ku, Nagoya 466-8555, Japan
e-mail: {yamamoto,ito}@itolab.mta.nitech.ac.jp

Q. Bai and N. Fukuta (Eds.): Advances in Practical Multi-Agent Systems, SCI 325, pp. 49–62.
springerlink.com

techniques of experimental markets that have been used in the field of experimental economics to make predictions.

A prediction market is a market in which participants make trades of securities predicting the result of a certain event that will be decided in the future. A security provides a dividend based on the result of the event. The price of the security serves to predict the event's realization probability. Prediction markets include markets of various types.

1. Type1
   In a game of baseball, it is assumed that Stock A corresponds to "Win by Team A" and Stock B corresponds to "Win by Team B". As a result, when Team A wins, there is a dividend of 100 cents per share of Stock A, and Stock B has a dividend of 0 cents.
2. Type2
   In an election, Stock A corresponds to "the number seats acquired by political party A" and Stock B correspond to "the number of seats acquired by political party B", and Stock C corresponds to "the number of seats acquired by political party C". The price per share of stock X and the dividend of the number of seats acquired by the political party X (cent) are determined as the result of an election.

A participant predicts a result from various sources of information related to the target phenomenon, and he trades to gain profits from his predictions. The market price of the result unifies all the members' predictions. When the market price is Type 1, the price predicts the likelihood that Team A will win. When the market price is Type 2, it predicts the likely number of seats that will be acquired by political party X.

Some prediction markets in the U.S. are the Iowa Electronic Market (IEM)[1] and the Hollywood Stock Exchange (HSX)[2]. Some of the prediction markets in Japan are at sites such as Shuugi.in[3] and Kounaru[4]. The IEM prediction market in the United States has been effective in predicting election outcomes. It has correctly predicted 75% of the results of elections traded on its exchange, a success rate that compares favorably with that of opinion polls. An example of previous research in Japan on prediction markets was an experiment by Suginoo [10] as to "whether the viewership of the Winter Sonata of televising would exceed 17% on August 13, 2005". However, in this study the problem of reconciling completely different opinions was not resolved through a prediction market. There were not enough transactions conducted, and poor results in a market, such as described in the above research, will dampen participants' motivation.

Thus, in this research, we studied what kind of influence the use of an agent had on a prediction market. For example, we studied how much influence

[1] http://www.biz.uiowa.edu/iem/index.cfm
[2] http://www.hsx.com/
[3] http://shuugi.in/
[4] http://kouna.ru/

an agent would have on predictive accuracy through an increase in trading volume.

The rest of this paper consists of the following four sections. Section 2 discusses prediction markets and collective intelligence. Section 3 considers current research and problems. Section 4 describes the experiment conducted in this research and discusses its results. Finally, Section 5 summarizes the paper.

2 Research Background

2.1 Prediction Market

A prediction market is a futures market for carrying out predictions in the future. The economist Friedrich A. Hayek [4] asserted, "The price in a free market is determined by the mechanism of communicating information about the possibility created of a prospective phenomenon". A prediction market is something made based on this opinion, and various prediction realization probabilities are reflected in its market transaction price. The opinion of the crowd about a phenomenon will be collected in one number in the future using the mechanism of a futures market and make the number a "predicted value". Since the structure of a futures market is used, the predicted value may change in real time, and may change greatly, depending on external information such as news events.

2.2 Structure of a Predicted Value

Suppose that a prediction market participant buys the claim (virtual contract) "100 yen will be gained if tomorrow's Nikkei stock average closing price is 13,000 yen or more". What would be an appropriate price then for this claim? If it is certain the Nikkei will close at 13,000 yen or more, the price of the virtual contract would be worth a maximum of 100 yen, the amount to be paid out if the Nikkei closes at 13000 yen or more. However, if it turns out that the possibility of the Nikkei's closing at 13,000 yen or more is much less certain, then the price of the virtual contract will be zero. That is, the price of this virtual contract can be expressed in terms of expected values, according to the occurrence probability of the phenomenon.

Expected value (E) = 100(yen)(Occurrence probability of the stock price of 13000 yen or more)*

If this becomes the case, then, to buy a right for himself as well, Participant A will to raise the proposed price and compete with participant B as a rival. On the other hand, if there is a participant C who thinks that the probability of the Nikkei of reaching 13,000 or more is lower, he will submit a lower bid and

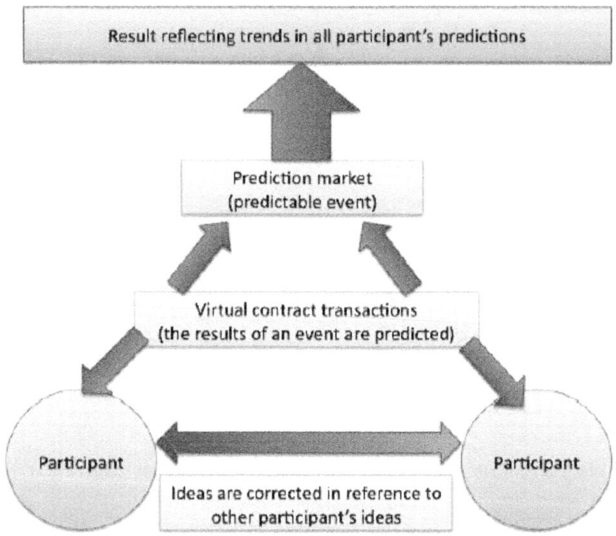

Fig. 1 Concentration of prediction

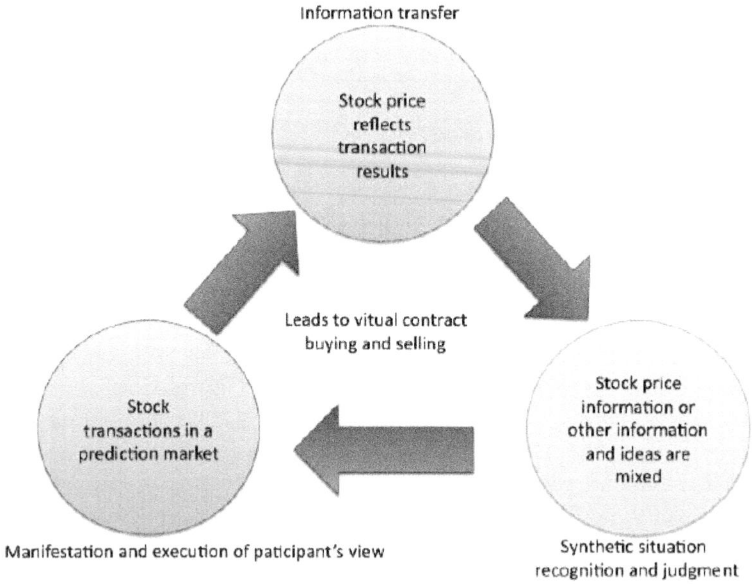

Fig. 2 Change of prediction

the price will fall (Table 1). Thus, according to the expectations (predictions) of the participants based on various kinds of information they have, deals will be made on prediction markets, and the price of bids submitted on these markets will fluctuate (Fig. 1 and Fig. 2).

Table 1 Structure of how a prediction results from participants' consensus

One's predicted value and the stock price of a virtual contract	A market participant's action	Change of the stock price in a market	The result predicted from participants' consensus
One's prediction > The present stock price	A virtual contract is bought	The price rises	Occurrence probability increases
One's prediction < The present stock price	A virtual contract is sold (It sells short)	The price falls	Occurrence probability falls
One's prediction = The present stock price	Nothing is done	Nothing is done	Nothing is done

The manager of a prediction market evaluates predictions by purchasing all virtual contracts according to the actual Nikkei stock average closing price after the defined end of a transactions period. The sum total of the trading difference at this time serves as the participants' profits.

Since the market is designed so that a profit results if a prediction is right, a participant focuses on making the most exact possible prediction. In this way, the regular price can be said to be that of the predicted value of the occurrence probability as arising from a consensus among participants in their total transactions. That is, the price of virtual contracts becomes synonymous at 60 yen with the market participants predict that there is a probability of 60% that a certain phenomenon will occur. This case is the expected value based on the occurrence probability of a phenomenon in the price of virtual contracts. There are also techniques that use concrete numerical values, such as the box-office receipts of a movie, as an event to be predicted value and traded on.

2.3 Comparison of the Prediction Method

Suginoo used a matrix to compare prediction markets with other prediction methods. He evaluated five prediction methodsC prediction markets, statistical analysis, questionnaire (choice), questionnaire (free answer), and the Delphi method according to four factorsC synthetic judgment, incentives, cost, and interaction. Table 2 shows this evaluation matrix.

- Synthetic judgment
  Future anticipation appears in the numerical value of the realization probability of the market price in a prediction market. Moreover, prediction markets can respond if futures contracts are made for various themes or predictable events. A market participant becomes something by which the market price that is decided also reflects consideration of various factors

Table 2 Comparison of the prediction method

	Prediction market	Statistical analysis	Questionnaire (choice)	Questionnaire (free answer)	Delphi method
Synthetic judgment	◎	×	×	○	○
Incentives	○	×	△	△	△
Cost	○	○	△	△	△
Interaction	○	×	×	×	×

and types of information that make possible judgment of the probability of an event.

Prediction markets can anticipate the data collected in statistical analysis and the likely possibility that there was a factor influencing the expectations of a future occurrence, but for which no statistical data was collected. In a questionnaire form with limited choices, the range of answers will be restricted, depending on how the the questionnaire was designed.

In a free answer questionnaire, various kinds of information can be collected as in a prediction market. However, it is a serious task to weigh and evaluate diverse information submitted on this type of questionnaire. On this point this type of questionnaire is inferior to prediction markets.

With the Delphi method one can carry out synthetic judgments. Yet, according to Surowiecki, a well-informed person has a well-informed personfs bias. It is said that the system is excellent for prediction markets. If fully different judgments can be collected, it is possible to offset extreme judgments. In addition, overall it will be possible to extract an exact judgment.

- Incentive

  Prediction markets offer real money and virtual money as an incentive. The existence of an incentive is not related to statistical analysis.

  Although remuneration is likely to become an incentive in a questionnaire, it is an incentive to the act itself in reply to an investigation. The questionnaire does not have the structure of an incentive for replying correctly to a question.

  In the Delphi method, remuneration by having participated in the Delphi method can be received. However, the remuneration is not fixed, and it does not depend on whether the Delphi method was able to make a correct judgment.

- Cost

  In a prediction market, if the system is already set up, costs will not be incurred. In the Inkling prediction market anyone can open a prediction market for free.

  In statistical analysis, if there is data, costs will not be incurred.

  In a questionnaire, in order to raise accuracy, it is necessary to enlarge the scale of the investigation, and this enlargement will incur more costs by requiring more time, money, and participants.

To raise accuracy with the Delphi method, it is necessary to assemble specialists and persons well informed on the theme being considered.

- Interaction
  The synthetic judgment in a prediction market is based on a spiral movement of interactions that leads to new discovery and evolution.
  Unlike judgments in a questionnaire, the overall judgment of the prediction market is based on the interactions of respondents and the market.
  A prediction market participant makes a synthetic judgment. After reading about delicate changes and the mood of a market, he can sense trends and influences in a timely manner and can reflect them in his predictions of the future.

As mentioned above, Suginoo has singled out the excellence of prediction markets compared with other prediction methods.

2.4 Collective Intelligence

James Surowiecki [11] has said that a prediction market totals various opinions comparatively simply and can collect them into one judgment as a group. This collecting of the opinions of many human beings is called collective intelligence.

It is thought that collective intelligence, as found in social networking services (SNS), social bookmarks (SBM), and Wikipedia[5], demonstrate the enormous power of having the right person in the right place [8]. Although it trends toward collective intelligence, decision-making by a group is said to bring about extreme deviations in many cases. Janis [5] studied the failures in U.S. foreign policy, such as the Bay of Pigs fiasco, in detail. Janis says decision-making persons are homogeneous and can lapse into group thinking easily. Moreover, there are problems that groups cause, such as "ochlocracy" in history, "group psychology" in social psychology, and "the tragedy of commons" in biology.

Surowiecki mentions four conditions in which collective intelligence functions appropriately to bring about desirable results.

1. Diversity of opinion
   If each participant has an original viewpoint, many proposed solutions can be listed up over the whole. When the search space is restricted, a suitable solution in this space may not be possible.
2. Independence
   Each participant's independence needs to be secured so that his/her opinion or proposal is not influenced by other participants. There are various dangers that people's opinions will be determined by the inclination of the group, especially in small group discussions.

[5] http://www.wikipedia.org/

3. Decentralization
   It is not necessary to abstract a problem, but each participant needs to judge based on information he or she acquires directly.
   Although it is expected that the kinds of information acquired by every participant will differ, to maintain diversity, one should not judge based only on attributes common to each participant.
4. Aggregation
   All participants share the knowledge they have acquired, taking advantage of the characteristics of the three above-mentioned points. Moreover, a structure that carries out comparisons and examinations of participants' various viewpoints and leads to a final conclusion is required.
   Although it is expected that the kinds of information acquired by every participant will differ, to maintain diversity, one should not judge based only on attributes common to each participant.
5. Aggregation
   All participants share the knowledge they have acquired, taking advantage of the characteristics of the three above-mentioned points. Moreover, a structure that carries out comparisons and examinations of participants' various viewpoints and leads to a final conclusion is required.

A prediction market is equipped with the structure of the four above-mentioned points. If seen from the viewpoint of utilizing collective intelligence, a prediction market will have a well-made, elegantly simple structure.

2.5 Examples of Prediction Markets

Some of the prediction markets in the U.S. are the Iowa Electronic Market(IEM) and the Hollywood Stock Exchange(HSX). Moreover, there are prediction markets, such as Shuugi.in and Kounaru, in Japan.

We introduce as a concrete example the "2004 presidential election vote tally rate prediction futures market" carried out by IEM. Yamaguchi analyzed [14] this prediction market. In this futures market, each candidates percent of the number of votes obtained serves as a price. That is, a price will be set to \$0.55 if a candidate tallies 55% of the vote. Therefore, the price at the time of the final vote tally will serve as an expected value about the end result.

Fig. 3 has a blue line of an original numerical value, a green line for the moving average for seven days, and a red line of 51.5% for the actual vote tally percentage. Naturally, the price converged on the actual vote tally percentage as of November 5. However, it did not separate from 51.5% as a whole, but it turned out that the circumference was moved. The average price from January 2004 at the beginning of the graph was \$0.52, and the standard deviation was \$0.017. The difference between the actual vote tally percentage and the average price was \$0.005. It was not separated from the average price by one third of one standard deviation. Moreover, the average price for seven days up to the election eve was \$0.512, and it was approaching closely to the

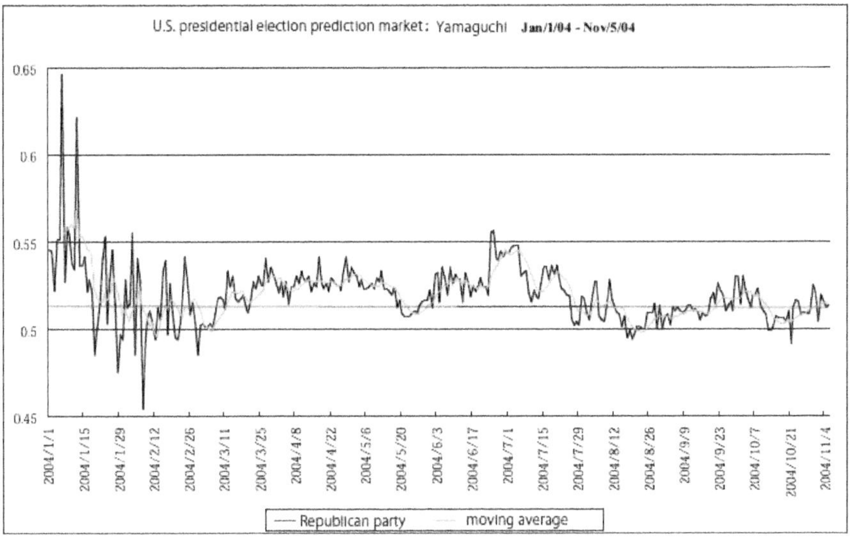

Fig. 3 U.S. presidential election prediction market 2004

actual vote tally percentage. At a stage in early 2004 a value below and above [$0.51 $0.52] was already being shown, and this value was not very much different from the end result. This tendency was mostly consistent through one year, although there were inaccuracies at some points. Although opinion poll results and analysis by specialists showed significant inaccuracies, most of the prediction results that prediction markets made did not deviate much from the actual final result. Of course, such excellent results by prediction markets are not seen in all elections. However, it can be concluded that predictions from prediction markets are more accurate than opinion polls areC at least that was the case for the prediction markets for the November 2004 election.

2.6 Related Work

In related work on this topic, Justin Wolfers and Eric Zitzewitz have introduced prediction markets [12]. J. Berg et al. have reported their 12-year research findings on the election futures market [1]. J.E. Berg and T.A.Rietz have proposed using a prediction market as a decision support system [2]. E. Servan-Schreiber et al. have considered the differences between the results of using actual currency and virtual currency in prediction markets [9]. Justin Wolfers and Eric Zitzewitz have provided relevant analytic foundations, describing sufficient conditions under which prediction markets prices correspond with mean beliefs [13]. C.F. Manski has presented the first formal analysis of price determination in prediction markets where traders have heterogeneous beliefs [7]. B. Cowgill et al. had illustrated how markets can be used to study how an organization processes information [3]. S. Luckner has studied the impact of different monetary incentives on prediction accuracy in a field experiment [6].

3 Current Research and Problems

In Japan Suginoo [10] has conducted previous research on a prediction market. Suginoo conducted an experiment on the prediction market "whether the viewership of the August 13, 2005 episode of the Winter Sonata TV drama would exceed 17%". In this experiment, the following three points about the design of a prediction market were examined and discussed.

1. It was not possible to gather people with completely different opinions in a prediction market. If only people with similar judgments gather in small numbers, it will not be easy to have active dealings in a prediction market.
2. Sufficient transactions will not be conducted unless participants have an interest in a market.
3. Poor results in a market will dampen participants' motivation. (The test of a market then becomes the way participants feel about it).

If a position is taken on a market mechanism principle, then it may be sufficient to derive this principle from a set of factual premises that, taken together, point to the truth of this position. However, the strong point of a prediction market is that in it a set of various opinions takes on the structure of an incentive, and an evolution of judgments through interactions occurs, so that a synthetic judgment that is not just an extrapolation from mere past events is made. Whether such a structure operates is also closely connected with the state of the culture or society where a prediction market takes place. Results attained from prediction markets carried out in the United States do not come out the same from simply transplanting these markets to Japan. The kind of reaction shown to information on a market will vary according to the types of societies and the characters of the people involved. Suginoo has said that this study related to this point about prediction markets must introduce research from psychology or sociology.

In the research we are reporting on, we considered what kind of changes introducing an agent would have on increases in trading volume and participants' motivation in relation to the above problem of how different people with different characters or from different societies react to information in a prediction market.

4 Experiment Outline and Discussion

4.1 Zocalo

We used Zocalo[6] the experiment we did in this research. Zocalo is a toolkit that can easily be used to set up a prediction market. In Zocalo, two kinds of prediction markets can be chosen. One is a binary market (The numerical value 0 to 100 is taken in the market decided by 'yes' or 'no') and the other is a multiple outcome (a market with two or more choices). The binary market was used in the experiment of this research.

Fig. 4 shows the Zocalo market. The time series data of the market price can be shown graphically, and the participants in a market can be confirmed. Moreover, the past transactions history of a participant and the amount and brand of goods a participant currently possesses can be checked. The participants make predictions based on the information from the time series data and public information on the transactions.

Fig. 4 The Zocalo market

[6] http://zocalo.sourceforge.net/

4.2 Experiment

The experiment in this research used Zocalo. The subjects (market participants) were students from a laboratory, and the number of subjects was 18. The period of the experiment was one week, from September 7, 2009. An experiment was conducted in regard to the following two prediction themes.

Theme 1. Will the weather be fine at 15:00 on September 14, 2009?

Theme 2. Will the "Chunichi Dragons" win in the game of "Chunichi Dragons" VS the "Tokyo Yakult Swallows" held on September 13, 2009?

For Theme 1, if the weather is clear, one dollar will be paid out for a 'yes' answer to the theme question. For Theme 2, if "Chunichi Dragons" wins, one dollar will be paid out for a 'yes' answer to the theme question.

For Theme 1, this experiment was conducted without an agent, and for Theme 2, it was conducted with an agent. Table 3 shows an agent's action.

Table 3 Agent's actions

Agent	Action
Buy or Sell	Random
Quantity	Random in the range of 1-10
Value	Random in the range of ± 20 from the latest value

4.3 Results and Discussion

Fig. 5 shows the result of predictions for Theme 1, and Fig. 6 shows the result of predictions for Theme 2. Regarding the correct answer, Themes 1 and 2 were set to 'yes' in this experiment.

We would like to discuss the accuracy of the predictions. In Theme 1, the result was "fine weather" and good accuracy was shown in anticipating this result, with a probability of 90%. We thought that this accuracy came about because the theme was one of weather forecasting, whose accuracy has comparatively improved. In Theme 2, the result was "Chunichi Dragons win" with the anticipated result being that the "Chunichi Dragons" would win by a probability of 70%. It is not easy to anticipate which team will win a baseball game. However, in the result of this experiment, one can say the accuracy was good.

Next, we will discuss trading volume. For Theme 1, the number of dealings was 3. For Theme 2, the number of dealings was 14. We consider this difference came about because the Theme 1 part of the experiment was conducted without an agent and the Theme 2 part of the experiment was conducted

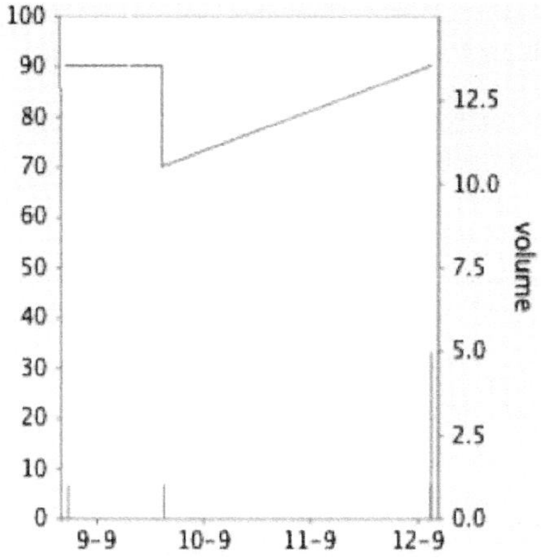

Fig. 5 Experiment:Theme 1

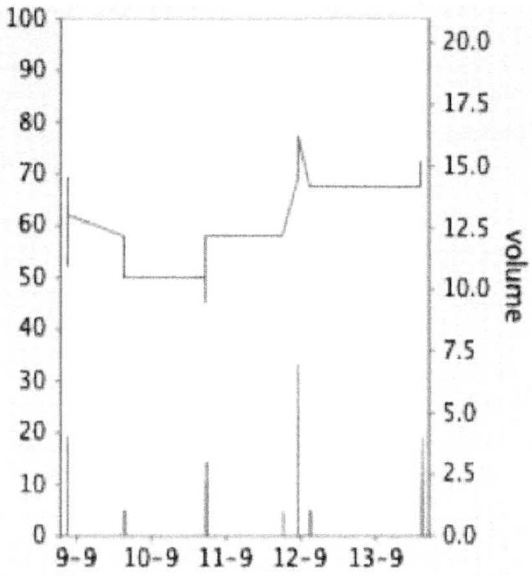

Fig. 6 Experiment:Theme 2

with an agent. For a trade, two persons, a "buyer" and a "seller", are required. Since there were few subjects in this experiment, we thought that there would be few dealings for Theme 1. However, for Theme 2, there was an increase in the number of dealings due to the introduction of am agent. Therefore, it can be said that an agent is useful for the improvement of dealings frequency. Moreover, the liquidity of a market can also be assured by an increase in the number of dealings.

5 Conclusion

In this research, we demonstrated the effect of introducing an agent in a prediction market. From the result of the experiment, we found that introducing an agent improved the number of dealings. Although we conducted the experiment on only two themes, in subsequent research we must increase the number of experiments done and carry out comparative studies.

References

1. Berg, J., Forsythe, R., Nelson, F., Rietz, T.: Results from a Dozen Years of Election Futures Markets Research, vol. 1, pp. 742–751. Elsevier, Amsterdam (2008)
2. Berg, J., Rietz, T.: Prediction Markets as Decision Support Systems. Information Systems Frontiers 5(1), 79–93 (2003)
3. Cowgill, B., Wolfers, J., Zitzewitz, E.: Using Prediction Markets to Track Information Flows: Evidence from Google (2008), `www.bocowgill.com/GooglePredictionMarketPaper.pdf`
4. Hayek, F.: The Use of Knowledge in Society. American Economic Review XXXV(4), 519–530 (1945)

5. Herek, G.M., Janis, I.L., Huth, P.: Quality of U.S. Decision Making during the Cuban Missile Crisis: Major Errors in Welch's Reassessment. Journal of Conflict Resolution 33, 446–459 (1989)
6. Luckner, S.: Prediction Markets: How Do Incentive Schemes Affect Prediction Accuracy? In: Jennings, N., Kersten, G., Ockenfels, A., Weinhardt, C. (eds.) Negotiation and Market Engineering, no. 06461 in Dagstuhl Seminar Proceedings. Internationales Begegnungs- und Forschungszentrum für Informatik (IBFI), Schloss Dagstuhl, Germany, Dagstuhl, Germany (2007)
7. Manski, C.: Interpreting the Predictions of Prediction Markets. Economics Letters 91(3), 425–429 (2006)
8. Oomukai, K.: Web2.0 and Collective Intelligence. Information Processing Society of Japan 47(11), 1214–1221 (2006)
9. Servan-Schreiber, E., Wolfers, J., Pennock, D.M., Galebach, B.: Prediction Markets: Does Money Matter? Electronic Markets 14(3), 243–251 (2004)
10. Suginoo, T.: A New Prediction Tool "Prediction Market". Communication Inquiries 8, 8–13 (2006)
11. Surowiecki, J.: The Wisdom of Crowds. Anchor Books (2005)
12. Wolfers, J., Zitzewitz, E.: Prediction Markets. Journal of Economic Perspectives 18(2), 107–126 (2004)
13. Wolfers, J., Zitzewitz, E.: Interpreting Prediction Market Prices as Probabilities. CEPR Discussion Paper No. 5676 (2006)
14. Yamaguchi, H.: It Looks Back Upon the U.S. Presidential Election (2004), http://www.h-yamaguchi.net/2004/11/post_31.html

A Cognitive Map Network Model

Yuan Miao

Abstract. This paper presents a cognitive map network model which models causal systems with interactive cognitive maps. Cognitive map is a family of cognitive models that have been widely applied in modeling causal knowledge of various systems like gaming and economic systems. Many causal systems have multiple components which causally evolve concurrently, interacting with each other. Modeling such a system as a whole (cognitive map) is not an ideal solution. Sometimes it is also not possible as the individual parties may not want to release their knowledge to other parties or the coordinating component. The cognitive map network model proposed in this paper represents a causal system as an ecosystem with individual components modeled as cognitive agents. It is a cognitive map ecosystem whose evolution is driven by the component cognitive agents. The cognitive ecosystem model is applied in a toy economic system to illustrate its power in study of hidden patterns.

1 Introduction

Cognitive map is a family cognitive models [14] representing causal knowledge and facilitating inference accordingly, modeling human cognition and reasoning. Cognitive Maps (CMs) [1], Fuzzy Cognitive Maps (FCMs) [9] and Dynamical Cognitive Networks (DCNs) [12] are three related models for cognitive modeling. Cognitive Maps are the first and fundamental models of the family. CMs have been applied in various application domains, for example analysis of electrical circuits [19], analysis and extension of graph-theoretic behavior, and plant control modeling [7]. CMs are easy to use and straightforward models; however, they do not compare the strength of different relationships. Rather, each node simply makes its decision based on the number

Yuan Miao
School of Engineering and Science, Victoria University, Australia
e-mail: Yuan.Miao@vu.edu.au

Q. Bai and N. Fukuta (Eds.): Advances in Practical Multi-Agent Systems, SCI 325, pp. 63–72.
springerlink.com

of positive impacts and the number of negative impacts; hence a CM is an over simplified model for many applications.

Fuzzy Cognitive Maps extended CMs by introducing fuzzy weights among factors to describe the strength of the relationships. FCMs have gained comprehensive recognition and been widely applied in entertainment [15], games [4], multi-agent systems [11], social systems [5], ecosystems [6], financial systems [21], and earthquake risk / vulnerability analysis [17]. However, FCMs are still not capable of handling complex causal systems [12]; several core modelling capabilities are missing in the FCM model. FCMs do not differentiate the strength of factors and FCMs do not describe the dynamics of the relationships. To address the limitations of CMs and FCMs, several analysis and extensions have been proposed [12] [16] [8] [20] [13]. These proposals, including Dynamic Cognitive Network (DCN), introduced more values to concepts including real valued concepts, nonlinear weight, and time delays. DCNs are powerful models in the family: they systematically address the modeling defects of other cognitive maps and are able to handle complex dynamic causal systems.

Most of the applications of the cognitive map models are applied to model a real world system as a whole. However, many systems have a number of concurrent evolving and interacting components. For example, an economic system have many players, each has his/her own cognition and decision making over various factors. These players interacting with each other, jointly drive the evolution of the whole systems. In such a case, a whole cognitive map is not proper. Using a group of cognitive agents to model the component will be more suitable. There have been several proposals in modeling software agents with cognitive maps [11] [10] [2] [18] [3], but not yet been a report in models of a cognitive agent ecosystem, which is provided in this paper. In such a cognitive ecosystem, cognitive agents carry cognitive knowledge, evolving and interacting with other agents. There are two types of agents: active agents and responsive cognitive agents. An active agent can take strategic actions to affect other agents. A responsive agent has internal laws regulating its behavior and respond to the impact from other agents. For example, a cognitive ecosystem model for a foreign exchange system can be consists of active agents like traders, and responsive agents like the market. The traders can actively trade in the system while the market responds according to collective trades from traders. Both active cognitive agents and responsive cognitive agents are modeled with cognitive models like fuzzy cognitive maps or dynamic cognitive networks. Inter-links among agents are used to model the interaction among agents.

The rest of the paper is organised as follows: Section 2 proposes the cognitive ecosystem model; Section 3 applies the cognitive ecosystem model on a toy economic system to study the hidden patterns and Section 4 concludes the paper.

2 A Cognitive Map Network Model

A cognitive map network model is a cognitive ecosystem model represents the real world causal system with a collection of agents and their inter-links. Each agent has cognitive map modeling its causal knowledge. An inter-link can be solid or informational. A solid link represents non deniable impact on the affected agent. However, the affected agent has choice whether or not to take into account of the impact from an informational link. The detailed definition is given in Definition 1.

Definition 1. *A* ***Cognitive Map Network****, denoted as W, is defined as a set tuple: $W = \{A, L\}$, where A is a set of cognitive agents and L is a set of inter-links. $A = \{A_1, A_2, ..., A_n\}$ is a set of agents $A_i, i = 1, 2, , n$. $L = \{L_1, L_2, ..., L_m\}$ is a set of inter-links $L_i, i = 1, 2, ..., m$. Each agent has a cognitive model and a tag indicating active or responsive type: $A_i = \{M_i, ar\}$, where $ar \in \{f_a, f_r\}$. f_a is the tag for active agents. f_r is the tag for responsive agents. M_i is the cognitive model of the agent; Each inter-link has a source agent an affected agent: $L_i = \{A_s, A_t, sv\}$, where A_s is the source agent creating the impact, and A_t is the agent receiving the impact. $sv \in \{f_s, f_i\}$, f_s is the tag for solid links, f_i is the tag for informational links. L_i is also denoted as $L(s, t, sv)$.*

A cognitive map network is also noted as a cognitive ecosystem.

2.1 A Market Cognitive Ecosystem

Fig. 1 is an economic system to illustrate how a cognitive ecosystem is applied. In this economic system, there are 7 agents, 6 are active agents and 1 is responsive agent. An active agent is represented by a double line circle, and a responsive agent is represented by a double line box.

The 6 active agents are players in the market: 3 are consumer players and 3 are producer players. The consumer agents are represented with webbed double circles and producer agents are represented with blank double line circles. Each active agent or player has its own cognitive model and based on which makes its decision of buying or selling. A consumer agent ($A_i, i = 1, 2, 3$) is more likely to buy because it consumes; but it may also sell when it has stock and the price is higher than the balanced price. Producer players ($A_i, i = 4, 5, 6$) are more likely to sell because they produces. However, a producer player may also decide to buy if it has space to store and the price is lower than the balanced price. It can sell out later to make profit by such tradings.

A_7 is a responsive agent representing the market pricing mechanism. If more players buy than those who sell, the price rises; otherwise, the price drops. A responsive agent does not have an ability to initiate impact on other agents. Instead, it responds according to the internal laws.

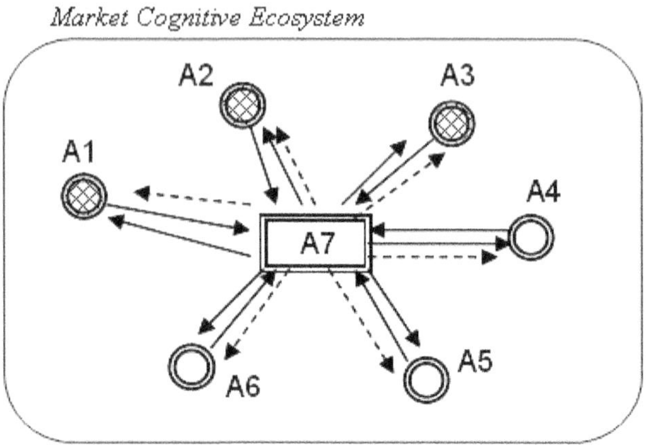

Fig. 1 A market Cognitive Ecosystem

In the Cognitive Ecosystem (CE) model, there are 18 inter-links:

- $L(A_i, A_7, f_s), i = 1, 2, ..., 6$ are solid links that an active agents providing purchase (buy or sell) instructions; Once an instruction is sent in, it cannot be withdrawn and the market agent has to process it: the impact is solid;
- $L(A_7, A_i, f_s), i = 1, 2, ..., 6$ are solid links that the market responsive agents providing trade (buy or sell) commitment instructions; Once an instruction is sent in, the active agent has to process it: the impact is solid;
- $L(A_7, A_i, f_i), i = 1, 2, ..., 6$ are informational links that the market responsive agents providing the information of current price; The active agents may or may not use this information for the trade decision making.

2.2 *Consumer Active Cognitive Agents*

Fig. 2 shows the cognitive model of a consumer active agent.

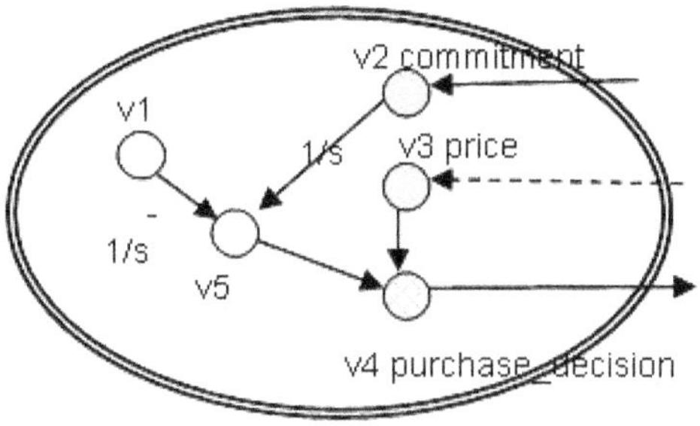

Fig. 2 Cognitive model of a consumer active agent

In the consumer active agent, the cognitive model involves 5 vertexes: $v_i, i = 1, 2, ..., 5$. Shaded nodes represent vertexes that receive impacts from other agents; webbed nodes represent vertexes that have impacts on other agents; blank nodes represent internal vertex which are not visible by other agents. There are two causal links in the cognitive model having $1/s$, or $-1/s$ as the weight. This is a dynamic link proposed in Dynamic Cognitive Network [12] (transfer function approach, with a integral type of transfer function), indicating the causal impact is accumulative. This cognitive model shows that in long run, the purchase decision shall be purchase because it consumes. However, in short term, the purchase decision is based on the storage and price. The decision function of the major vertex v_5 is:

$$x_5(k+1) = f_5(x_5(k), y_{51}(k+1), y_{52}(k+1)) = x_5(k) - x_1(k) + x_2(k),$$

where f_5 is the decision function of v_5, y_{51} is the impact from v_1 to v_5 (defined by the accumulation dynamic weight), and y_{52} is the impact v_2 from to v_5.

As to v_4, the possible decisions are

$$x_4(k+1) = \begin{cases} n_4 \\ 0 \\ -n_4 \end{cases},$$

where n_4 is the purchase decision with amount of n_4, 0 is the no action and $-n_4$ is the sell out decision with amount of n_4. Different decision strategies will be compared in Section 3.

2.3 Market Responsive Cognitive Agent

Fig. 3 shows the cognitive model of a producer active agent.

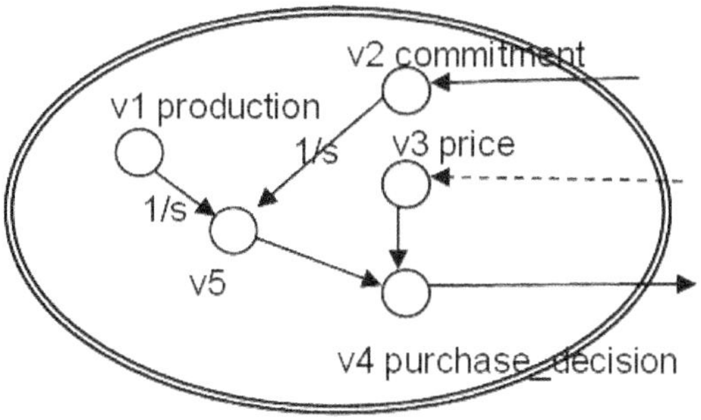

Fig. 3 Cognitive model of a producer active agent

A producer active agent is similar to that of consumer agents but v_1 is production and have positive accumulative impact on the storage. The decision function of the major vertex v_5 is:

$$x_5(k+1) = f_5(x_5(k), y_{51}(k+1), y_{52}(k+1)) = x_5(k) + x_1(k) + x_2(k),$$

where f_5 is the decision function of v_5, y_{51} is the impact from v_1 to v_5 (defined by the accumulation dynamic weight), and y_{52} is the impact v_2 from to v_5.

As to v_4, the possible decisions are

$$x_4(k+1) = \begin{cases} n_4 \\ 0 \\ -n_4 \end{cases},$$

where n_4 is the purchase decision with amount of n_4, 0 is the no action and $-n_4$ is the sell out decision with amount of n_4. Different decision strategies will be compared in Section 3.

2.4 Market Responsive Cognitive Agent

Fig. 4 is the cognitive model of A_7. It is a market responsive cognitive agent. A_7 has 14 vertexes. $v_{11}, v_{12}, ..., v_{16}$ are vertexes receiving trade decisions from the solid inter-links connecting from active agents; $v_{21}, v_{22}, ..., v_{26}$ are vertexes sending trade commitment through solid link to other active agents; Here the decision making is probabilistic (when buyers and sellers are not equal, randomly match possible buyers and sellers); v_3 is an internal vertex based on the market law worked out the market move (up or down), and the corresponding price, which passed to v4. v4 has informational links to other agents, informing the new market price; The market price rises if there are more purchase than sells, and vice versa. At a same time, only the matched trades are committed, which the market randomly select matching parties. Two different market pricing mechanisms are given in Section 3.

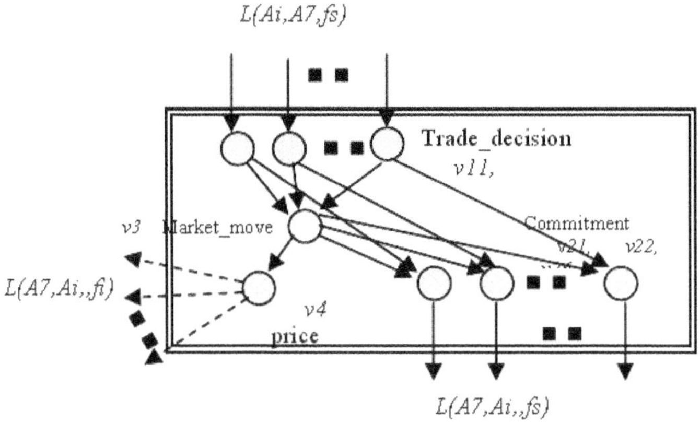

Fig. 4 Cognitive model of the responsive market agent

3 Analysis of the Economic System with the CM Network Model

This section presents a number of inference patterns on different market mechanisms and cognitive agent strategies.

Fig. 5 shows the pattern that the market moves its price by maximum 1 at each time. The balance price is 10. Each agent will buy low and sell high unless the storage does not allow (stop purchase if no storage space; no borrowed selling; or naked short in the stock market.). Fig. 5 shows that the market is able to slowly stablise itself on disturbance.

Fig. 6 shows the pattern of a different market pricing mechanism: the market moves its price according to the purchase and sells difference: the higher the difference between the buy power and sell power, the stronger of the move. The rest setting is the same as that in Fig. 5. It shows that the market gradually falls in an oscillation cycle, because of the over correction to the disturbance.

Normally, the trading itself costs. Traders won't jump into the market when the price slightly moves from the balance price. Fig. 7 shows the pattern when

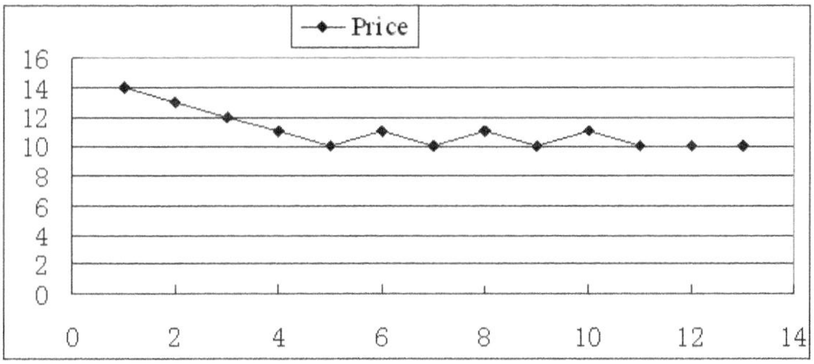

Fig. 5 Market slowly stabilise after disturbance

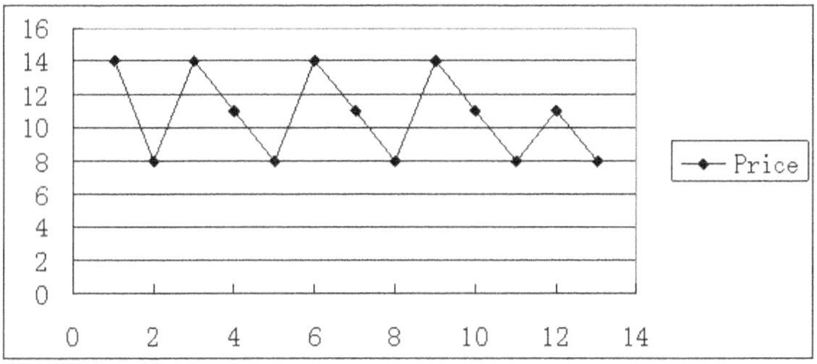

Fig. 6 Market falls in an oscillation dues to the over correction

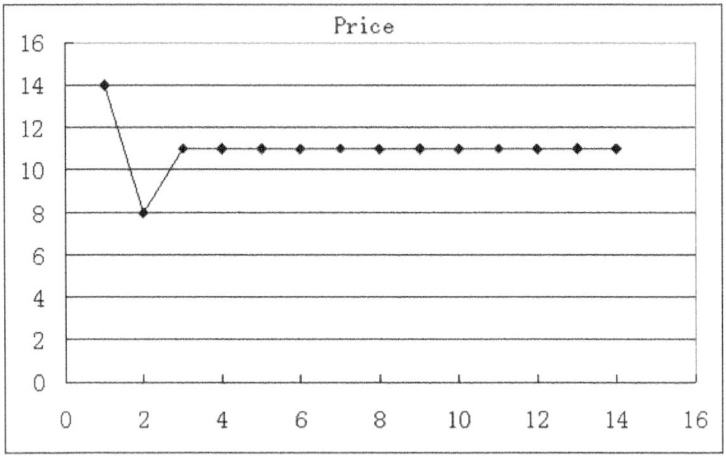

Fig. 7 Market stabilise with proper trading costs or spreads

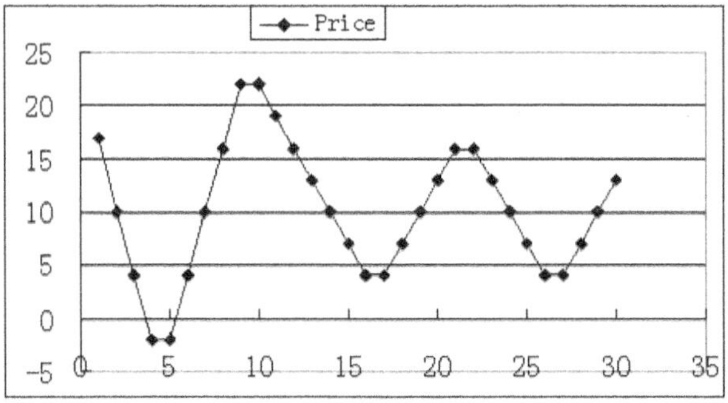

Fig. 8 Market falls in a severe oscillation with agents' decision on delayed market move

the active cognitive agents set a gap threshold. If the price is away from the balance price farther than the gap, then trading decision is made. The rest is the same as that in Fig. 6. It shows that the market can be pulled back to the balance area (instead of a balance price) quickly. This pattern also proves that the setting of a "spread" is important to avoid non-necessary sentiment in the market, and reduce the sensitivity.

In some cases, a trading decision could not be made immediately following the market move. Each trading strategy takes time to be implemented. Fig. 8 shows the pattern when the trading decision can only be altered after 2 time slots of the market move. The rest setting is the same as that in Fig. 7. It shows that the existing market is in a big oscillation, and the pattern is a composition of a few periodic oscillations. This is not a desirable pattern. Large oscillation can be damaging to the economy. However, many large companies could not make swift decisions. Therefore, the speculation should be allowed and encouraged in the market. So the move can be smoothed out. The patterns in Fig. 7 and Fig. 8 indicate that speculation should be well managed: not too less and not too much.

4 Conclusions

Cognitive map is a family of cognitive models that have been widely applied in modeling causal knowledge. To date, all the applications of cognitive maps model a whole causal system with a whole cognitive map. As many causal systems have multiple components which causally evolve concurrently, interacting with each other, this paper proposes a cognitive ecosystem model that collectively models these components and allows them to interact within a cognitive ecosystem. When such a system contains a large amount of components, the traditional analysis approaches have difficulties to provide reliable analysis. In the recent economic crisis, many AAA products from a large number of major banks failed. Those AAA products are supposed to fail only in a very small probability (e.g. 1 out of a million chances). Therefore, these banks and products fail in a same crisis should be almost impossible. That fact shows that the economic models were not reliable. Many economic systems are so complex that traditional analysis tools have to omit too many factors to come up with a mathematically manageable model, which becomes unreliable. Cognitive ecosystem takes a different approach. Each component in the model is not too complex thus can be modeled with cognitive agents reasonably well. The whole ecosystem patterns can be revealed with simulations, with affordable computation. Such analysis is also able to give the probability of each result pattern, which is very useful for pricing and policy making. Such probability is also likely to be more reliable than those derived from the existing models. This paper uses a small number of cognitive agents to illustrate different patterns. However, a large number of agents do not have essential difference in term of the analysis but with a larger amount of pattern data to process.

References

1. Axelrod, R.M.: Structure of Decision: The Cognitive Maps of Political Elites. Princeton University Press, Princeton (1976)
2. Borrie, D., Isnandar, S., Ozveren, C.S.: The Use of Fuzzy Cognitive Agents to Simulate Trading Patterns within the Liberalised UK Electricity Market. In: Proceedings of Universities Power Engineering Conference, vol. 3, pp. 1077–1081 (2006)
3. Cai, Y., Miao, C., Tan, A.H., Shen, Z.: Creating an Immersive Game World with Evolutionary Fuzzy Cognitive Maps. IEEE Computer Graphics & Applications 29 (2009)

4. Chen, M.E., Huang, Y.P.: Dynamic Fuzzy Reasoning Model with Fuzzy Cognitive Map in Chinese Chess. In: Proceedings of SICE Joint Symposium of 15th System Symposium and 10th Knowledge Engineering Symposium, vol. 3, pp. 1353–1357 (1995)
5. Craiger, P., Coovert, M.: Modeling Dynamic Social and Psychological Processes with Fuzzy Cognitive Maps. In: Proceedings of IEEE World Congress on Computational Intelligence, vol. 3, pp. 1873–1877 (1994)
6. Eccles, J., Dickerson, J., Shao, J.: Evolving a Virtual Ecosystem with Genetic Algorithms. In: Proceedings of 2000 Congress on Evolutionary Computation, vol. 1, pp. 753–760 (2000)
7. Gotoh, K., Murakami, J., Yamaguchi, T., Yamanaka, Y.: Application of Fuzzy Cognitive Maps to Supporting for Plant Control. In: Proceedings of SICE Joint Symposium of 15th System Symposium and 10th Knowledge Engineering Symposium, pp. 99–104 (1989)
8. Hagiwara, M.: Extended Fuzzy Cognitive Maps. In: Proceedings of IEEE International Conference on Fuzzy Systems (1992)
9. Kosko, B.: Fuzzy Cognitive Maps. International Journal of Man-Machine Studies (24), 65–75 (1986)
10. Miao, C.Y., Goh, A., Miao, Y.: A Multi-agent Framework for Collaborative Reasoning. CRC Press, Boca Raton (2002)
11. Miao, C.Y., et al.: DCM: Dynamical Cognitive Multi-Agent Infrastructure for Large Decision Support System. International Journal of Fuzzy Systems 5(3), 184–193 (2003)
12. Miao, Y., Liu, Z., Siew, C., Miao, C.: Dynamical Cognitive Network - an Extension of Fuzzy Cognitive Map. IEEE Transaction on Fuzzy Systems 10, 760–770 (2001)
13. Miao, Y., Liu, Z.Q.: On Causal Inference in Fuzzy Cognitive Maps. IEEE Transactions on Fuzzy Systems 8(1), 107–119 (2000)
14. Miao, Y., Tao, X., Shen, Z., Liu, Z.Q., Miao, C.Y.: Transformation of Cognitive Maps. IEEE Transaction on Fuzzy Systems (to appear)
15. Parenthoen, M., Tisseau, J., Morineau, T.: Believable Decision for Virtual Actors. In: Proceedings of 2002 IEEE International Conference on Systems, Man and Cybernetics, vol. 3, p. 6 (2002)
16. Park, K.S., Kim, S.H.: Fuzzy Cognitive Maps Considering Time Relationships. International Journal of Human-Computer Studies 42(2), 157–168 (1995)
17. Rashed, T., Weeks, J.: Assessing Vulnerability to Earthquake Hazards through Spatial Multicriteria Analysis of Urban Areas. International Journal of Geographical Information Science 17(6), 547–576 (2003)
18. Rodin, V., Querrec, G., Ballet, P., Bataille, F., Desmeulles, G., Abgrall, J., Tisseau, J.: Multi-Agents System to Model Cell Signalling by Using Fuzzy Cognitive Maps- Application to Computer Simulation of Multiple Myeloma. In: Proceedings of the Ninth IEEE International Conference on Bioinformatics and Bioengineering, pp. 236–241 (2009)
19. Styblinski, M., Meyer, B.: Fuzzy Cognitive Maps, Signal Flow Graphs, and Qualitative Circuit Analysis. In: Proceedings of the 2nd IEEE International Conference on Neural Networks (ICNN 1987), vol. 2, pp. 549–556 (1988)
20. Tsadiras, A.: Comparing the Inference Capabilities of Binary, Trivalent and Sigmoid Fuzzy Cognitive Maps. Information Sciences 178, 3880–3894 (2008)
21. Zhang, W., Liu, L., Zhu, Y.C.: Using Fuzzy Cognitive Time Maps for Modeling and Evaluating Trust Dynamics in the Virtual Enterprises. Expert Systems with Applications 35(4), 1583–1592 (2008)

Coordination Strategies and Techniques in Distributed Intelligent Systems – Applications

Abdeslem Boukhtouta, Jean Berger, Ranjeev Mittu, and Abdellah Bedrouni

Abstract. This paper is devoted to describing the broad range of application domains which implement many of the coordination strategies and techniques from the field of multi-agent systems. The domains include defense, transportation, health care, telecommunication and e-business, emergency management, etc. The paper describes the diversity of the applications in which multi-agent coordination techniques have been applied to overcome the challenges or obstacles that have existed with regard to performance, interoperability and/or scalability. While the number of application domains is steadily increasing, the intent of this paper is to provide a small sampling of domains which are applying coordination techniques to build intelligent systems. This paper will also describe an emerging and important problem

Abdeslem Boukhtouta
Defence Research & Development Canada - Valcartier, 2459 Pie-XI Blvd. North, Val-Belair, G3J 1X5 (Quebec), Canada
and
Concordia University, Concordia Institute for Information Systems Engineering (CIISE), 1455 de Maisonneuve, Montreal, Quebec, H3G 1M8, Canada
e-mail: abdeslem.boukhtouta@drdc-rddc.gc.ca

Jean Berger
Defence Research & Development Canada - Valcartier, 2459 Pie-XI Blvd. North, Val-Belair, G3J 1X5 (Quebec), Canada
e-mail: jean.berger@drdc-rddc.gc.ca

Ranjeev Mittu
US Naval Research Laboratory, Washington, 4555 Overlook Avenue, SW, Washington, DC 20375, USA
e-mail: mittu@itd.nrl.navy.mil

Abdellah Bedrouni
Concordia University, Department of Industrial Engineering, 1455 de Maisonneuve, Montreal, Quebec, H3G 1M8, Canada
e-mail: abedrouni@gmail.com

Q. Bai and N. Fukuta (Eds.): Advances in Practical Multi-Agent Systems, SCI 325, pp. 73–93.
springerlink.com

domain which requires the coordination among many entities across the civil-military boundary, and can benefit from multi-agent coordination techniques.

1 Introduction

It is commonly recognized that many different disciplines can contribute to a better understanding of the coordination function as a way to build and provide appropriate tools and adequate approaches for designing complex organizations and systems. Agents incorporated into these self-regulating entities represent "communities of concurrent processes which, in their interactions, strategies, and competition for resources, behave like whole ecologies". We describe in this paper applications requiring coordination or distributed decision-making. A common characteristic of these applications is the highly distributed nature of them. These applications can be also seen as challenging problems for multi-agent coordination (See [20]).

The research community working in the area of Distributed Artificial Intelligence (DAI) unanimously endorses the idea that coordination - a fundamental paradigm - represents a challenging research area. A clear consensus has thus emerged around a forged, well-articulated, and strongly advocated common vision that coordination is a central issue to agent-based systems engineering research (See Bedrouni et al. [2] for more details).

This paper provides an overview of the application domains in which multi-agent systems have been developed. The inherent distributed nature of these application domains reveals that benefits in performance or efficiency can be derived through the coordination between multiple agents. In other words, the collective behaviour of the system is improved through the coordinated interaction of their parts. In this paper, we will cover applications from defence, transportation, health care, telecommunication, E-business, emergency management, and ambient intelligence.

From the defence sector, we will describe how multi-agent systems have been deployed to support course of action analysis through their ability to monitor the battlefield environment as depicted in command and control systems, and additionally by tapping into simulations as ground truth are able to reason about deviations occurring in the movement of entities represented in the command and control systems. We also describe a large multinational experiment in which hundreds of agents were federated in a simulated coalition scenario. From the transportation industry, multi-agent coordination techniques have been applied to manage and improve the overall efficiency of incoming aircraft. From the health care industry, distributed multi-agent system coordination concepts have been prototyped to support the monitoring and treatment of diabetic patients. In the area of communication networks, coordination techniques have been applied to diagnose faults in such networks through agents that are able to communicate and exchange local information in order to understand non-local problems. In another setting, coordination techniques have been applied to more efficiently route information in a communication network, utilizing auction mechanisms to purchase or sell commodities such as bandwidth on links and paths, comprised of multiple links from source to destination. E-business application is described in the area of supply chain management, which incorporates Coloured Petri nets as a coordination mechanism between agents. Coordination techniques are also described for emergency management. Lastly, coordination techniques for the ambient intelligence are discussed.

The paper will also describe an emerging problem domain requiring the coordination of large teams that span the civil-military boundary, specifically during stability, security, transition and reconstruction operations. These operations require the coordination of the military with the civilian sector and non-governmental organizations in response to either natural or man-made disasters. The goal of these operations is to bring stability, rebuild or provide security in order to maintain order during such crisis situations. We will describe how multi-agent coordination techniques can provide tremendous value in these situations.

The remainder of this paper is organized as follows. Section 2 describes broad range of military applications which implement many of the coordination strategies and techniques from the field of multi-agent systems. Section 3 describes a set of problems drawn from transportation domain and using the distributed artificial intelligence paradigm. Section 4 discusses the healthcare applications. On the other hand, Section 5 describes various applications related to the communication networks. Furthermore, Section 6 is specifically devoted to the coordination in E-business applications. Section 7 discusses the coordination for emergency management and disaster relief. Application of the coordination for ambient intelligent in Section 8. Finally, some concluding remarks and recommendations are given in Section 9.

2 Defence

2.1 Control of Swarms of Unmanned Aerial Vehicles (UAVs)

The ability to coordinate military operations with the aid of information technology has become an imperative for command and control systems for the past several years. One of the most important military problems (in vogue) where coordination mechanisms should be used is in the control of swarms of UAVs (unmanned aerial vehicles). The UAVs are considered in this case as highly mobile agents that can be used for performing reconnaissance in a combat environment. In this case the communication between the UAVs is too costly because it would reveal their location to the enemy. The UAVs are sent out to collect information from certain locations and then return. The information sought by a UAV may depend on other information that it has collected by another UAV. There are many other cooperative control problems of the UAVS, we enumerate: cooperative target assignment, coordinated UAVs intercept, path planning, feasible trajectory generation, Etc (See [1]). A recent study of performance prediction of an unmanned airborne vehicle multi-agent system has been developed by Lian and Deshmukh [13]. The Markov Decision Processes (MDP) techniques and the Dynamic Programming approach have been widely used to solve the problems cited above (See Goldman and Zilberstein [11]).

2.2 Coordination for Joint Fires Support (JFS)

Coordination techniques are also useful for the systems that address the Joint Fire Support (JFS) problem. The mandate of Joint fires is to assist air, land, maritime, amphibious, and special operations forces to move, manoeuvre, and control territory, populations, and airspace. The JFS and coalition operations require shared approaches and technologies. The challenge is large, however, with a legion of command-and-control computer systems all developed along slightly different lines of approach. Some of these systems operate jointly within their spheres of influence. The bigger challenge is tying these and other efforts together in a shared global information network, linking ground, air and sea forces.

2.3 Simulation of C4I Interoperability

As the complexity of modern warfare increases, managing and interpreting operational data will continue to be one of the greatest challenges to commanders and their staffs. The wealth of data collected and distributed via Command, Control, Communications, Computers and Intelligence (C4I) systems during battlefield operations is staggering. The ability to effectively identify trends in such data, and make predictions on battlefield outcomes in order to affect planning is essential for mission success. Future commanders will be forced to rely upon new information technologies to support them in making decisions.

Simulations have been used by analysis and planning staffs for years during exercises and operations. Typically, combat simulations are used most heavily during the planning stages of an operation, prior to execution. However, simulations are increasingly being used during operations to perform course of action analysis (COAA) and forecast future conditions on the battlefield. Recent efforts by the Defense Modeling and Simulation Office (DMSO) to improve the interoperability of C4I systems with simulations have provided a powerful means for rapid initialization of simulations and analysis during exercises, and have made simulations more responsive and useable during the execution portion of an exercise. Real-time interfaces between C4I systems such as the Global Command and Control System (GCCS), and the

Integrated Theater Engagement Model (ITEM) simulation have provided command staffs with the capability to perform faster, more complete, COAA.

As can be seen in Figure 1, intelligent agents, coupled to C4I systems and simulations, offer another technology to help commanders manage information on the battlefield. The Defense Advanced Research Projects Agency (DARPA) has sponsored the development of the Control of Agent-Based Systems (CoABS) Grid. The Grid is middleware that enables the integration of heterogeneous agent-based systems, object-based applications and legacy systems. The CoABS grid was used to develop the Critical Mission Data over Run-Time Infrastructure (CMDR) that allows dynamic discovery, integration and sharing of High Level Architecture (HLA) Run-Time Infrastructure (RTI) compliant simulation objects with legacy C4I systems and grid-aware software agents. The bridging of the CoABS grid and the HLA RTI using CMDR makes it possible to leverage the power of agent technology with the ability to tap into multiple C4I sources and simulation systems simultaneously. This synergy could lead to profound benefits in situation assessment and plan-execution monitoring using agents.

The key idea behind the capability as depicted in Figure 1 was to present the intelligent agents with real (notional) and simulated battlefield information, so that these agents can analyze both streams of data in order to understand and provide alerts when the notional battlefield information has changed as compared to the information contained in the data stream coming from the simulation [18]. Several types of agents were developed to check for positional deviations in the battlefield entities using the simulated data as ground truth, both using extrapolation and interpolation techniques.

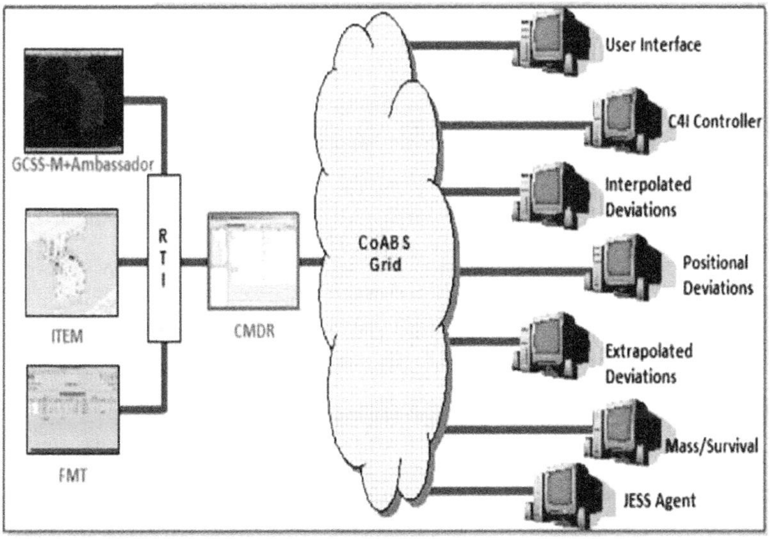

Fig. 1 Federation of intelligent agents, C4I systems and simulations

Extrapolation techniques - to project the entity's real reported position - were needed when the entity's reported time was less than the reporting time of that entity in the simulation in order to compare any changes to position at the same instance of time. Similarly, interpolation techniques were used when the entity's real reported position was greater than the reporting time of that entity in the simulation. An additional mass monitoring agent was responsible for detecting deviations in combat worth based on thresholds defined in the original plan in the simulation. This example demonstrates end-to-end system level coordination, facilitated by various middleware technologies such as HLA RTI, CMDR and CoABS grid. While the agents did not specifically communicate with each other in order to achieve some desired level of coordinated activity within the federation, the infrastructure does allow the agents to send and receive messages in order to enable their ability to communicate.

In [3], while the author's specifically describe agents to monitor plans, additional agents that are described for future development in the paper include plan understanding agents. Through the development of appropriate plan understanding agents, such agents would be able to determine critical events and relationships within a plan so that this information could be provided to the monitoring agents, which would then focus on monitoring the important aspects of a plan. The authors suggest that techniques such as natural language processing and sublanguage ontologies may be useful to extract key events from military operational orders or through leveraging XML-based techniques such the Battle Management Language [3].

2.4 Coalition Interoperability

Coalition operations are characterized by data overload, concerns who to share information with, dealing with security issues and the need to overcome the stovepipe nature of systems – all of which contribute to the challenges in achieving coalition interoperability. The goal of the Coalitions Agents Experiment (CoAX) was to demonstrate that coordination through multi-agent techniques could provide the mechanism to improve interoperability between systems. The CoAX [18] was an international effort between the Defense Advanced Research Projects Agency (DARPA), Defense Science Technology Office (DSTO), Defense Scientific Technical Laboratory (DSTL), and The Technical Cooperation Program (TTCP). The CoAX participants included the Department of Defense (DoD) laboratories, industry as well as academia, which were brought together with the purpose of demonstrating the value of multi-agent systems to support the ability to rapidly construct and maintain a coalition Command and Control structure. The experiment leveraged investment from the DARPA Control of Agent-based Systems (CoABS) program, specifically the CoABS grid framework.

In the context of the scenario, the goals of CoAX were to demonstrate that stovepipe barriers could be overcome across systems through techniques such as policy enforcement to bind the behavior of the agents in accordance with the coalition partner's requirements so that some form of trust could be achieved. For instance, services provided through the Knowledgeable Agent-Oriented System (KaOS) [8] assured that the agents would follow certain operating policies. The CoABS grid provided the infrastructure for the agents by providing registry, communication and logging services. There were dozens of agents that were developed and integrated using the CoABS grid in support of the scenario. In one particular instance within the scenario, agents must deconflict air plans, and this is accomplished through a multi-level coordination agent. This agent is built upon the notion of sharing plans that are represented hierarchically, and through the upward propagation of the conditions that hold true and the timing constraints between subplans, relationships between subplans at different levels of abstraction can be identified. Hence, agents can share these various subplans with each other at the appropriate level of detail in order to coordinate their activities.

2.5 Tiered Systems

A key enabler of a sustainable military force is the notion of a tiered system. A tiered system is an integrated, multi-tier intelligence system encompassing space and air-based sensors linked to close-in and intrusive lower tiers [6]. The lower tiers (e.g., UAVs) are not only the critical source of intelligence; they can also serve as a key cueing device for other sensors. It should be noted that tiered-system components such as UAVs or space-based assets are not only useful for ISR activities supporting more traditional combat operations, but may also enable effective SSTR operations. Given the diversity of the assets, and the fact that coordination must be achieved both in the horizontal and vertical planes, and the environments in which the components of a tiered system will operate; it is not likely that a single coordination approach or even a family of coordination approaches will work well from a static perspective. It is more reasonable to expect that systems should learn which approaches work well and under which circumstances, and adapt appropriately.

It is commonly recognized that many different disciplines can contribute to a better understanding of the coordination function as a way to build and provide appropriate tools and adequate approaches for designing complex organizations and systems. Agents incorporated into these self-regulating entities represent "communities of concurrent processes which, in their interactions, strategies, and competition for resources, behave like whole ecologies". This paper has described a diverse set of application domains in which coordination of multi-agent systems has improved the behavior of the overall system. A common characteristic of these applications is the highly distributed nature of them.

3 Transportation

3.1 Air Traffic Flow Management

Air traffic congestion is a world-wide problem leading to delays as well as subsequent monetary loss to the commercial flight industry. The ability to either manage the flow of traffic more efficiently or improve the infrastructure (e.g., adding more runways) to increase capacity are two available mechanisms to improve the efficiency of commercial aviation. Given the fact that increasing the capacity through improvements to the infrastructure is a costly and very time consuming solution, information technologies may be useful in helping to efficiently manage the traffic flow.

The Optimal Aircraft Sequencing using Intelligent Scheduling (OASIS) [14] is an intelligent agent-based system which is able to ingest data from Sydney Airport at the rate of 65 aircraft arriving in 3.5 hours. It is designed to assist the Flow Director (who is responsible for coordinating the landing sequences for aircraft) in arranging the sequencing of incoming aircraft through intelligent agents that represent the aircraft and supporting global agents. The aircraft agents are responsible for predicting aircraft trajectories, monitoring the actual movement versus predicted to understand discrepancies, and engaging in planning activities. The global agents consist of a coordinator agent, sequencer agent, trajectory manager agent, wind model agent and user interface agent. The coordinator agent is the central node for interacting with all of the other agents in the system. The sequencer agents uses search techniques based on the A* search algorithm to determine the arrival sequence which minimizes delay and overall cost. The trajectory manager agent ensures that aircraft maneuvers do not violate statutory separation requirements. The wind model agent is responsible for computing and providing the forecast winds based on the reported wind by the aircraft agents. Lastly, the user interface agent is the mechanism by which the flow director receives information from the system.

The planner agents accept a schedule time from the sequencing agent and decide how to execute that plan to meet that schedule, and these agents must monitor the actual movement against the plan and be able to predict the trajectory. If there are discrepancies, the aircraft agent must notify the global agents so that another sequence may be computed. The agents coordination is facilitated through communicate of their activities using the Procedural Reasoning System (PRS). The PRS system is an agent-based framework for agents to coordinate their activities with each other based on the Belief, Desires and Intention (BDI) model. The PRS framework is comprised of knowledge bases that contain agent beliefs as represented through first order logic, agent's goals or desires, a library of plans and a reasoning engine. Depending on the agent's beliefs and desires, the appropriate plans are triggered for execution, which in turn affect the environment.

The OASIS system is compared against COMPAS, MAESTRO and CTAS. The COMPAS system is developed for the German Civil Aviation Authority to provide flow control. It has been pointed out that the key difference between COMPAS and OASIS is that the former only handles single runway sequences and the aircraft's progress is not monitored. The MAESTRO system is developed for the French Civil Aviation Authority. Similarly to COMPAS, MAESTRO only considers a single runway, but does provide aircraft monitoring capabilities to provide a comparison of the progress against the plan. The CTAS is being developed by NASA-AMES for the US Federal Aviation Administration, and is described as more complex than either MAESTRO or COMPASS.

3.2 Other Transportation and Network Management Applications

Agents are deployed in this problem domain to cooperatively control and manage a distributed network. Agents can also handle failures and balance the flows in the network. Various approaches are developed in [4] to route packets in static and ad-hoc networks. DAI offers suitable tools to deal with the hard transportation problems [9]. Fischer describes in [9] the modeling autonomous cooperating shipping companies system (Mars), which models cooperative order scheduling within a society of shipping companies. Three important instances for DAI techniques that proved useful in the transportation application, i.e., cooperation among the agents, task decomposition and task allocation, and decentralized planning are presented in this paper. Indeed, the problem tackled in this paper is a complex resource allocation problem. The auction mechanism is used for schedule optimization and for implementing dynamic replanning.

Coordination mechanisms are also used for distributed vehicle monitoring problems or traffic management. The objective in such problems is to maximize the throughput of cars through a grid of intersections or a network. Each intersection is equipped with a traffic light controlled by an agent. The agents associated to different intersections need to coordinate in order to deal with the traffic flows [19,7]. Another application concerns the Air fleet control or airspace deconfliction. A multi-agent approach for air fleet control is reported in [22]. In this application, the airspace used by the traffic controllers is divided into three-dimensional regions to guide the airplanes to their final destination. Each region can hold a limited number of airplanes. The multi-agent solution is used to guide the planes from one region to another along minimal-length routes while handling real-time data.

4 Healthcare - Monitoring Glucose Levels

The ability to carefully monitor blood sugar levels is critical in the essential care of diabetes, which can lead to serious health problems and even death if not treated in a timely manner. A prototype application called the Integrated Mobile Information System (IMIS) for diabetic healthcare is described in [17]. The IMIS demonstrates how multi-agent coordination can be leveraged to support critical collaboration between healthcare providers in the Swedish health care industry in support of monitoring diabetic patients.

Two types of problems are identified in [21] with regard to the proper care of diabetic patients, that of accessibility and interoperability. The accessibility problem is further categorized as physical service accessibility and information accessibility. The former implies that medical resources should be available to all healthcare providers while the latter implies that the appropriate information must be available from those resources. The key challenge identified in the diabetic healthcare system in one specific province, Blekinge, is information accessibility across the municipality and county council, each of which is responsible for various aspects of healthcare. For example, the municipality is responsible for the elderly, disabled or non-medical healthcare while the county council is responsible for medical and surgery related care. The various medical records cannot generally be shared between county council and municipality. The interoperability problem is one of overcoming the various forms in which the information is stored; in other words, there is a need for an information sharing standard to enable semantic interoperability.

The IMAS system is a multi-agent system that supports coordination between agents through the act of communication, in order to provide decision support to the human actors to support their collaborative activities. It was developed from the IMIS system and supports the collection and manipulation of healthcare information, and subsequent presentation of that information to the human actors based on their preferences in order to help them better collaborate. Three collaboration levels are identified in this specific problem domain: Glucose management, Calendar arrangement and task delegation. Glucose management involves providing the healthcare actors, perhaps through a centralized database, with the right information so that they can take the appropriate actions. Calendar arrangement includes the ability to schedule patient visits, while task delegation involves the ability of the healthcare providers or software agents to take certain actions (and the ability to report on the completion of those actions) based on the diagnosis.

There are four types of agents implemented in the IMAS system: PatientAgent, ParentAgent, Hospi-talNurseAgent and SchoolNurseAgent. These agents act on behalf of their user's preferences in order to better coordinate their activities through the communication of appropriate information (e.g., glucose levels). For example, in the scenario described in the paper [21], if a young boy (named Linus) has an elevated sugar level which is detected by the PatientAgent through readings from a database (which is updated by Linus via the IMAS patient control panel installed on his mobile device), then a message will be sent to the ParentAgent and SchoolNurse Agent to give Linus an insulin shot. When the alarm is received by the ParentAgent, this agent will search for additional information or charts to present to the mother, who will take the appropriate action. The IMAS system was implemented using the Java Agent Development Environment (JADE) and uses an ontology and uses the Foundation for Intelligent Physical Agents (FIPA) compliant messages for communication, and more specifically, the FIPA Agent Communication Language (ACL). The ACL is based on speech act theory, which is based on the fact that when one makes a statement, it is implied that one is doing something.

Future areas for investigation include techniques to ensure the privacy of the information so that only the right individuals or agents have access, and also the necessity to properly deal with the security of the information that is being transmitted so that it cannot be altered by third parties.

5 Communications Networks

5.1 Fault Diagnosis

As with any communication network, telecommunications networks are inherently distributed with expertise and data residing throughout various nodes of the network, hence, the management of the network requires access to, and analysis of, data that resides in various parts of the network. Due to the distributed nature of these networks, monitoring and diagnosing faults can benefit from a distributed approach as can be offered through multi-agent system techniques. The Distributed Intelligent Monitoring and Analysis System (DIMAS) system is described in [12], which builds upon the more centralized Intelligent Monitoring and Analysis System for monitoring and diagnosing faults. Specifically, the paper describes a distributed architecture based on a multi-agent approach for monitoring and detecting faults in cellular communication networks.

The problem which is described here consists of monitoring and detecting faults across cellular base stations. Since multiple base stations may use the same frequency for communication due to the limited number of frequencies available, this has the effect of increasing the capacity of the network. Interference of frequencies across base stations is minimized by limiting the coverage and strength of the radio signal at each base station. However, in certain cases, natural phenomena such as winds or storms can cause the base station antenna to misalign. Such as misalignment can cause the frequencies to interfere, thereby affecting other nearby base stations by causing "noise" on the frequencies which overlap.

The DIMAS system is applied to monitoring this problem and diagnosing which base station is causing the frequency interference. The DIMAS system is comprised of a data store, data filter and expert system which is interfaced to a network simulator. The data store contains information obtained from the network, the data filter enables the processing of a large amount of information, and the expert system is the engine used to provide the diagnostic capabilities based on information obtained from the data store. The expert system in DIMAS is built on the AT&T C5 language, and its workflow processes through a series of steps: monitor symptoms, reply to queries, do actions, analyze results, analyze evidence and decide next step.

The authors also differentiate the DIMAS system with other related work. For example, they compare DIMAS to Distributed Big Brother (DBB). The DBB uses the Contract Net Protocol to distribute the monitoring task to Local Area network managers. However, within DIMAS, rather than offering tasks to potential bidders and then awarding the task to one bidder, the DIMAS agents are asked whether they can suggest possible actions to diagnose a problem. The suggested actions are then evaluated by the temporary owner of a problem and then contracted out to the agent that suggested the action. For example, each agent that detects a symptom of a problem may suggest further diagnostic tests, such as checking for changes to base station parameters or monitoring the signal strength of call attempts. However, for each action there is only one agent that can execute it. The decision that must be made by the coordinating agent is which action to choose, rather than which agent will execute that action. The subtle difference is that in DIMAS, the contracting is action-centric and not agent-centric, as a series of actions may be spread over multiple agents.

The Large-internetwork Observation and Diagnosis Expert System (LODES) is an expert system for diagnosing faults in a LAN that consist of multiple networks connected via routers. Each LAN has its own LODES system, which is used to detect remote faults based on local data. The primary difference between LODES and DIMAS is that the latter is a passive approach while the former is both passive as well as active.

In DIMAS, the agents can only observe local symptoms and/or faults. Coordination between the agents is facilitated through the use of communication using KQML performatives such as ask, tell and achieve to gather non-local information. These performatives are used in various phases of the agent's process such as assign-responsibility, gather-evidence and allocate-blame. In the assign-responsibility phase, the problem is detected using the ask and tell performatives. The base station at which the symptom is detected uses the ask performative to query the other base station agents about possible causes for increase in calls with poor transmission quality, while the other base stations use the tell performative to report back as to the fact that the cause may be due to a misaligned antenna or a frequency change. The affected base station agent then uses the achieve performative to ask the other base station agents to check for changes in frequency within their logs (gather-evidence). When this is ruled out, then more expensive requests are made such as checking for conditions which are indicative of a misaligned antenna. The allocate-blame phase assigns responsibility to the offending base station, and communicates its findings to other base stations.

5.2 Routing in Telecommunications

A market-based approach to routing in telecommunication networks using a multi-agent system model is described in [10]. It is argued that a market-based approach requires no a priori cooperation between agents and provides a robust mechanism for routing. Specifically, the approach relies on an adaptive price setting and inventory strategy based on bidding in order to reduce call blocking. The approach relies on several layers of agents, specifically, link agents that interact in link markets to purchase slices of bandwidth, path agents that interact in path markets to sell paths and on link markets to buy necessary links in order to offer paths on the path market, and lastly the call agents. Each of these agents buys or sells their commodity on the appropriate market. The author compares the market based approach to routing with other approaches. For example, in a centralized scheme where traffic predictions are used to compute the paths there are issues with scalability particularly as the size of the network grows, the amount of information to monitor and process also increases and there is a single point of failure. Similarly, the authors argue that in optimal network flow algorithms there are performance issues in heavily loaded networks and unpredictable oscillations between solutions. Hence, the author suggests that a decentralized solution may be appropriate. However, the authors also describe a few challenges in decentralized solutions. For instance, they indicate that such solutions may have non-local effects on the entire network and it would be difficult to understand these effects due to the lack of complete information about the entire network. Complete information is infeasible since the network is dynamic and there is a delay in propagation and therefore the solution is prone to error and there may also be scaling issues.

As previously mentioned, the system architecture is comprised of link agents, path agents and call agents, each operating on one or several markets. A link agent and path agent is associated with each link and path, respectively, in the network. A call agent is associated with each source and destination pair in the network. Link and path markets are sealed bid double blind auctions in order to minimize any delays in the system, as callers will not want a delay in establishing a circuit in the network. The agents buy and sell the appropriate resource for which they are responsible, and depending on what they have sold and/or purchased, they use a simple function in order to either increase or decrease their buying/selling price. Similarly, path agents maintain an inventory that is profitable. Lastly, call agents trigger the auctions at the path and link agent level.

The authors present various results from their experiments. In the first experiment, they compare Vickery and First price auction strategies in their market based approach to static routing on a 200 node network with link capacities for 200 channels. They show that they can do at least as well as static routing, but with the added advantage of achieving the performance level through an open architecture without the need to rely on static network policies. They demonstrate improved scalability through their approach, since the entire network topology need not be known in advance and, furthermore, using their approach are able to provide a quicker response as information is not needed about the broader network. The authors also provide results that demonstrate that about 61% of the time, the call was routed through the most efficient route and 46% of the time the call was routed through the least congested route. Research areas that are left for future work include the ability of the callers to request different types of services, which might be routed differently through the network and the ability for market-based routing to support such activities.

6 E-Business

6.1 Supply Chain Management

A supply chain is described in [5] as "a network of suppliers, factories, warehouses, distribution centers and retailers through which raw materials are acquired, transformed, reproduced and delivered to the customer". A Supply Chain Management System (SCMS) manages the interaction of these processes in order to achieve some end goals.

In [5] the authors describe a SCMS based on the coordination of multi-agent systems, as represented through information agents and functional agents. The information agents are a source of information to the functional agents, while the functional agents are responsible for specific roles in the dynamic supply chain. The key notion behind these agents is that there is no assumption that a certain number of agents be available in the supply chain. The agents are assumed to enter and leave the system through a negotiation process based on the negotiation performatives that have been developed based on the FIPA ACL. For example, accept-proposal, CFP, proposal, reject-proposal and terminate are used for pairwise negotiation between agents while a third performative, bid, has also been included to capture third party negotiations, for example, through an auctioneer. The paper [5], further describes the use of Colored Petri Nets (CPN) as a modeling tool for managing the negotiation process between agents. A simple example of the interaction between buyer and seller agents is described through the use of CPNs. The buyer agent initiates the negotiation process by sending a message indicating a call-for-proposal from the seller agent. The buyer agent then waits, while the seller agent receives the message for consideration. When the seller agent sends the reply (proposal, accept-proposal, reject-proposal or terminate) to the buyer agent, the buyer agent then enters the waiting or thinking/inactive state. If the thinking transition is triggered, the cycle repeats and it is up to the buyer agent to send the next round of message. The paper also describes the internal logic of the functional agents in making decisions by providing an example of an agent that represents a company's need for certain quantity of supplies in a given time period. This company is represented through an agent that sends a CFP message to other agents. The other agents evaluate the CFP and provide the quantity of supplies that can be provided, their cost and lead time. After receiving the other agents offers the initiating agent constructs a search tree for evaluating the constraints. If a solution can be found, then the initiating agent will accept one or more offers. If a solution cannot be found, either the initiating agent will relax its constraints, or ask the other agent to relax their constraints.

Several key issues are described for future research, including understanding the convergence behavior of the network and strategies for deciding how and when to relax the agent constraints.

6.2 Manufacturing Systems

Modern and big manufacturing depends heavily on computer systems. In fact, in many manufacturing applications the centralized software is not as effective as distributed networks. Indeed, to provide competitiveness in global markets, manufacturers must be able to implement, resize, design, reconfigure, respond to unanticipated changes, and maintain manufacturing facilities rapidly and inexpensively. These requirements are more easily satisfied by distributed small modules than by large monolithic systems. Moreover, small modules allow system survivability. Multi-agent systems offer a way to build production systems that are decentralized rather than centralized, emergent rather than planned, and concurrent rather than sequential. Besides, the emergence of internet-based computing and communication to execute business processes has emerged as a key element in the supply chain domain. By using the internet, businesses gain more visibility across their network of trading partners, and it helps them to respond quickly to customer demands, resource shortages, etc. Min and Bjornsson present a method based on computer agents' technology for building virtual construction supply chains [16].

7 Emergency Management and Disaster Relief

The emergence of new doctrine is enabling Security, Stabilization, Transition and Reconstruction (SSTR) operations to become a core U.S. military mission. The immediate goal in SSTR is to provide the local populace with security, restore essential services, and meet humanitarian needs. Large scale disasters are an example where military support can improve SSTR operations by providing a much needed boost to foreign governments and non governmental organizations (NGOs) which may already be under great stress to respond in a timely and effective manner. However, without the means to effectively coordinate the efforts between many diverse groups across the civil-military boundary during SSTR operations, basic assistance and relief operations may be severely impeded. There are many operational challenges in the ability to coordinate such a large group across the civil-military boundary during SSTR operations. Usually, in the civil sector, coordination is the result of voluntary effort; hence coordination by "directing" is rarely effective. Generally, relief agencies partly function within a framework of self-interest. In other words, they assist their beneficiaries in such a way that their good works are seen and valued by the donor community and the "profile" of their agency is enhanced. In many cases, farther down on the list is the goal of recognizing the contribution of others or admitting someone else can do the job better and therefore coordination is not necessarily an agencies first priority. With regard to coordination across the civil-military boundary there are also a number of challenges. The military tends to be a highly structured organization with a hierarchical command and control structure, while the civil sector in many cases is more loosely structured and tend to be less formal.

Understanding emerging social networks and using the information that is provided through such networks is important for effective coordination. Social networks provide a visual cue as to the relationships between people or groups and can be used to identify "mutual" friends. This is particularly important when there is a requirement by one or more individual or groups to locate other individuals or groups that might be able to provide support in order to achieve a desired end result. Research and tools from the Social Network Analysis (SNA) community may allow users to understand who the experts are in the SSTR community and to whom and how they are linked. As a simple example, social maps that depict connectivity between users in the context of their discussion threads, and ability to filter the content within a social map based on specific keywords are likely to provide the foundation to enable the community to identify service providers or those that may offer similar services or capabilities. The ability to rate individuals within the social network may also be an important aspect in building trust within the community of users. This is particularly important during pre-disaster situations so that some level of trust and common understanding can be achieved. Furthermore, pre-disaster interactions can help in the development of concept of operations or doctrine to provide guidance during real life situations by helping people or organizations form the bonds of working together. One of the challenges, however, will be to effectively visualize such a network or efficiently filter through the various dimensions of information contained in the social network. There is also a lack of automated coordination tools to support the ability of people or groups to actively coordinate their activities. There are processes in place but most coordination is manual. The ability to manage tasks across the diverse actors involved in these operations has the opportunity to improve the overall efficiency of these operations. Due to the complexity of SSTR operations, multi-agent coordination techniques may provide benefits.

There is already ongoing research into developing information sharing environments to be used during SSTR operations within which automated coordination tools could be developed. These environments leverage content management systems and web technologies to integrate various collaboration mechanisms such as weblogs, wiki's and forums. One of the key uses of such sites would be to enable service requestors to find service providers during crisis situations. As potential service providers are identified from the information sharing site's content, additional techniques could be developed and applied to transform and map the services that are being requested into appropriate tasks. These tasks would be defined by appropriate users of the system, and would be defined such that when executed, would enable the services providers to meet the needs of the service requestors. The tasks may be hierarchical or peer-to-peer. For instance, hierarchical tasks may be appropriate for the military as protocols tend to be very structured and organized, while tasks for the civilian side may be peer-to-peer. However, the determination and selection of the appropriate task representation is best handled locally and there is no strict requirement that organizations use one representation or the other.

In order to automate task scheduling and assignment to a degree, the tasks and potential task performers might be organized into an assignment matrix, and the entries in the matrix would contain numerical values that represent the cost for doing each of the tasks across each of the performers (Figure 2). The costs may be monetary or may be based on a more complex cost function and it is up the participants to agree upon and choose the appropriate cost model. The computation of the cost functions across performers may vary, so some normalization techniques may need to be applied. Once these costs are determined, the goal of the matching algorithm would be to find a set of assignments between tasks and performers in the matrix that globally minimizes the total cost of doing the tasks across all of the performers. Alternatively, numerical values for preference could be included instead of cost in the assignment matrix, and the objective would be to find a set of assignments that globally maximize the preference.

The task network would provide a visualization of the tasks and would allow the user to modify the tasks or include new tasks, which would be subsequently reflected in the assignment matrix. The task network would also allow the specification of preconditions that need to be true before a task or subtask can be executed. The assignments that are calculated based on the algorithm would update the task network to show which performers have been assigned to various tasks. Multi-agent coordination techniques could be employed to support a negotiation process to allow the performers to interact with each other to change their tasks assignments for various reasons that might not have been accurately reflected in the cost functions. The negotiation process could be facilitated, for example, through various coordination techniques such as an auction mechanism. Such a negotiation would be pursued until task equilibrium is reached. Once equilibrium is reached, the passing of appropriate messages using any method chosen by the user would permit the transmission of the assignments to potential responders for action or renegotiation.

The ability to coordinate a large and diverse group of first responders begins with the ability to communicate guidance or orders, while receiving situation reports from those in the field. The lack of a stable communications infrastructure will negatively impact the efforts of those that need to coordinate and share information. Recent technological advances in Mobile Ad-Hoc Networks (MANET) are key enablers in the deployment of net-centric cooperative multi-agent systems in disaster areas. MANET technology holds the promise of enabling communications between first responders when the local communications infrastructure is unusable. These networks support mobile entities, connected through a wireless network which supports discovery and self-organization through peer-to-peer message exchanges, leading to an increase in the robustness of the overall network.

The concept of "NETwork-Aware" Coordination and Adaptation is a potential area worthy of exploration. In such an approach, the users or applications are aware of the state of the network, thereby allowing the applications to adapt in order to "work around" network constraints, while the network is aware of the state of the applications or mission needs in order to better handle traffic flows. Such cross-layer information exchange is important to enable a more robust communication strategy for the first responders in order to support their coordination activities. To the extent possible, coordination strategies also have to be robust against message loss and equipment failures.

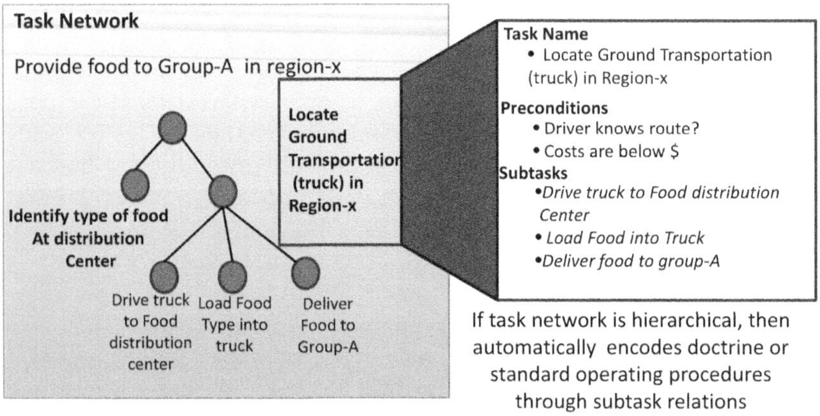

Performers

Tasks	NGO-1	NGO-2	NGO-3
Drive Truck	0.1	0.2	0.3
Load Food Truck	0.2	0.4	0.6
Deliver Food	0.3	0.6	0.9

Fig. 2 Task Assignment Concept

A few of the research issues in network-aware coordination include defining measures for determining network congestion or other types of failures such as loss of connectivity within the network, in order to provide such measures and parameters to the application layer. The key challenges for the application layer include how to best utilize that information in order to adapt communication strategies (e.g., sharing images that are smaller in size, prioritizing certain information, or identifying certain nodes to act as communications relays). Such a feedback loop may be continuous, so that the network could support larger bandwidth exchanges as congestion is proactively alleviated in the network.

8 Ambient Intelligence

The early developments in Ambient Intelligence took place at Philips in 1998. The Philips vision of Ambient Intelligence is: "people living easily in digital environments in which the electronics are sensitive to people's needs, personalized to their requirements, anticipatory of their behavior and responsive to their presence". The main purpose of Ambient Intelligence applications is to coordinate the services offered by small smart devices spread in a physical environment in order to have a global coherent intelligent behaviour of this environment. From a computing perspective, ambient intelligence refers to electronic environment that is sensitive and responsive to the presence of people. Ambient intelligence paradigm builds upon several computing areas and human-centric computer interaction design. The first area is ubiquitous or pervasive computing. Pervasive computing devices are very tiny (can be invisible) devices, either mobile or embedded in almost any type of object imaginable, including cars, tools, clothing and various consumer goods. All these devices communicate through increasingly interconnected networks (ad hoc networking capabilities). The second key area is intelligent systems research. Learning algorithms and pattern matchers, speech recognition and language translators, and gesture classification belong to this area of research. A third area is context awareness; research on this area lets us track and position objects of all types and represent objects' interactions with their environments. See [15] and [23] for more details about Ambient Intelligence applications.

9 Conclusion

Information technologies and information systems have become the cornerstone that shapes the present and the future of our society and its underlying affairs. They have penetrated almost every aspect of our lives. This omnipresence is stemming from significant advancements in computer systems, software, middleware, telecommunications infrastructure, etc. Also, enterprises, organizations and governments usually use a large variety of networked information systems, interconnected software, etc. From other side, military organisations as well as military coalitions share more information than before, such that decision making occurs at all levels within the chain of command. Indeed, current information technology gives organisations the opportunity to take advantage of all available information and coordinates the different actions. Consequently, coordination is becoming very important to consider and to take into account.

Emerging application domains such as Ambient Intelligence, Grid Computing, Electronic Business, Semantic Web, Bioinformatics, and Computational Biology will certainly determine future agent-oriented research and technologies. Over the next decade, it is thus expected that real world problems will impose the emergence of truly open, fully scalable agent-oriented systems, spanning across different domains, and incorporating heterogeneous entities capable of reasoning, learning, and adopting adequate protocols of communication and interaction.

References

1. Beard, R.W., McLain, T.W., Goodrich, M.A., Anderson, E.P.: Coordinated Target Assignment and Intercept for Unmanned Air Vehicles. IEEE Transactions on Robotics and Automation 18(6), 911–922 (2002)
2. Bedrouni, A., Mittu, R., Boukhtouta, A., Berger, J.: Distributed Intelligent Systems: A coordination Perspective. Springer, New York (2009)
3. Carey, S., Kleiner, M., Heib, M., Brown, R.: Standardizing Battle Management Language- Facilitating Coalition Interoperability. In: Proceedings of the Euro Simulation Interoperability Workshop, SIW (2002)
4. Chang, Y.H., Ho, T., Kaelbling, L.: Multi-agent learning in mobilized ad-hoc networks. In: Proccedings of AAAI Fall Symposium, Technical Report FS-04-02 (2004)
5. Chen, Y., Peng, Y., Finin, T., Labrou, Y., Cost, S., Chu, B., Yao, J., Sun, R., Wilhelm, B.: A Negotiation-based Multi-agent System for Supply Chain Management. In: Workshop on Agents for Electronic Commerce and Managing the Internet-Enabled Supply Chain, Seattle, WA (1999)
6. Dahlburg, J.: Developing a Viable Approach for Effective Tiered Systems. NRL Memorandum Report 1001-07-9024. USA (2007)
7. Dresner, K., Stone, P.: A Multiagent Approach to Autonomous Intersection Management. Journal of Artificial Intelligence Research 31, 591–656 (2008)
8. Feltovich, P., Bradshaw, J.M., Jeffers, R., Uszok, A.: Order and KAoS: Using policy to represent agent cultures. In: Proceedings of the AAMAS 2003 Workshop on Humans and Multi-Agent Systems, Melbourne, Australia (2003)
9. Fischer: Cooperative Transportation Scheduling: An application Domain for DAI. Applied Artificial Intelligence 10(1), 1–34 (1996)
10. Gibney, M.A., Jennings, N.R., Vriend, N.J., Griffiths, J.M.: Intelligent Agents for Telecommunication Applications. In: Proceedings of the 3rd International Workshop IATA (1999)
11. Goldman, C.V., Zilberstein, S.: Decentralized Control of Cooperative Systems: Categorization and Complexity Analysis. Journal of Artificial Intelligence Research 22, 143–174 (2004)

12. Leckie, C., Senjen, R., Ward, B., Zhao, M.: Communication and Coordination for Intelligent Fault Diagnosis Agents. In: Proceedings of the Eighth IFIP/IEEE International Workshop DSOM 1997, Sydney, Australia, pp. 280–291 (1997)
13. Lian, Z., Deshmukh, M.A.: Performance prediction of an unmanned airborne vehicle multi-agent system. European Journal of Operational Research 172, 680–695 (2006)
14. Ljungberg, M., Lucas, A.: The OASIS Air Traffic Management System. In: Proceedings of the Second Pacific Rim International Conference on Artificial Intelligence, PRICAI 1992 (1992)
15. McCullough, M.: Digital Ground: Architecture, Pervasive Computing, and Environmental Knowing. MIT Press, Cambridge (2005)
16. Min, J.U., Bjornsson, H.: Construction Supply Chain Visualization through Web Services Integration, Technical Report, Civil & Environmental Engineering Dept., Stanford University (2004)
17. Mittu, R., Abramson, M.: Walters. J.: Improving Simulation Analysis through C4I Systems and Use of Intelligent Agents. In: Proceedings of the Spring Simulation Interoperability Workshop, SIW (2004)
18. Mittu, R., Segaria, F., Guleyupoglu, S., Barber, S., Graser, T., Ross, R.: Supporting the Coalition Agents Experiment (CoAX) through the Technology Integration Experiment (TIE) Process. In: Proceedings of the 8th International Command and Control Research and Technology Symposium (ICCRTS), July 17-19, National Defense University, Washington (2003)
19. Nunes, L., Oliveira, E.: Learning from multiple sources. In: AAMAS Proceedings of the Third International Joint Conference on Autonomous Agents and Multiagent Systems, New York, NY, USA, pp. 1106–1113 (2004)
20. Panait, L., Luke, S.: Cooperative Multi-Agent Learning: The State of the Art. Autonomous Agents and Multi-Agent Systems 11, 387–434 (2005)
21. Peng, Z., Guohua, B., Bengt, C., Stefan, J.J.: Applying Multi-agent Systems Coordination to the Diabetic Healthcare Collaboration. In: Proc. The Int. Conference on Information & Communication Technologies: from Theory to Applications- ICTTA, Damascus, Syria (2008)
22. Steeb, R., Cammarata, S., Hayes-Roth, F., Thorndyke, P., Wesson, R.: Distributed intelligence for air traffic control. In: Bond, A., Gasser, L. (eds.) Readings in Distributed Artificial Intelligence, pp. 90–101. Morgan Kaufmann Publishers, San Francisco (1988)
23. Wang, X., Dong, J.S., Chin, C., Hettiarachchi, S.R., Zhang, D.: Semantic Space, A Semantic Web Infrastructure for Smart Spaces. IEEE Pervasive Computing 3(3), 32–39 (2004)

Multi-Agent Area Coverage Using a Single Query Roadmap: A Swarm Intelligence Approach

Ali Nasri Nazif*, Alireza Davoodi*, and Philippe Pasquier

Abstract. This paper proposes a mechanism for visually covering an area by means of a group of homogeneous reactive agents through a single-query roadmap called Weighted Multi-Agent RRT, *WMA-RRT*. While the agents do not know about the environment, the roadmap is locally available to them. In accordance with the swarm intelligence principles, the agents are simple autonomous entities, capable of interacting with the environment by obeying some explicit rules and performing the corresponding actions. The interaction between the agents is carried out through an indirect communication mechanism and leads to the emergence of complex behaviors such as multi-agent cooperation and coordination, path planning and environment exploration. This mechanism is reliable in the face of agent failures and can be effectively and easily employed in cluttered environments containing narrow passages. We have implemented and evaluated the algorithm in different domains and the experimental results confirm the performance and robustness of the system.

1 Introduction

The area coverage problem in robotics deals with the use of one or more robots to physically sweep [8] or visually sense [5]the free space of an area. The aim of this research study is to propose a robust mechanism to cope with the problem of multi-agent visual environment coverage.

Ali Nasri Nazif · Alireza Davoodi · Philippe Pasquier
Metacreation, Agents and Multiagent Systems (MAMAS) Group, School of Interactive Art and Technology, Faculty of Communication, Arts and Technology, 250 - 13450 102nd Avenue, Surrey, BC, Canada, V3T 0A3
e-mail: anasri@aut.ac.ir, {alireza_davoodi,pasquier}@sfu.ca
http://www.siat.sfu.ca

* The first and second authors have equally contributed in the paper.

Q. Bai and N. Fukuta (Eds.): Advances in Practical Multi-Agent Systems, SCI 325, pp. 95–112.
springerlink.com

To realize this purpose, we apply an emergent coordination approach and introduce a bio-inspired technique which utilizes a roadmap called WMA-RRT and employs a group of multiple autonomous agents to cover a given environment. To this end, agents should be capable of performing in the area and cooperating with each other to achieve the mutual objective of covering an area.

In accordance with the *Swarm intelligence* principles originally introduced by Beni and Wang [3], the proposed system is made up of several simple autonomous agents. These agents locally interact with the environment and pursue their own goals by following some simple and explicit condition-action rules and performing the their corresponding actions. In such a scenario, the multi-agent cooperation is an invaluable by-product of a special communication mechanism. In the other words, the direct agents' interaction with the environment leads to the emergence of indirect communications and consequently teamwork among the agents.

In this research study, we assume that the structure of the environment and a 2D map of the area are available. Furthermore, the robots are considered to be point robots for the sake of simplicity, and it does not affect the solution. If the robots are circular robots of radius r, Minkowski sum of the static obstacles and the robots are computed and considered instead of the original static obstacles in the environment [11]. Moreover, the robots are assumed to have 360 degree field of view and they are able to observe part of the area which is not further than a pre-defined visibility range.

The approach to the environment coverage problem, presented in this paper, can be divided into two processes. Firstly, a special roadmap called WMA-RRT is constructed and embedded in the environment to discretely represent the free space. Secondly, WMA-RRT is distributed among all the agents. Each agent independently bears the responsibility of locally traversing its associated part of the WMA-RRT roadmap.

The agents employ the WMA-RRT both as a roadmap and an information exchange channel. This channel is used by the agents to interact with the environment and indirectly communicate to one another. In the other words, the agents change the situation of the environment by altering the states of the edges and nodes of the roadmap while traversing the roadmap tree. Meanwhile, this information is used by the agents in order to decide which action to take, where to go next, how to return to the origin, how to support other working agents and handle the agents' failures during the operation. We have implemented our approach and evaluated it considering some criteria such as the number of robots, different visibility ranges and maps.

The rest of the paper is organized as follows. First, in Section 2, we review some preliminary concepts and related works. The WMA-RRT roadmap is introduced in Section 3. The agents architecture is addressed in Section 4 and followed by the WMA-RRT traversing algorithm in Section 5. Also, the implementation of the mechanism and experimental results are presented in

Section 6. Finally, Section 7 concludes with a short summary of the research and introduces some future works.

2 Preliminaries and Related Works

2.1 Roadmaps

A roadmap is a graph constructed in the environment to capture the connectivity of the free space of the area. Generally, two types of roadmaps have been introduced by the researchers. The first group of roadmaps requires and utilizes the exact structure of the environment. `Visibility Graphs` and `Generalized Voronoi Diagram` belong to this group. Kai and Wurm [19] introduce an approach for environment exploration using `Voronoi Diagram`. They use `Voronoi Diagram` to decompose the environment into several regions and assign each of them to a robot to explore. Also, Jiao and Tang [17] employ a visibility-based decomposition approach for multi-robot boundary coverage in which the robots are supposed to collectively observe the boundary of the environment.

On the contrary, there are sampling-based roadmaps which do not need the explicit geometrical representation of the workspace. Instead, they use some strategies for generating sample points throughout the free space and connecting them to build the roadmap. Basically, two types of sampling-based approaches have been introduced by the researcher: `Multiple Query` and `Single Query`. In multiple-query planners [9], a roadmap is constructed by simultaneously expanding some trees from several randomly distributed starting points and merging them to prepare the entire roadmap. It is likely that the roadmap is not connected in a cluttered environment. Probabilistic roadmaps, `PRM`, introduced by Kavraki is the most popular multiple-query planner.

Single-query planners, on the contrary, have been introduced to construct a connected tree which spreads over the environment. They use an incremental approach to build the roadmap by expanding the tree from a random starting point. `RPP`, `Ariadnes Clew`, `2Z`, `EST`, `Lazy RPM` and `RRT` are single-query planners. Since the planner introduced in this paper is a variation of Rapidly-Exploring Random Tree `RRT` [10], this section briefly explains how `RRT` algorithm works.

The roadmap constructed by the `RRT` planner is a tree which expands throughout the free space of a given environment. This planner is probabilistically complete meaning that the probability of covering every location of the environment converges to one, if the algorithm keeps running for a long enough time.

In order to construct `RRT` roadmap, first a uniform distribution is used to insert a sample point q_{init} in the environment. This point is added as a node to the `RRT` tree if collision-free. In each iteration another sample point,

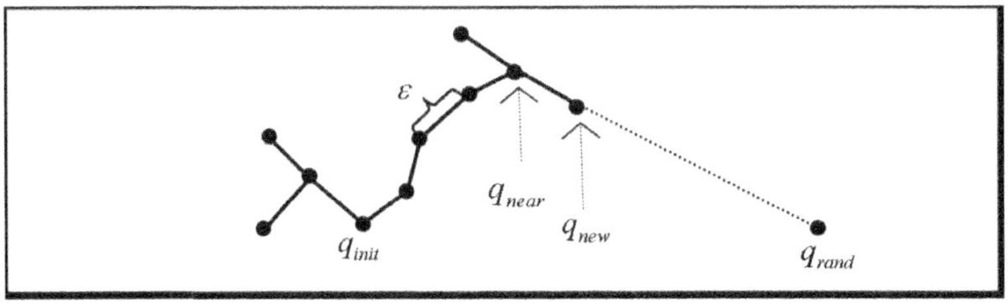

Fig. 1 This figure illustrates how a RRT tree is expanded given the current state of the tree.

q_{rand}, will be placed randomly in the free space and its nearest neighbor node, q_{near}, among previously added nodes will be selected for further expansion. A new node, q_{new}, is produced as a result of moving q_{near} by ϵ, a predefined value called `step size`, toward q_{rand} and it will be added to the roadmap if collision-free. Figure. 1 illustrates a snapshot of the expansion process.

2.2 Multi-Agent Environment Coverage

The Swarm Intelligence model, SI, relies on emergent coordination and is inspired by the collective behaviors observed in a multitude of species which exhibit social life such as ants and honey bees. As an AI technique, SI discusses the systems consisting of a population of simple agents interacting locally with one another or the environment [1, 15, 18] and these interactions often lead to the emergence of complex and global behaviors.

In one of the oldest attempt, Gordon and Wagner [6] employed honeybees' foraging behavior to coordinate a group of agents with no elaborate language and advanced communication device. Svennebring and Koenig [16] introduce a bio-inspired algorithm for covering an environment decomposed to a number of cells in which the robots use an artificial pheromone to mark the visited cells. Also Bayazit [2] discusses a multiple query roadmap used as a means of indirect communication between agents for covering an environment. Another application of indirect communications between agents is the work of Halasz [7] in which he deals with the dynamic redistribution of a swarm of robots among multiple sites to cover the environment.

Moreover, the swarming behavior model observed among birds flocks, fish schools and sheep herds motivated Reynolds [14] to introduce a bio-inspired technique to steer a group of rule-based autonomous agents, called Boid *birds-like object*. The Boids are endowed by some simple behaviors like seeking, fleeing, obstacle avoidance, wandering etc. The Boids make use of these actions to navigate the environment and represent more complex actions.

On the contrary, in intentional cooperation approaches, the agents deliberately communicate or negotiate with each other in order to cooperatively achieve a mutual goal. Zlot [20] addresses a market-based mechanism for environment exploration in which they use an auction based method to assign exploration targets between the agents. Furthermore, Choset [4] surveys different approaches such as approximation, grid-based and decomposition approaches. In one of the recent attempt Packer [12] applies a computational geometry approach in order to compute multiple Watchman routes which can cover inside a given polygon. In this paper, she first solves the art gallery problem and finds a set of guards which can collectively cover the entire area. Then, the visibility graph of the obstacles and static guards is constructed and decomposed into several routes. Finally, each route is allocated to one agent to traverse.

Grid-based approaches are also used in environment coverage. Hazon and Kaminka [8] propose a grid-based mechanism in which the environment is decomposed into several same size cells on the basis of the size of the robots. Considering the initial positions of the robots, they find a spanning tree of the free cells which can be decomposed to some balanced sub-trees. Finally each sub-tree is assigned to a robot.

3 Weighted Multi-Agent RRT

`WMA-RRT` is an extension of the `RRT` planner and consequently, categorized as a single-query planner. Moreover, `WMA-RRT` roadmap is an infrastructure which supports simultaneous movements of many agents throughout the cluttered environments which contain narrow passages.

Construction of `WMA-RRT` begins by figuring out the location of the root of the roadmap which is called `Main Root` and continues to specify the nodes that are adjacent to the `Main Root` and located at the next level of the tree. These nodes are called `Secondary Roots`. The `WMA-RRT` is then constructed by expanding all the edges between the `Secondary Root` and `Main Root` in all directions throughout the free space.

In the construction of `WMA-RRT`, it is assumed that all the agents in the environment are situated close to each other and consequently can be surrounded by a simple polygon. A simple polygon is a polygon whose boundary does not cross itself. In order to construct this polygon, the convex hull algorithm [13] is applied to build a polygonal boundary enclosing all the agents. The convex hull algorithm considers the locations of the agents as points and constructs the minimal convex hull of the points regardless of the obstacles in the environment. Then the intersection point of the two longest diameters of the polygon is calculated and considered as the starting point for the construction of the `WMA-RRT` and called `Main Root` of the roadmap. If the intersection point is not collision-free, the intersection points of other next longest diameters of the polygon are considered. If the convex hull is a triangle, the intersection point of two longest edges of the triangle is considered as the `Main Root`.

In the second step, the goal is calculating and assigning one particular node of `WMA-RRT` to each agent. To this end, the algorithm considers all the

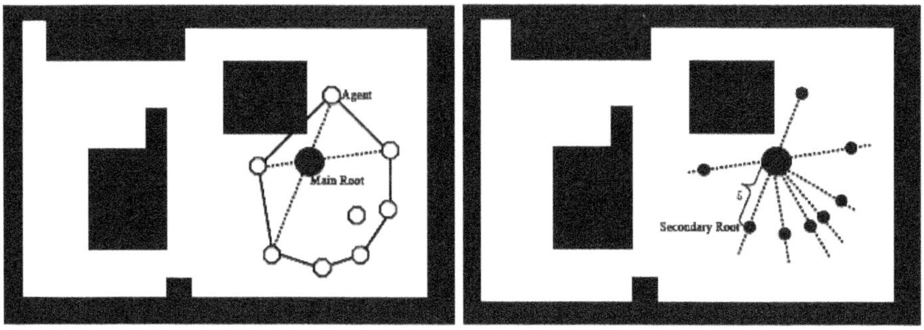

Fig. 2 Illustrates a situation in which eight agents (small white polygons) are available. The figure on the left represents the polygonal boundary enclosing the agents, the diameters and the `Main Root` (the black circle). The figure on the right illustrates the corresponding `Secondary Roots`

edges, which connect the agents initial locations to the `Main Root`. A set of candidate nodes is produced as a result of moving out from the `Main Root` by the value of ϵ towards the agents' locations along the edges. The value of ϵ is specified by the user and must be equal or less than the value of the `step size` as defined in the previous section. If a candidate node is collision-free, it is called a `Secondary Node` and assigned to the corresponding agent whose location point is an end-point of the edge. If the candidate node is not collision-free, the algorithm finds the nearest agent which has been assigned a `Secondary Node` and assign it to the agent. Figure. 2 illustrates the `Main Root` and `Secondary Roots` of a particular configuration.

From this point on, the algorithm is similar to the `RRT` planner algorithm with two differences. The first one is in the number of initial points, q_{init}. While `RRT` there is only one q_{init} node, which is randomly located in the free space, the `WMA-RRT` includes as many q_{init} nodes as there are `Secondary Roots`. The other difference is observed in the initial expansion of the roadmap. In `RRT` the roadmap expands through the root of the roadmap, q_{init}, while in `WMA-RRT` the `Main Root` is exempted from expansion and instead, the `Secondary Roots` are considered for expansion. The initial roadmap expands very rapidly throughout the free space of the environment and yields a tree with one `Main Root`, which has as many sub-trees as there are `Secondary Roots`. Each sub tree is called a `Branch`.

The edges' weight of the roadmap is initialized to 0 so the roadmap is considered as a weighted tree. During the covering operation, at-least one agent is assigned to each Branch. The weights of the edges are updated while traversing the roadmap. The agents use `WMA-RRT` roadmap to move throughout the free space, interact with the environment and indirectly communicate with one another. In the other words, while traversing the `WMA-RRT` roadmap, the agents perceive information available through the roadmap and update their knowledge about the environment. Further, the agents change the state of the environment by depositing information into the nodes of the `WMA-RRT`.

Table 1 Represents all the information available in the nodes of the WMA-RRT roadmap.

information	Main Root	Secondary Root	Intermediate Nodes
e_o: label of each outgoing edge	No	Yes	Yes
e_i: label of the incoming edge	No	Yes	Yes
w_i: weight of the incoming edge	No	Yes	Yes
w_o: weight of the outgoing edge	No	Yes	Yes
c: shows all the sub-trees of a node have been completely explored	Yes	Yes	Yes
l_o: label of the outgoing edge which is temporarily locked and is not accessible	No	Yes	Yes
c_o: label of the first edge of a sub-tree which has been completely explored	No	Yes	Yes

There is no information available through the edges. Table 1 represents the information that may be deposited by the agents and available to them in the WMA-RRT roadmap.

In some situations, it is impossible to have as many branches as there are agents. This is specially the case when the environment is highly cluttered and contains many narrow passages since some of the Secondary Roots may not be collision free and consequently, have to be deleted from the roadmap. In such situations, in order to make use of the maximum capabilities of the available agents, the algorithm assigns more than one agent to each branch. Figure. 3 represents an environment and its corresponding WMA-RRT roadmap containing two branches and there are four agents available in the environment. As illustrated, the algorithm assigns two agents to each branch.

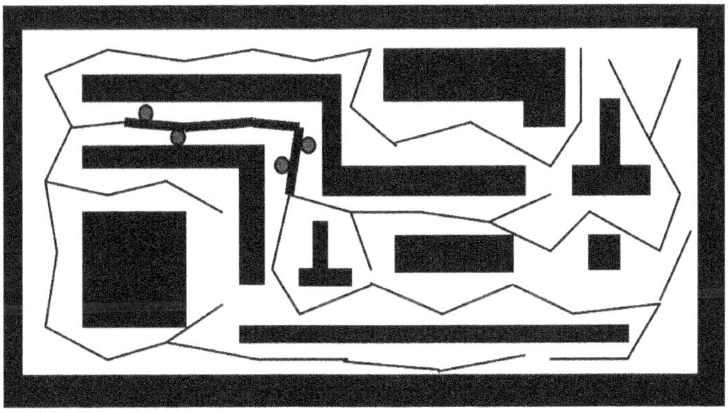

Fig. 3 A cluttered environment with narrow passages. Two agents have been assigned to each branch.

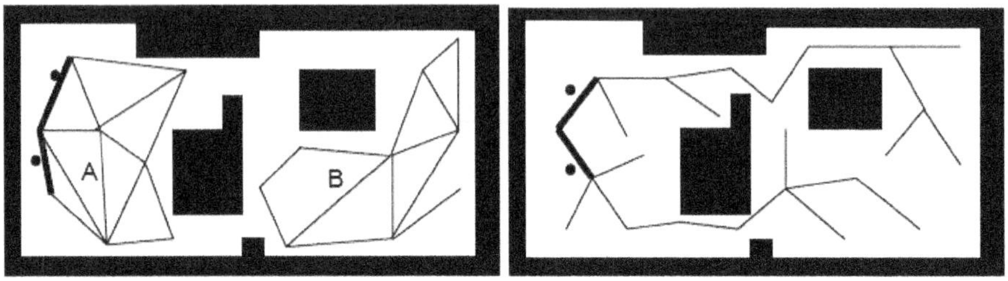

Fig. 4 The figures on the left and right represent the results of running PRM and WMA-RRT algorithms respectively on an environment.

Finally, we compare WMA-RRT tree as a single query roadmap with the Probabilistic Road Map, PRM, as a multiple query roadmap. As mentioned above, a multiple query planner might lead to a disconnected roadmap and consequently, some parts of the environment are not accessible by any agents. Figure. 4 represents the results of running PRM and WMA-RRT algorithms on a sample environment.

4 Agent Architecture

As mentioned in previous sections, the agents in our system are simple autonomous individuals capable of following some explicit condition-action rules and performing the corresponding actions. To this end, our agents are modeled based on reactive architecture and are implemented in such a way that supports `beliefs`, `actions` and `Plans` of the agents. Furthermore, they are utility based entities, which try to maximize their own utilities independently. The utility function of an agent is defined as the average over the weight of all the edges, the agent traverses. The utility function of an agent motivates it to select the edges with minimum weight.

Additionally, the agents perceptions are represented as beliefs in the agents' belief bases and each agent contains some pre-defined plans stored in its plan library. Each plan is an instruction which tells the agent what to do in a particular circumstance. Each agent has a goal and uses a plan to approach the goal. In the rest of this section, we discuss some basic beliefs, actions and plans used by the agents during the covering operation. We use Agent Speak, a multi-agent programming languages standard syntax and format in order to explain the agents beliefs, actions and plans. As discussed in previous section, the agents update their knowledge base about the world when they reach a node and perceive the information available at that particular node. In the other words, while traversing the WMA-RRT tree, an agent detects the information available in the nodes of the roadmap and updates its belief base accordingly. The followings are some beliefs of the agents:

1. `subtreeState(e, w)`: Standing at each node, an agent could detect some information about all the outgoing and incoming edges, including the edges' labels ,`e`, and weights, `w`.
2. `currentEdge(e,w)`: When an agent decides which outgoing edge to take, it can remember the edges label and weight until it reaches the end-points of the edge and updates the related information about the traversed edge.
3. `completed(e)`: It means that the agent believes the sub-tree of the current node that starts with edge `e`, has been completely explored.
4. `currentBranch(b)`: It represents the current branch, b, which is being explored.
5. `agentName(ag)`: Each agent has a unique name which is stored in the belief base of the agent.

Furthermore, each agent is capable of performing some basic external and internal actions to change the state of the environment and its internal state. The followings are some of the actions:

1. `allocateBranch`: Standing at a `Secondary Root`, an agent updates the state of the current roadmap's node, by adding information about the label and weight of the incoming edge to the current node, which is a `Secondary Root`
2. `traverse`: An agent can traverse the selected edge by moving from the start node of the edge and reaching the end point of it.
3. `departNode`: When an agent decides which edge to traverse next, it increases the weight of the selected edge by one and updates its corresponding information, including the label and new weight of the selected edge, in the current node.
4. `reachNode`: As an agent reaches a node, it updates the state of the current node to represent the label and weight of the incoming edge to the current node.
5. `comeback`: When an agent is to return one level back to the parent of the current node, it should set the current node as completed. It means that the entire sub-tree of the current node has been completely explored. Then, as soon as the agent arrives in the parent node, it updates the state of the parent node to indicate that one of its sub trees has been completely explored.
6. `lock(Edge)`: When an agent enters an edge, it locks the edge temporarily to prevent other agents enter the edge simultaneously. In order to do this, the agent adds some information in the start node of the edge representing that the edge with label `Edge` is locked. Each locked edge, gets unlocked after a specific period of time automatically. This period of time equals to the twice the maximum amount of time an agent needs to completely traverse an edge, plus some amount of time the agent requires to detect the information available and update its belief base accordingly. This time can be specified at the beginning of the covering operation. Remember

that the maximum length of each edge of the WMA-RRT equals to the step size, ϵ,which has been already defined in WMA-RRT construction algorithm.

7. detect: The agents use their sensors in order to percept all the information available in the current node. Standing at a node, an agent also can detect the outgoing edges of the current node.
8. scan: Standing at a node, each agent can access a local roadmap containing the current node, incoming edge and outgoing edges of the current node. The agents cannot store these local roadmaps in their limited memories but an agent can temporarily employ the local roadmap in order to enumerate the outgoing edges and assign them temporary labels.
9. enumerate: Standing at a node, an agent changes the state of the current node by updating information relevant to the incoming and outgoing edges of the current node. To differentiate the outgoing edges, the agents locally enumerate them and assign each of them a label. To this end, all the agents assume that, the labels of all the edges going out of the Main Root, are zero. As mentioned in previous sections, the roadmap is locally available to the agents so they can detect all the outgoing edges and label them by enumrating all the outgoing edges in a clock-wise order starting from the incoming edge and incrementally assigning each of them a label.To this end, the algorithm assums that the label of all the Secondary Roots are zero. In other words, suppose that the weight of the incoming edge of the current node is w and there are two outgoing edges. Then, by doing a clock-wise search starting from the incoming edge, the label of the first and second edges are $w + 1$ and $w + 2$ respectively. Figure 5, illustrates how the edges of WMA-RRT are locally enumerated by the agents.

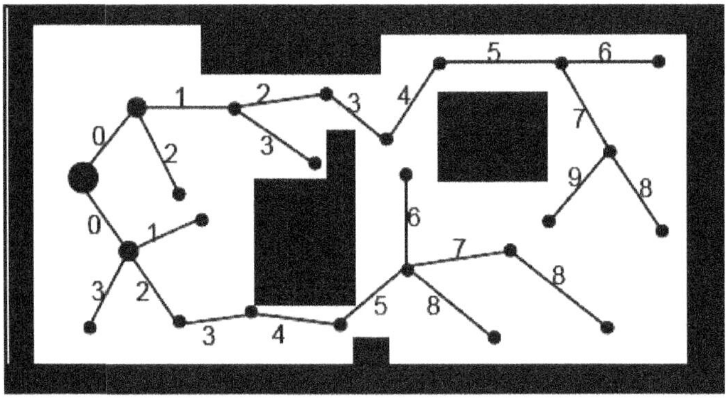

Fig. 5 This figure represents how the edges of the WMA-RRT are locally enumerated and labeled by the agents. Here, the biggest black circle is the Main Root and two second biggest are the Secondary Roots.

Furthermore, each agent maintains a plan library including some built-in plans. Some of the simplified versions of the plans available in the plan library of the agents are as followings:

1. `findingBranch`: Each agent applies this plan to find a branch to explore. This plan starts by performing the action of `scanning` around the `Main Root` to build a local roadmap. The local roadmap contains the `Main Root`, `incoming edge` and `outgoing edges`. The `incoming edge` to the `Main Root` is exclusive to each agent since the agent considers the edge it takes to reach the `Main Root` from its initial position as mentioned in Section 3. The local map is then used by the agent to locally enumerate and label the outgoing edges. Then, the agent traverses the outgoing edges sequentially in order to find the first outgoing edge with minimum weight. This is done through detecting the information available in the `Secondary Roots`.

```
!+findingBranch : constructed(wma-rrt) <-
scan;
internalAction.enumerate;
internalAction.getTheUnCheckedEdgeInOrder(Edge);
traverse(Edge);
detect;
!allocation(Edge).

!+allocation(Edge) : isMinimum(Edge) <-
!findingBranch.

!+allocation(Edge) : not isMinimum(Edge) <-
allocateBranch(Edge);
!nextDestination(Edge).
```

2. `nextDestination`: This plan is used when an agent reaches a node. Each agent is simply able to temporarily remember the label and weight of the incoming edge to the current node. This plan tells the agent to scan around the current node, locally enumerate the outgoing edges to assign each of them a label, detect all the information available in the current node and find the next best edge to traverse. The best edge is an edge which is not currently locked and has not been set as completed and also has the minimum weight among all other outgoing edges of the current node.

```
+!nextDestination : true <-
scan;
internalAction.enumerate;
detect;
internalAction.findNextEdge(NextEdge);
!traversing(NextEdge).
```

3. `traversing`: This plan is invoked when the agent has made a decision on which outgoing edge to traverse. This plan makes the agent lock the selected edge, depart from the current node, traverse the edge and reach the other endpoint of the selected edge.

```
+!traversing(Edge) : true <-
departNode;
lock(Edge);
traverse(Edge);
reachNode.
```

4. `return`: This plan is used when an agent reaches a leaf or blind node, a node whose sub-trees have been traversed and set as completed, so the agent cannot move forward and has to return one level back to the parent of the current node.

```
+!return(Node) : leaf(Node) | blind(Node) <-
internalAction.nextEdge(Edge);
comeback(Node);
traverse(Edge);
detect.
```

5 Exploration of WMA-RRT Roadmap

As discussed earlier, the `WMT-RRT` roadmap, is considered as a set of several sub-trees called `Branch`. The ultimate goal of our algorithm is to assign the branches of the `WMA-RRT` roadmap to the agents effectively and let the agents traverse the tree completely. To achieve this objective, each agent is supposed to independently find a branch and commit itself to traversing it. An agent start performing the coverage mission as soon as it arrives at the `Main Root`.

When an agent arrives at the `Main Root` the only things it can detect are the edges going out of the `Main Root`. Using the `findingBranch` plan the agents find a branch and commit to traversing it. Since there is no information maintained in the `Main Root`, the agents have to check the `Secondary Roots` to figure out which one has not been yet assigned to one agent. Using the same enumeration approach each agent can exclusively enumerate all the outgoing edges from the `Main Root` and visit them sequentially. The information available at the `Secondary Roots` tells the agent whether the current branch has been already assigned to any other agent.

Each agent continues to search for a branch to eventually find a branch and commit to exploring it. The utility function of an agents motivates it to select unallocated branches. If there is no remaining unallocated branch, they look for a branch in which the minimum numbers of agents are working. Running the `findingBranch` plan directs each agent to reach its corresponding

`Secondary Root`. The agents then update the state of the environment accordingly by changing the states of the nodes.

At each node, the agent executes the `nextDestination` plan to find the next edge to traverse. Basically, The outgoing edge with minimum weight is selected for exploration since edges with higher weights have been already visited more so it is reasonable to select those edges with fewer weights. Then the agent performs the `trversing` plan to update the state of the start node of the edge, traverse the edge and update the state of the end node of the edge. If an agent reaches a leaf or blind node of the current branch, it applies the `return` plan in order to return to the parent node and the leaf or blind node is marked as completed. An agent sets a node as completed if and only if the node's entire sub-trees have been marked as completed. When an agent is going to perfom the `return` plan, it should know which edge is the incomming edge. Since the states of the incoming edge and outgoing edges of the current node are available, the agent can figure out which edge is the incoming edge since the incoming edge is the only edge which has not been marked as completed.

Eventually, when an agent returns to its corresponding `Secondary Root`, it marks the `Secondary Root` as completed and returns one level back to the `Main Root` to contribute in exploration of other branches if there is any remaining. As soon as an agent realizes that all the branches have been set as completed, it marks the Main Root as completed to indicate that the exploration process is done.

It is very likely that one agent completes the navigation of its corresponding branch and returns to the origin while some other agents are still working in other branches. In such situations, an agent manages to maximize its contribution by helping some other agents in their exploration missions. As an agent returns to its corresponding `Secondary Root` and consequently, reaches the `Main Root`, it applied its findingBranch plan to find the branch in which the minimum number of agents are working in and commit to exploring it. Consequently, the agent could help other agents working in the same branch to finish the exploration task. When an agent arrives in a node, which has been marked as completed, it does not go further and therefore, returns to the parent node by executing the `return` plan. This process guarantees the exploration of the environment even in the case of occurring massive failures in the agents. This feature makes our system reliable and robust.

6 Implementation and Experimental Results

6.1 Implementation

Since the agents are modeled according to the `BDI` architecture, we have applied the same design pattern used in the Agent-Speak multi-agent programming language to implement the system. According to this pattern, there is an environment which is shared between all the agents in the environment and is implemented in the class `SharedEnvironment`. All the required data structures including the data structures for `WMA-RRT` have been defined and implemented in a class called `EnvironmentModel` which models the environment.

Each agent is a thread which is started in the `SharedEnvironment` class as soon as the offline process of constructing `WMA-RRT` is completed. Since all the agents are homogeneous, each agent is implemented as an object of the `Agent` class. In this class, variables such as max-speed, max-force, current-active-branch, current-position, current-velocity and target-position are defined. This class also includes all the required data structures for representation of beliefs, belief base, plans library and all necessary methods like deliberation, plan selection and belief update. Also, the actions of each agent have been implemented as methods in the Agent class.

Each plan is also implemented as an object of the Plan class. This class includes some data structures in order to maintain the plans pre-conditions, event triggers and a list of references to the actions listed in the plans.

6.2 Experimental Results

We have evaluated the mechanism discussed in this paper, in a variety of environment in order to capture the effect of the number of sample points, step size (distance between two adjutant nodes of `WMA-RRT`) and the sensor range, which is the visibility range of the agents. All the experiments have been conducted in a Pentium 4 CPU 3.00GHz with 1.00 GB of RAM machine. Figure 6 represents a sample environment including three agents initially located near each other and its corresponding `WMA-RRT` tree and areas covered by the agents.

The Tables 2 and 3 summarize the results of running the proposed area coverage mechanism in the presence of three agents on two environments

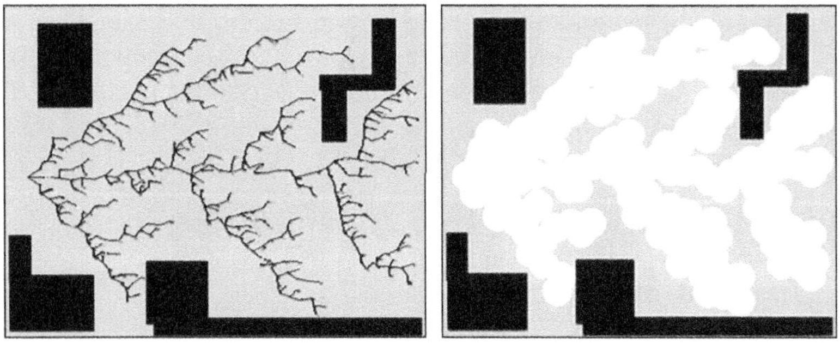

Fig. 6 Shows the results of running the `WMA-RRT` tree and the coverage mechanism in the presence of three agents.

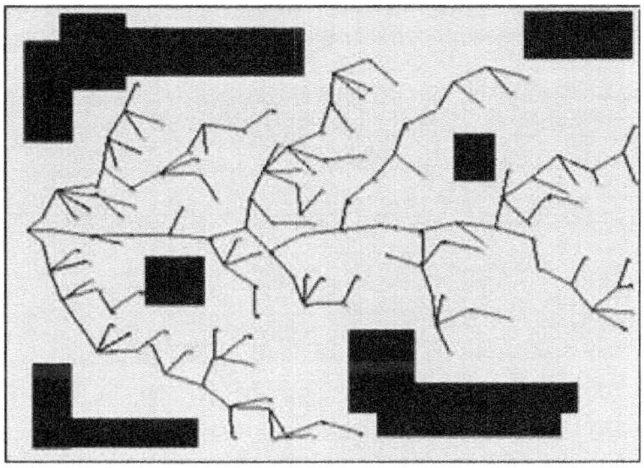

Fig. 7 Represents another sample environment and the corresponding WMA-RRT tree in the presence of three agents.

Table 2 Represents all the information available in the nodes of the WMA-RRT roadmap.

number of nodes	step size	run/per each group of agents	visibility range	coverage percent-age
400	20	10	r	74
400	20	10	$2r$	87
400	20	10	$4r$	90
400	20	10	$infinite$	99
800	20	10	r	83
800	20	10	$2r$	88
800	20	10	$4r$	92
800	20	10	$infinite$	99
200	40	10	r	88
200	40	10	$2r$	93
200	40	10	$4r$	95
200	40	10	$infinite$	99
400	40	10	r	93
400	40	10	$2r$	95
400	40	10	$4r$	97
400	40	10	$infinite$	99

represented in Figure 6 and 7 respictively. According to the results represented in Tables 2 and 3, it can be concluded that as the number of samples, step size and sensor range increase, the coverage percentage of the environment improves. As the results show, in cluttered environments the step-size should be as small as possible.

Table 3 Represents all the information available in the nodes of the `WMA-RRT` roadmap.

number of nodes	step size	run/per each group of agents	visibility range	coverage percentage
150	20	10	r	68
150	20	10	$2r$	80
150	20	10	$4r$	88
150	20	10	$infinite$	97
300	20	10	r	83
300	20	10	$2r$	92
300	20	10	$4r$	95
300	20	10	$infinite$	99
600	40	10	r	93
600	40	10	$2r$	95
600	40	10	$4r$	97
600	40	10	$infinite$	99
150	40	10	r	68
150	40	10	$2r$	80
150	40	10	$4r$	88
150	40	10	$infinite$	97

7 Future Works and Conclusion

In this paper, we discuss the problem of visual environment coverage by means of a group of many simple agents. In the proposed mechanism, a special roadmap called `WMA-RRT` is constructed in the environment and used by the agents both as roadmap and a communication channel. We use an emergent coordination mechanism in which the agents cooperate with each other to achieve a final goal while pursuing their own individual objectives. The proposed approach is specially effective in situations in which the exact structure of the environment is not known and it is possible to use many numbers of simple agents each of which is only capable of performing some simple actions and there is no communication and negotiation devises. While the `WMA-RRT` can be easily used in an environment with narrow passages but the covering time increases as the number of narrow passages increases. This is due to the effect of the value of `step size` in `WMA-RRT`. In order to cover the narrow passages of environments, the value of `step size` should be small enough to let the `WMA-RRT` tree expand throughout the narrow passages. As further works, we are interested to extend our study in order to apply a group of heterogeneous agents with a different visibility range to cover the environment. Also, we are interested in finding a more appropriate roadmap which decreases the total time of the environment coverage in the presence of many narrow passages.

References

1. Abraham, A., Guo, H., Liu, H.: Swarm intelligence: Foundations, perspectives and applications. In: Swarm Intelligent Systems. Studies in Computational Intelligence, pp. 3–25. Springer, Heidelberg (2006)
2. Bayazit, O.B.: Roadmap-based flocking for complex environments. In: Proc. 10th Pacific Conference on Computer Graphics and Applications, PG 2002, pp. 104–113 (2004)
3. Beni, G., Wang, J.: Swarm intelligence. In: Proceedings Seventh Annual Meeting of the Robotics Society of Japan, pp. 425–428. RSJ Press, Tokyo (1989)
4. Choset, H.: Coverage for robotics - a survey of recent results. Ann. Math. Artif. Intell. 31(1-4), 113–126 (2001)
5. Davoodi, A., Fazli, P., Pasquier, P., Mackworth, A.K.: On multi-robot area coverage. In: 7th Japan Conference on Computational Geometry and Graphs, JCCGG 2009, pp. 75–76 (2009)
6. Gordon, N., Wagner, I.A., Bruckstein, A.M.: Discrete bee dance algorithm for pattern formation on a grid. In: IEEE International Conference on Intelligents Agents Technologies, Toronto, CA, pp. 545–549 (2003)
7. Halász, Á.M., Hsieh, M.A., Berman, S., Kumar, V.: Dynamic redistribution of a swarm of robots among multiple sites. In: International Conference on Intelligent Robots and Systems, IROS, pp. 2320–2325 (2007)
8. Hazon, N., Kaminka, G.: On redundancy, efficiency, and robustness in coverage for multiple robots. Robotics and Autonomous Systems (2008)
9. Kavraki, L.E., Svestka, P., Latombe, J.C., Overmars, M.H.: Probabilistic roadmaps for path planning in high-dimensional configuration spaces, vol. 12, pp. 566–580 (1996)
10. Lavalle, S.M., Kuffner, J.J.: Rapidly-exploring random trees: Progress and prospects. In: 4th Workshop on the Algorithmic Foundations of Robotics, Algorithmic and Computational Robotics: New Directions, pp. 293–308 (2000)
11. Oks, E., Sharir, M.: Minkowski sums of monotone and general simple polygons. Discrete & Computational Geometry 35(2), 223–240 (2006)
12. Packer, E.: Computing multiple watchman routes. In: McGeoch, C.C. (ed.) WEA 2008. LNCS, vol. 5038, pp. 114–128. Springer, Heidelberg (2008)
13. Preparata, F.P., Hong, S.J.: Convex hulls of finite sets of poin ts in two and three dimensions. Commun. ACM 20(2), 87–93 (1977)
14. Reynolds, C.: Steering behaviors for autonomous characters. In: Game Developers Conference (1999)
15. Sahin, E., Labella, T.H., Trianni, V., Louis Deneubourg, J., Rasse, P., Floreano, D., Gambardella, L., Mondada, F., Nolfi, S., Dorigo, M.: Swarm-bot: Pattern formation in a swarm of self-assembling mobile robots. In: Proceedings of the IEEE International Conference on Systems, Man and Cybernetics, Hammamet, pp. 6–9. IEEE Press, Los Alamitos (2002)
16. Svennebring, J., Koenig, S.: Trail-laying robots for robust terrain coverage. In: ICRA, pp. 75–82 (2003)
17. Tang, L.J.: A visibility-based algorithm for multi-robot boundary coverage. International Journal of Advanced Robotic Systems 5, 63–68 (2008)

18. Tarasewich, P., McMullen, P.R.: Swarm intelligence: power in numbers. Commun. ACM 45(8), 62–67 (2002)
19. Wurm, K.M., Stachniss, C., Burgard, W.: Coordinated multi-robot exploration using a segmentation of the environment (2008)
20. Zlot, R., Stentz, A., tony Stentz, A., Dias, M.B., Thayer, S.: Multi-robot exploration controlled by a market economy. pp. 3016–3023 (2002)

An Approach for Learnable Context-Awareness System by Reflecting User Feedback

InWoo Jang and Chong-Woo Woo

Abstract. As the ubiquitous computing becomes popular, the context awareness computing also becomes more interesting reasearch issue. Despite of many important research results on this issue, there still are some limitations that can be enhanced to provide more reliable solution to the user. In this paper, we are describing a design and development of agent based context awareness system that can improve such limitations. The system composed of mainly three layers; the hardware layer receives numerical raw data from sensors and converts it into meaningful semantic data, the middleware layer takes care of ontology modeling, and finally the application layer makes adaptive inference and provides personalized solution to the user. Our approach focuses on the following two main issues. First, we have built ontology based context modeling using fuzzy data that can provide more reliable solution. Second, our CBR based inference engine can provide more personalized and adaptive service by interacting with users feedback. The simulated experimentation has been made and result shows some significant importance.

1 Introduction

Recently, as the ubiquitous computing becomes popular, context awareness computing is also becoming interesting research issue. The context in context awareness system is any entities that affect interaction between human and computing devices in the environment. An entity can be a human, place, or any object relevant to the interaction between human and application [1] [14].

Among the many research results on this issue, the most prominent systems are the CoBrA (Context Broker Architecture) [2], SOCAM (Service Oriented Context-Aware Middleware) [3], Context-Toolkit [4]. Some of the

InWoo Jang · Chong-Woo Woo
School of Computer Science, Kookmin University
e-mail: `zx1018@nate.com`, `cwwoo@kookmin.ac.kr`

Q. Bai and N. Fukuta (Eds.): Advances in Practical Multi-Agent Systems, SCI 325, pp. 113–127.
springerlink.com

systems were not flexible for providing formalized expression and also for the extensibility and interoperability with other systems. To handle this issue, CoBrA and SOCAM used ontological model that supports easy sharing in different domains, and reusing the related knowledge. Therefore, the ontological modeling approach becomes popular solution to represent the data in the recent context awareness systems. But, when we try to build the system based on the ontological modeling, we still have some problems as follows. First, we often run into unreliable results that come from converting numerically sensed data into symbolic expression. Second, the system always responds with the same context to different users. And this could be much effective if we could support personalized solution based on the each individuals preference. And finally, if the system could learn the users behavior, then it will perform much adaptively by providing a personalized service to the user.

In this paper, we are describing a design and development of agent based context awareness system that can handle such limitations of the current context awareness system. Our system is designed as three main layers, namely, hardware layer, middleware layer, and application layer. The hardware layer receives numerical raw data and makes data conversion using Fuzzy inference mechanism. Middleware layer takes care of ontology modeling, which has been developed but will not be discussed in this paper. The application layer makes inferences using case-based reasoning (CBR) engine, and generates final solution to the user. Our approach focuses on the next two main issues; building ontology based context modeling using fuzzy data, and second, reusing adapted rules by interacting with users feedback. The simulated experimentation has been done, and results show some significant importance.

2 Related Studies

2.1 Context-Awareness System

Recently, the context awareness computing has been studied explosively, and CoBrA [2], SOCAM [3], Context-Toolkit [4][6], GAIA [5][9], CoolTown [7], are the major research results in this area. These studies showed advent of some context models for representing a context, such as, Attribute-Value, Web Based model. And later, this has been developed into Entity-Relationship model, UML model, and ontology model. This development is due to the need for sharing not only the context information, but also the need for inference of the users intention. Therefore, the context awareness computing can be implemented in many different ways, including more formalized representation of the context information, and the use of ontology becomes main stream in this paradigm. The general process of the system that based on the ontology model includes next few steps [2][3][5][9].

- First, the system collects numerical raw data from the physical sensors in the distributed environments. Since the ubiquitous sensor network has been built successfully, we could receive information from the sensors through the network.

- Second, the collected data from the sensors needs to be further processed to become ontology data that requires semantic value. For instance, if measured illumination value is 1000 lux, then this value must be converted into semantic data, bright, and saved on the ontology model. This numeric value is called low-level context and the semantic value is called as high-level context.

- Third, the data inference mechanism runs into some complicated ontological process. For instance, if the processed data from the above stage is *room*, then we need to get related context from the ontology model, and *home* is the upper class from the ontology model.

- Fourth, we then need to generate appropriate solutions for the context by applying domain specific rules. For example, *'if Tony comes into a room, then turn on the light'* is the final solution for the current situation.

This is the most general procedure for the recent context awareness system, which has strong advantages, such as, reusing data, sharing data, interoperability, and so on. But the system using the ontology, still needs to be studied further for its evaluation on the provided services. And also the sensed data needs to be investigated further for its reliability, which will be discussed in the next section.

2.2 Agent in Previous Context-Awareness System

Some of the previous context awareness systems employed multi-agent structure. For instance, COBRA utilized multi-agent structure, which has capabilities of knowledge sharing based on the negotiation policy. The agent in this system basically manages information for user context, and decides whether to share it. In this case, the inference engine exists in the server, and the role of the agent is the management of user context and sharing it with other agents. In GAIA system, the agents role is to inference for the stored context. The system comprised of multi-agents with multiple inference engines, and decides which inference engine to use for the given situation. OCASS [7] uses agent capability for acquiring, sharing, and making inference for the context data. But, these systems still has some limitation of personal record management, self-learning capability, and uncertainty in context data. Therefore, we are going to design and develop a system that could overcome these problems.

3 System Architecture

The main structure of our system consists of three layers as in the Fig. 1, namely, hardware layer, middleware layer, and application layer. And each layer of the system works as follows.

- First, the hardware layer is organized as physical sensor and software Context Management Agent (CMA). The physical sensor receives raw data from the environment, processes through the CMA, and sends the processed data to the middleware of the system. To make this work, the CMA controls the physical sensor and converts the raw data into the semantic data through the fuzzy reasoning engine.

- Second, the middleware layer is in charge of sharing data through ontology model, resource managing, and managing data input and output. Basically, it connects hardware layer and application layer through data I/O, and generates intermediate data.

- Third, the application layer is composed of Situation Reasoning Agent (SRA), and some computing devices for the user. This layer receives context from the middleware, analyzes it, and provides solution to the user. The SRA recognizes the context through the knowledge base and CBR engine, and generates solution for the situation.

In this paper, we are not going to describe middleware layer, since it is out of focus in this paper. The detailed descriptions for the agents of the system are in the below.

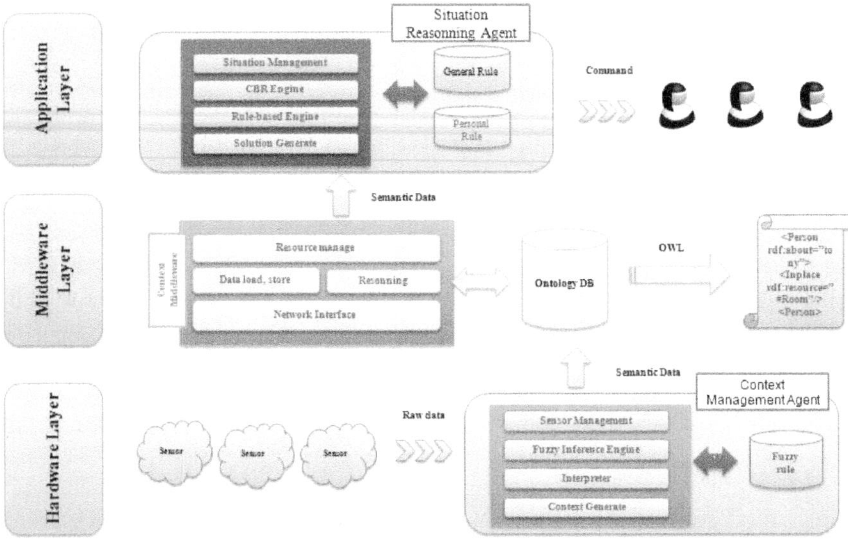

Fig. 1 The General System Architecture

3.1 *Context Management Agent (CMA)*

The structure of the CMA can be seen as in the Fig. 2. The context awareness system receives numeric raw data from the physical sensors that needs to be converted into semantic data. The previous systems use the if-then rules for

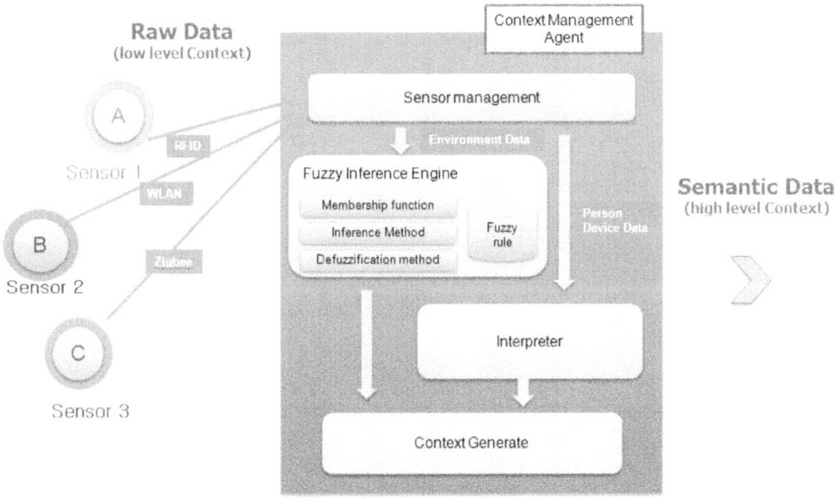

Fig. 2 Structure of CMA

data conversion, which showed some unreliable results. Therefore, we are going to use the Fuzzy rule for the conversion of data in this CMA.

The agent does not treat a single data, but it processes information from the multiple sensors with internal knowledge base and Fuzzy inference engine. After analyzing the processed information, it sends the result to the middleware. The agent receives data from the sensors at fixed hour, and it can control the physical sensors. It generates two different types of semantic data; first, it generates actual semantic data directly like temperature, and second, it also generates dependent semantic data like discomfort index that is dependent to the other semantic data. The dependent semantic data is generated according to the Mamdanis Min-Max algorithm [11].

Generation of Actual Context. The raw data from the physical sensor passes through the Fuzzy membership function first. The number of Fuzzy membership function exists as many as we express the situations. For instance, if we express the state of the humidity as 7 stages, then the number of Fuzzy membership function becomes 7. And the numeric value of each function ranges between 0 1 that reflects the reliability for the current status. For example, let us assume that we received the raw data value 35 from the humidity sensor. Then this value gets converted as, F1(x)=0.0, F2(x)=0.2, F3(x)=0.4, F4(x)=0, and so on (see Fig. 3). This kind of data will be used for getting dependent context, and also saved in the middleware as numeric data value 0.4 along with semantic data value of humidity function.

Generation of Dependent Context. The CMA generates the dependent context by using the Mamdanis min-max Fuzzy reasoning algorithm. The fuzzy rules are saved in the internal knowledge base, and there are some examples in the Table 1 below.

Table 1 Fuzzy rule example

Index	Rules
Rule1	IF x is A1 and y is B1 Then z is C1
Rule2	IF x is A2 and y is B2 Then z is C2
...	...

From the above table, the A1 stands for very dry, A2 is dry, B1 is warm, and B2 is very warm. With this Fuzzy rules, the CMA generates dependent context. Lets assume that fuzzy inference engine generates the discomfort index by using the sensed humidity and temperature. Then the engine goes through the next four steps as follows.

First, we seek for the suitability for the rules. This can be done by deciding the membership using the membership function, and take the smallest between value the A1 and B1, and find the result W1 as in the following equations.

$$\textit{Suitability for Rule 1: } W_1 = \mu_{A_1}(x) \wedge \mu_{B_1}(y)$$
$$\textit{Suitability for Rule 2: } W_2 = \mu_{A_2}(x) \wedge \mu_{B_2}(y)$$

Second, we get the inference result for each rule by applying the generated suitability value onto the Fuzzy sets. The fuzzy set indicates set for the discomfort index; C1 is very discomfort, C2 is discomfort. This kind of information also must be saved in the internal knowledge base. The inference results for the each rule gets smaller value between the suitability value W1 and value for Ci.

$$\textit{Inference Result for Rule1: } \mu_{c_1'}(z) = W_1 \wedge \mu_{c_1}(z).\forall z \in \mathbf{Z}$$
$$\textit{Inference Result for Rule2: } \mu_{c_2'}(z) = W_2 \wedge \mu_{c_2}(z).\forall z \in \mathbf{Z}$$

Third, find final inference result from the biggest value from acquired suitability values. In other words, we will take the max value if C1 (very discomfort) exists with more than a single value. Generally, for N rules, we can set the equation as follows.

$$\textit{Inference Result for Rule1:}$$
$$\mu_c(z) = \mu_{c_1'}(z) \vee \mu_{c_2'}(z) \vee \mu_{c_3'}(z) \cdots \vee \mu_{c_n'}(z).\forall z \in \mathbf{Z}$$

Finally, the result from the previous steps is expressed as fuzzy set, and it needs to go through the Defuzzification to get the final result. Among many methods, we are going to use the Center for Gravity method as follows.

$$Z_0 = \frac{\int \mu_c(z) \cdot z\mathrm{d}z}{\int \mu_c(z)\,\mathrm{d}z} \tag{1}$$

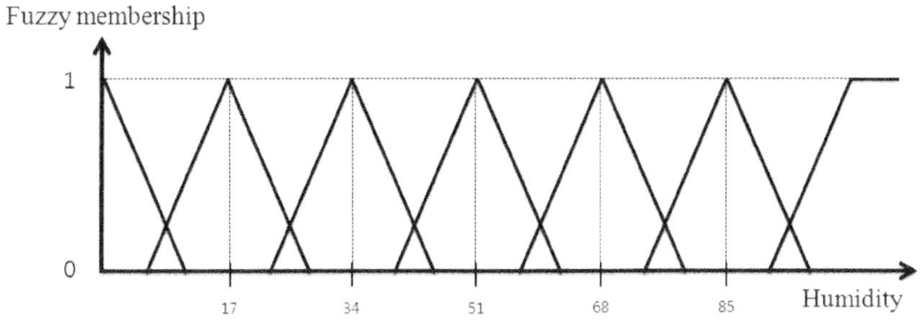

Fig. 3 Fuzzy membership function

Therefore, the data type for the semantic data that the CMA sends to the middleware layer is the highest membership status, and it is expressed as normalized real value (0 1).

3.2 Situation Reasoning Agent (SRA)

SRA works in the application layer. It extracts context from the middleware, and provides a solution for the user. The agent could solve the disadvantages of previous context awareness system, by providing personal profile and adaptive learning capability. The general structure of the system is in the Fig.4.

SRA is composed of CBR engine and knowledge base. Basically it extracts the semantic data from the middleware and aware the context. Therefore, it provides the command to the user who can control the devices. The work for the SRA recognizes the context from the CBR engine can be further divided into following stages: extraction and conversion of semantic data, searching for similar cases, adaptation of cases, and evaluation and evolution.

Extraction and Conversion of Semantic data. The SRA manages personal profile to communicate with middleware layer. When an event occurs that belongs to a certain place, the agent requests context surrounding the place. Since the data receives from the middleware has been extracted from the ontology model, it is rather semantic data type. The SRA converts the data into a context which will be a case. For instance, information about who, when, where will be represented as a string type, and the rest of the environmental information will be changed into Float type (see Fig 5).

Search for the Similar Case. The SRA uses history of previous context for the current context awareness. This context record composed of status of context and appropriate solution for the context. We assume that record has been built for some time, and this record will be accumulated continuously as time goes by. The agent will try to find proper solution for the current

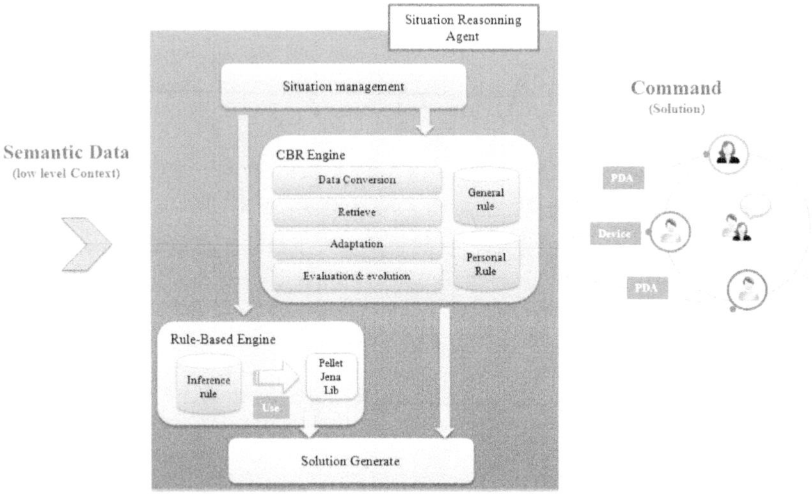

Fig. 4 Structure of SRA

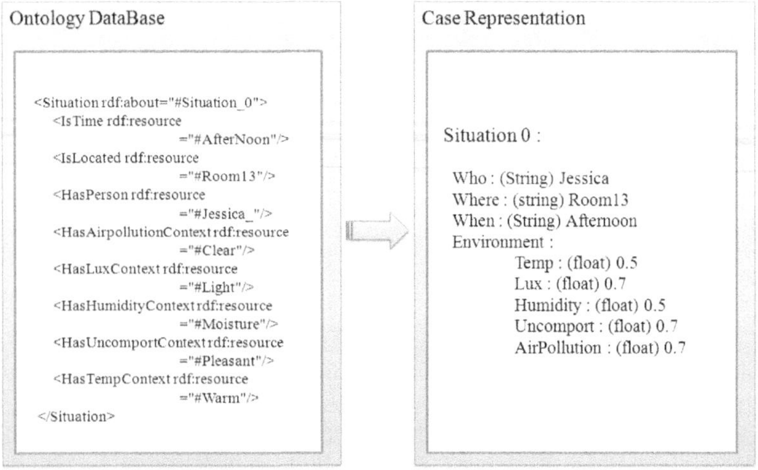

Fig. 5 Structure of a Case

context by using this previous record, and use K-nearest algorithm to do this. We are going to use the following equation to search for the most similar case.

$$Similarity = \frac{\sum W_i \cdot f(d)(C_{k_i}, C_{c_i})}{\sum W_i} \tag{2}$$

The parameter Wi is the weight for the context, and the function f represents the distance value between the previous context and the current context. When the function f gets implemented, the environmental information will be calculated as distance value. And the information for when, and where will be calculated through the similarity matrix. There are many ways to compute the weight, and one of the methods could be the following equation.

$$Weight_i = \frac{|f(d)(C_{k_i}, C_{c_i})|}{\sum |f(d)(C_{k_i}, C_{c_i})|} \quad (3)$$

From the above procedure, we can compute the similarities for the every case, and will choose one from them. But, if the selected one is the most similar case, and there still exists differences from the current case. Then we need to solve the differences by going through the adaptation stage.

Case Adaptation. The SRA performs the adaptation stage to resolve the differences between the current context and the selected case, by applying adaptation rules [12]. There are two types rule bases; one is the match rule base, and the other is range rule base (see Table 2). The match rule represents exact matches between current situation and similar case. The range rule allows the temporary boundary values. These Two types of the rules are pre-defined in the general knowledge base, and the learned result from the feedback will be stored in the personal knowledge base. When the adaptation is being done, then both the rule base will be searched. If more than one rule is triggered, then the most recent rule gets fired. The rule adaptation process is as follows.

Table 2 Two Type of Rules

Num	Type	Rule	Priority
Rule1	Match	IF Current Temperature is warm and Similar Temperature is Cold Current Humidity is dry Then set the Air conditioner level 2 down	0
Rule2	Range	IF Current Lux is very dark and Similar Lux level less than normal Then set Lux level at 4	1

Let us assume that we have a case as follows.

- Case = {Location, Person, Temp, Humidity, Lux, Discomport, air-Pollution}

Case example for Rule1 gets fired:

Current Case = {Room, Kim, emphwarm, *dry*, somewhat light, Pleasant, Normal}
Similar Case = {Room, Kim, cold, *dry*, somewhat light, Pleasant, Normal}

Case example for Rule2 gets fired:

Current Case = {Library, Kim, warm, dry, *very dark*, Normal, Very Clear}
Similar Case = {Library Kim, cold, dry, *dark*, Normal, Clear}

If the Rule1 fires, the solution changes as follows. We can generate more flexible rules like in table 3, and this approach could be worthwhile to produce more appropriate solution.

Table 3 Generated Solution before and after adaptation

Before adaptation	After adaptation
Air conditioner level is 4	Air conditioner level is 2
Moisture level is 2	Moisture level is 2
Open the Window	Open the Window
Play Ballade Music	Play Ballade Music

Evaluation and Evolution. The SRAs another job is to process learning based on users feedback to provide personalized context. The users feedback can be saved in the personal knowledge base as a new rule that does not affect the general rules. This strategy could make the system to evolve by adapting users evaluation [13]. If the acquired new rules have conflict with the previous rules, then the system can resolve it by using the meta-rules. The detailed process can be seen in the following Fig. 6.

The SRAs self-learning capability may differ based on the users evaluation and feedback. If the user satisfies the suggested solution, then the context and solution will be saved in the case library for the future use. If the user does not satisfy the solution, then the system generates new solution and it will be saved in the personal rule base. This situation happens when the system does not interpret the current situation, and there are two reasons. First, it happens when the system cant resolve the differences between the current context and the previous case. Second, the general adaptation rule cant resolve the differences between the previous context and current context, which can be resolved by the personal rule. Therefore, the personal rule has higher priorities over general rule in case of conflict. This situation can be explained in the following Fig. 7.

From the figure, we first can see that the agent compares the differences between the current context and similar case, and represent it as a rule. The result of the process will be saved into IF part of the rule. And then, compare between the suggested solutions with user feedback, and saves the differences into the THEN part of the rule. By going through this procedure, the agent generates a personal rule, which will be saved into the personal knowledge base.

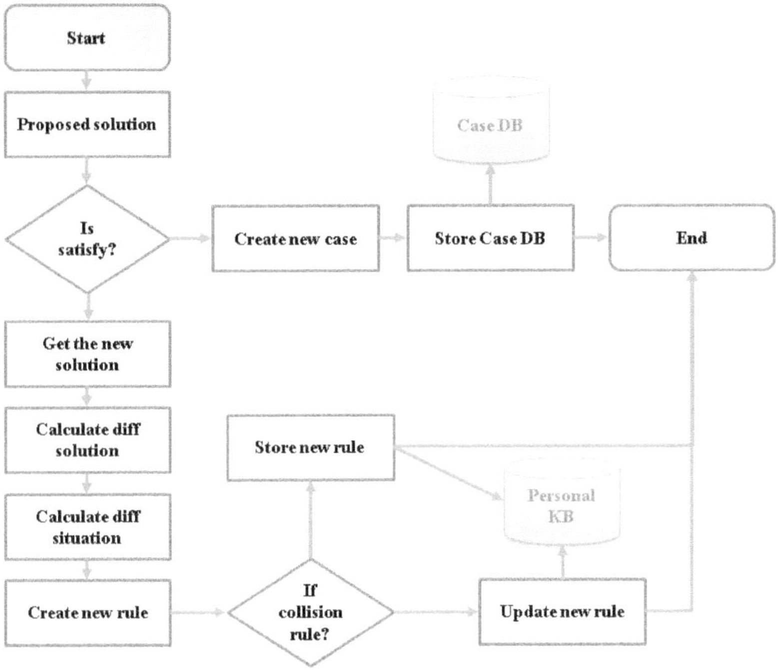

Fig. 6 Flowchart for User Feedback and Learning

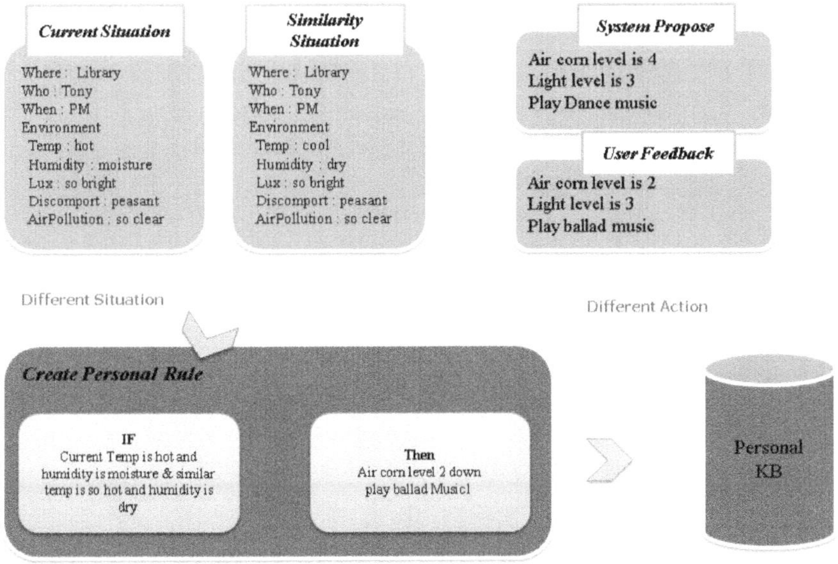

Fig. 7 Generation of User Adaptation Rule

4 Simulated Experimentation

We first designed the experimentation window as in the Fig. 9. In this figure, the upper left corner of the window shows extracted case from the ontology, which has semantic data type. The upper right corner of the window represents the result of comparison between the current case and past context in Case DB. The last column of the table shows the similarity, and we can see that second case is most similar one. The lower right corner of the window shows the systems adaptation rule, and we can find out the last column fired shows that which rule has been applied with the mark x and o. Finally, lower right corner of the window shows the system log as the system progresses, and the suggested solution.

The procedure for our simulated experimentation can be explained as follows. It first provides solution through the SRA, proceeds evaluation and learning based on the solution, and finally shows the result after the learning. The scenario for this experimentation is as follows.

- Tony comes home from the school after finishing exercise.
- When Tony goes to his room, the CMA recognizes him and sends this information and surrounding context information to the system.
- The system recognizes the status of Tony, and turns on the air-conditioner and dehumidifier.
- Tony move to the living room to study, and the CMA recognizes this and provides proper illumination for studying
- But Tony does not satisfy the illumination and temperature of air-conditioner.
- Therefore, Tony controls the illumination and temperature of air-conditioner.
- The system learns this feedback, and saves it as personal rule.

From the example, using the five environmental information, four actual contexts and one dependent context is generated through the CMA. If we apply it into real world application, additional attributes should be added based on the characteristics of a domain.

From the Fig 8, the left window represents the Humidity among the actual context and the right window shows Uncomfort, which is dependent context by using the Temp and Humidity.

Now, the simulation procedure works as follows. For the first step, when Tony comes into the room, the system recognizes and provides a solution as the following three commands, Play ballad music, set the air-conditioner level as 3, and set the dehumidifier level as 2, which can be seen in lower right corner window in the Fig.9. In this simulation, since there is no adaptation rule exists, the temperature context was not considered. Therefore, Tony set the air-conditioner level at 5 by himself, and the system recognizes the change and generates a new rule to solve this problem automatically.

From the Fig. 10 (extracted only the necessary sub-windows from the original window), we can see that a new rule is added in the bottom left

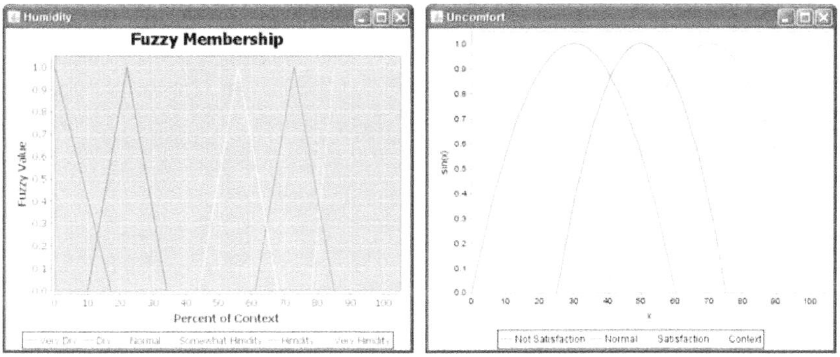

Fig. 8 Fuzzy Membership Representation

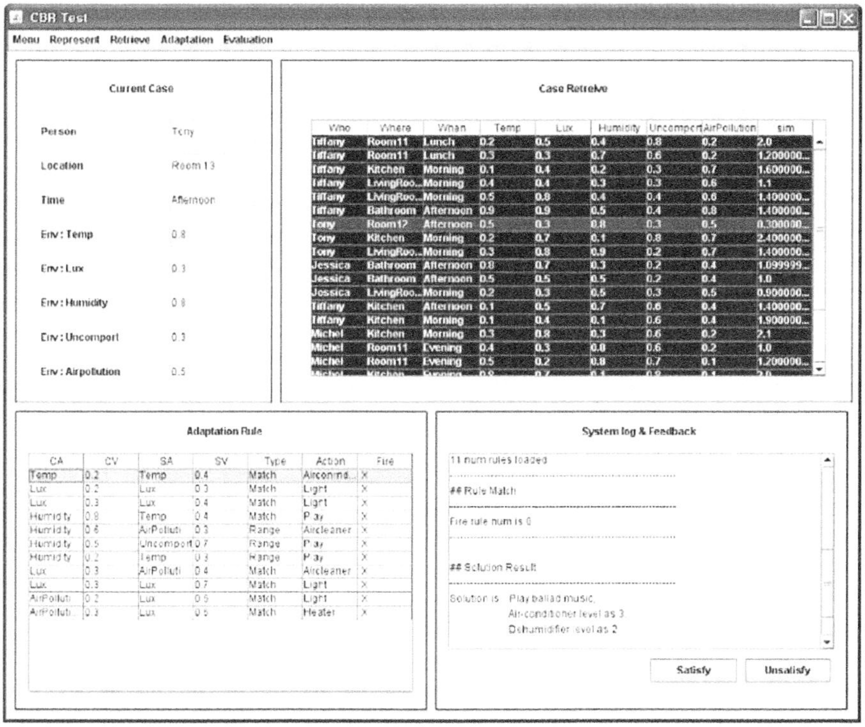

Fig. 9 Simulation with initial solution

of the window, which will be saved in Tonys personal knowledge base. The stored solution is the air-conditioner level 5, which has been set by Tony as a feedback. When the same situation happens for the next occasion, then the system will apply the personal rule first, as in the Fig. 11. We can see the system provides revised solution by adapting the differences from the current situation.

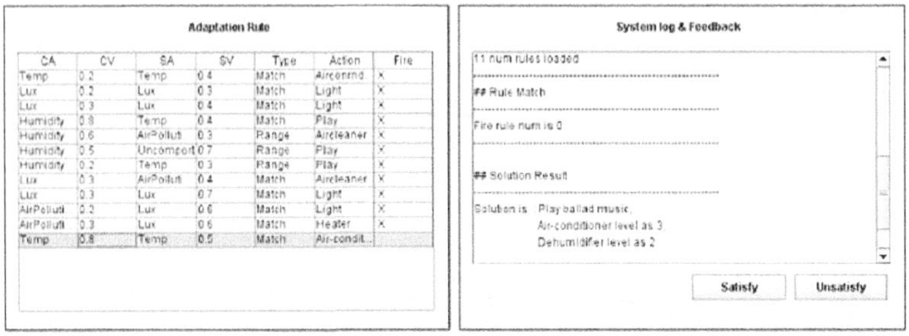

Fig. 10 Adaptation of User Feedback

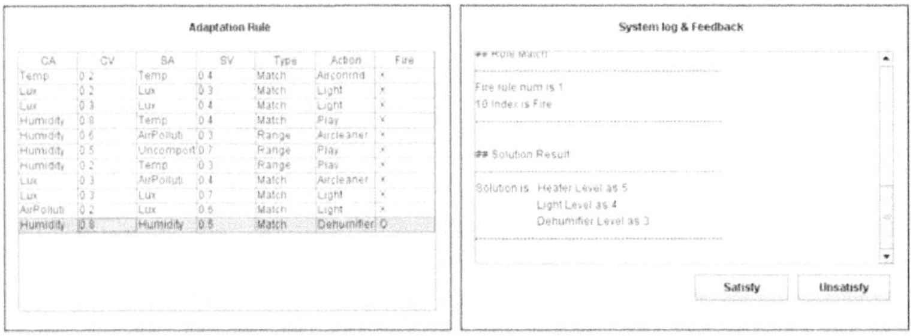

Fig. 11 Learning New Rule

5 Conclusion

In this paper, we have designed and developed agent based context awareness system that can handle some limitations of the previous context awareness systems. The significance of our study is as follows. First, out system architecture does not need to be modified for any environmental change, and it could be adapted into any application domains easily. Second, the system employed agent based architecture to enhance the intelligence of the system. For instance, the CMA treats the conversion of numerical data into semantic data using Fuzzy inference mechanism. This approach can provide more reliable solution to the user. Third, the SRA utilizes CBR engine to make inference, and also reflects user feedback as adaptation rule to enhance the inference capability. This approach brings advantage of providing personalized service to the user. Fourth, we have performed our approach as a simulated experimentation with a virtual scenario. Even if our approach shows some successful results, it still needs to be enhanced in few issues. First, the Fuzzy data conversion of the CMA needs to be tested with more sophisticated Fuzzy theorem, and the SRA needs to handle conflicts of personalized rules as the learning progresses.

Acknowledgements. This research was supported by the Seoul R & BD program (10848) of Korea.

References

1. Dey, A., Abowd, G.: Towards a Better Understanding of Context and Context-Awareness. In: Gellersen, H.-W. (ed.) HUC 1999. LNCS, vol. 1707, pp. 304–307. Springer, Heidelberg (1999)
2. Chen, H., Finin, T., Joshi, A.: An Ontology for Context-Aware Pervasive Computing Environments, Special Issue on Ontologies for Distributed Systems. Knowledge Engineering Review, 18, 197–207 (2003)
3. Gu, T., Pung, H., Zhang, D.: A Middleware for Building Context-aware Mobile Services. In: Proceedings of Vehicular Technology Conference (VTC 2004), vol. 5, pp. 2656–2660 (2004)
4. Dey, A., Abowd, G.: The Context Toolkit: Aiding the Development of Context-Aware Applications, In: Proceedings of the SIGCHI Conference on Human Factors in Computing Systems, pp. 434–441 (1999)
5. Ranganathan, A., Campell, R.: A Middleware for Context Aware Agents in Ubiquitous Computiong Environments. In: Proceedings of ACM/IFIP/USENIX International Middleware Conference, Rio de Janeiro, Brazil (2003)
6. Dey, A., Salber, D., Abowd, G.: A Conceptual Framework and a Toolkit for Supporting the Rapid Prototyping of Context-Aware Applications. Human Computer Interaction 16(2-4), 97–166 (2001)
7. Kindberg, T., et al.: People, Places, Things: Web Presence for the Real World. In: Proceedings of the 3rd IEEE Workshop on Mobile Computing Systems and Applications, Monterey, California, USA, pp. 19–28 (2000)
8. Gu, T., Wang, X., Pung, H., Zhang, D.: An Ontology-based Context Model in Intelligent Environments, In: Proceedings of Communication Networks and Distributed Systems Modeling and Simulation Conference, San Diego, California, USA (January 2004)
9. Kim, Y., Uhm, Y., et al.: Context-Aware Multi-agent Service System for Assistive Home Applications. In: Ma, J., Jin, H., Yang, L.T., Tsai, J.J.-P. (eds.) UIC 2006. LNCS, vol. 4159, pp. 736–745. Springer, Heidelberg (2006)
10. Roman, M., Compbell, R.: GAIA: Enabling Active Spaces. In: Proceedings of the 9th ACM SIGOPS European Workshop, Kolding, Denmark (2000)
11. Mamdani, E., Assilian, S.: An Experiment in Linguistic Synthesis with a Fuzzy Logic Controller. International Journal of Human-Computer Studies 51(2), 135–147 (1999)
12. Aamodt, A., Plaza, E.: Case-Based Reasoning: Foundational Issues, Methodological Variations, and System Approaches. AI Communications 7(1), 39–59 (1994)
13. Hanney, K., Keane, M.: The Adaptation Knowledge Bottleneck How to Ease it by Learning from Cases. In: Leake, D.B., Plaza, E. (eds.) ICCBR 1997. LNCS, vol. 1266, pp. 359–370. Springer, Heidelberg (1997)
14. Baldauf, M., Dustdar, S.: A Survey on Context-aware systems. International Journal of Adhoc and Ubiquitous Computing 2, 263–277 (2004)

A Hybrid Multi-Agent Framework for Load Management in Power Grid Systems

Minjie Zhang, Dayong Ye, Quan Bai, Danny Sutanto, and Kashem Muttaqi

Abstract. In order to cope with load management in power grid systems, this paper presents a hybrid multi-agent framework. This framework integrates the advantages of both centralized and decentralized architectures to achieve both accurate decisions and quick response, and avoid the single point of failure as well. The development of various agents and the behaviors of each agent in the framework are described. Moreover, an example is also introduced, which demonstrates the interaction among agents when a fault happens in a power grid system. The contribution of this paper is to combine local intelligence with global coordination in multi-agent system design to satisfy the challenging requirements in a power grid system.

1 Introduction

In power grid systems, faults at certain nodes can lead to cascading blackouts in the whole system. When a power system is facing the threat of a widespread power outage, the system will be in a vulnerable state [5]. Catastrophic failures in power grid systems bring great inconvenience to people's living and are very harmful for both power industries and the social economics. To avoid

Minjie Zhang · Dayong Ye
School of Computer Science and Software Engineering, University of Wollongong, Wollongong, NSW 2522 Australia
e-mail: {minjie,dy721}@uow.edu.au

Quan Bai
Tasmanian ICT Centre, CSIRO, Hobart, TAS 7001 Australia
e-mail: quan.bai@csiro.au

Danny Sutanto · Kashem Muttaqi
School of Electrical, Computer and Telecommunications Engineering, University of Wollongong, Wollongong, NSW 2522 Australia
e-mail: {danny,kashem}@elec.uow.edu.au

Q. Bai and N. Fukuta (Eds.): Advances in Practical Multi-Agent Systems, SCI 325, pp. 129–143.
springerlink.com

such failures, it is necessary to include protection mechanisms in power grid systems. The general principle of current power protection mechanisms are similar. When a fault occurs on a/some particular node(s), protection procedures are executed to isolate the fault from remaining power grids. Ideally, effective isolation of failure nodes can minimize the loss of the whole system and avoid cascading blackouts.

Cascading failures are normally happen in a very short time, e.g. less than 30 seconds. This requires the power grid system to make accurate decisions within a very limited period. Intelligent agent as a powerful artificial intelligence technology has been used for power grid systems. An intelligent agent is able to make rational decisions in an autonomous and dynamic environment, namely, blending pro-activeness and reactiveness, showing rational commitment to decision-making, and exhibiting flexibility in the face of an uncertain and changing environment [13]. A multi-agent system is a system that is composed of several intelligent agents, and each agent performs different functionalities. The agents in a multi-agent system can work autonomously, make decisions independently and interact with each other in order to achieve global objectives. Currently, most multi-agent based power grid management systems have hierarchical architectures and central controllers which are in charge of various activities of the system, e.g. [11]. However, especially in some large scale power grid systems, it is very hard to take system-wide sequential actions, which include communication, analysis, prediction and decision making, within a very short time. To overcome this limitation, some researchers proposed decentralized methods which allow nodes (grids) to communicate only with their neighbors, such as the approach proposed in [15]. Nevertheless, the information obtained by nodes may be incomplete if only communicating with neighboring nodes, and, therefore, nodes might not make accurate decisions. In this paper, we introduce a hybrid multi-agent framework which can help power grid systems to detect and response node faults quickly and avoid catastrophic failures. This framework, which combines the benefits of both centralized and decentralized manners, can avoid the single point of failure and provide sufficient information for nodes to make accurate decisions. In addition, this framework is topology independent and, hence, suitable for any structures of power grid systems.

The rest of this paper is arranged as follows. In the next section, an overview of current related research is provided. In Section 3, our framework and its detailed design are proposed. After that, an example regarding operation of our framework is demonstrated in Section 4. Finally, we conclude this paper and present several future research directions in Section 5.

2 Related Work

In the last decade, intelligent agent technology has been adopted for various aspects of power systems management, such as restoration [15], relaying [2], maintenance [6], substation automation [1] and state estimation [12]. In this paper, we focus on power system restoration when one or several generators in a power grid system are out of order.

Jung and Liu [5] presented a multi-agent framework that provides real-time information acquisition and interpretation, quick vulnerability evaluation of both power and communication systems, and preventive and corrective self-healing strategies to avoid catastrophic failures of a power system. However, their framework is only a preliminary work, and much details have to be done.

Nagata and Sasaki [10] developed a multi-agent framework for power system restoration. Their framework is in centralized design, which consists of several bus agents and a single facilitator agent. Bus agents are used to decide suboptimal target configuration after a fault occurrence, while the facilitator agent acts as a manager in the decision process. Srivastava et al. [16] proposed a similar method to use a coordinating agent with global information was used for reconfiguration of the shipboard power system. Nagata et al. [9] improved the restoration methodology proposed in [10], but facilitator agents were still required for coordination of the agents. Momoh and Diouf [8] refined the work done in [9]. They utilized power generation agents, bus agents, and circuit breaker agents to distribute the reconfiguration functionalities. However, the system proposed in [8] still needs facilitator agents. The methods proposed in [10] [16] [9] and [8] are centralized multi-agent systems. The centralized manner might bring the framework the potential of the single point of failure.

To overcome the drawback of centralized methods, some decentralized ones have also been presented. In [11], Nordman and Lehtonen proposed a new agent concept for managing an electrical distribution network. Their concept consisted of three aspects, which include secondary substation object, decentralized functionality and an information access model. However, because all secondary substations are copies of the secondary substation objects, all secondary substations are homogeneous with the same type of agent intelligence. This homogeneous feature might limit the adaptation of this concept. Moreover, according to their decentralized functionality, when a primary substation wants to deliver a permission message to a specific secondary substation, the primary substation has to pass the permission message along the communication path instead of directly to the target secondary substation. This communication feature might rise communication costs, and delay the decision making in emergency situations.

Solanki et al. [15] provided a multi-agent framework with detailed design of each agent, which is used to restore a power system after a fault. The framework is decentralized and topology independent which can overcome the scalability of limitations of existing restoration techniques. Although the decentralized manner can avoid single point of failure, the decision accuracy usually cannot be guaranteed. This is because each node in the decentralized architecture only has a limited view of the whole working environment and makes decisions based only on its incomplete information.

A multi-agent system based reconfiguration approach for mesh structured power systems was introduced in [4]. Although the authors claimed that the architecture of the multi-agent system is decentralized, some global information is still employed, e.g. the net power of the power system. Moreover, their work overlooked negotiation process of the multi-agent system, which is very important for a multi-agent system to perform properly.

In this research, we propose a Hybrid Multi-Agent Framework (HMAF) for load management in power grid systems, which deploys intelligent agents in different power grids, e.g. buses, generators and transformers, to execute monitoring, analysing and maintaining activities. Agents in different nodes can also interact with each other to exchange their local information and decisions. Through binding intelligent agents to power grids, we can decompose and allocate complex problems to local agents. In this case, a power grid system can be considered as a Multi-Agent System (MAS), and many agent coordination techniques could be borrowed to solve load balancing and system protection problems in the power system domain. Compared with centralized methods, e.g. [10] and [16], our framework can avoid the single point of failure. In contrast to current decentralized approaches, such as [11] and [15], our framework can provide sufficient information for nodes to make accurate decisions.

3 Framework Architecture and Detailed Design

A load cutting action will cause black-out in a power grid system in some particular areas, which will bring inconvenience to residents. Accurate decisions and actions are very important for load management in order to support the recovery of power grid systems. On the other hand, a widespread power outage will be caused if there is no quick response made within a very limited time. Therefore, for a load management system, speed and accuracy are both in high priorities. To this end, we introduce a hybrid multi-agent management framework for load management in power grid systems, which pays attention on both speed and accuracy. We first compare the benefits of both centralized and decentralized architectures in Subsection 3.1, and, then, provide an overview of the Belief-Desire-Intention (BDI) agent and our motivation in Subsection 3.2. Finally, the framework will be elaborated in Subsection 3.3.

3.1 Centralized Architecture vs. Decentralized Architecture

As described in Section 2, in the current stage, most power grid systems are in centralized architectures, i.e. using a central controller to monitor and manage different power grids. Centralized architectures can promise high accuracy, since a central controller has a global view of the whole system. However, such architectures may have low response speed especially when the system scale is large. For large scale power grid systems, there are a large number of nodes which need to be monitored and managed. This will bring huge computation and communication overheads to the central controller of the system. In this case, a central controller will not be able to give quick response to different nodes in the system within a limited time.

To overcome the above limitation, decentralized system architecture is a potential solution. In a decentralized architecture, a complex problem is decomposed into a number of sub-problems and assigned to different computing nodes. As each node only needs to process a small amount of data and handle a "simple" sub-problem, it can solve the assigned task in shorter time. Decentralized architecture enables a system short response time, but the accuracy cannot be guaranteed. This is because each node in the system only have a limited view of the whole working environment, and needs to provide solutions based on its incomplete vision.

3.2 Design Consideration

An agent can be defined as an intelligent entity, which performs given tasks by using its knowledge and information gleaned from the working environment [17]. It can act in a suitable manner toward achieving the given tasks successfully based on the following common properties [3] [18], particularly in power grid systems:

- *Autonomy.* An agent has some level of self-control ability. It can exist and execute tasks in an environment without external directions. In power grid systems, it is necessary for each node to make decisions autonomously in order to achieve quick response when there are some faults occurring. In addition, autonomy could take the pressure off system operators who form the last line of defence.
- *Adaptivity.* An agent has the ability to learn and improve its performance with previous experience. In power grid systems, a node should be able to make precise decisions based on its previous experience and current states.
- *Reactivity.* An agent can perceive its environment and respond in a timely fashion to changes that occur in the environment. In power grid systems, a node should have the ability to perceive the change of its environment and act in real time in order to reduce the loss when a fault happens.
- *Sociality.* An agent has the ability to interact, communicate and work with other agents. When a fault occurs in a power grid system, it might be inevitable for some nodes to cooperate together to deal with the fault. Therefore, the nodes in the power grid system should be able to communicate and negotiate with each other.

In order to realize our framework, we choose Belief-Desire-Intention (BDI) agent architecture for the framework design. The main reason to use BDI agent architecture is that a BDI agent [14] is able to continuously reason about beliefs, goals and intentions, and acts accordingly. There are four major concepts in the BDI architecture:

- *Beliefs* of an agent are information about the environment. Beliefs can also include inference rules, allowing forward chaining to lead to new beliefs.
- *Desires* are goals assigned to the agent. They represent objectives or situations that the agent would like to accomplish or bring about.
- *Intentions* are commitments by an agent to achieve particular goals. Intentions represent the deliberate state of the agent: what the agent has chosen to do.
- *Plans* are sequences of actions that an agent can perform to achieve one or more of its intentions.

It is now ready to describe the HMAF in detail.

3.3 Hybrid Architecture

To satisfy the requirements of power grid systems, we propose a framework which deploys a layered architecture, i.e. the Scheduler Layer and the Coordination Layer, to ensure accurate and quick response. In each layer of the system, a number of intelligent agents are used to monitor and manage loads in different nodes. Fig. 1 shows the architecture of the framework.

From Fig. 1, it can be seen that the power grid system is divided into a number of small grids, and the dashed arrows demonstrate the communication directions between different agents. A Local Scheduler Agent (LSA) is assigned to each grid to monitor and manage loads within the grid. To facilitate each LSA making accurate decisions, we allow interactions among different Local Scheduler Agents (LSAs). Namely, different LSAs can exchange their local information in order to make more accurate decisions. In addition, the framework includes a Local Coordinator Agent (LCA) in each area to coordinate the corporation of different LSAs, and the LCA has a global view of its associated area. Similarly, above the LCA, there is a Higher Level Coordinator Agent (HLCA) that coordinates the corporation of different Local Coordinator Agents (LCAs). The functionalities of LCA and HLCA are the same. This paper concentrates on the interactions between LSAs and LCA, since the interactions between LCAs and HLCA are analogous as those between LSAs and LCA. Thus, it can be found that there are two types of relations in HMAF. One is *Peer-Peer* relation which exists between LSA-LSA and LCA-LCA. The other is *Superior-Subordinate* relation which exists between LCA-LSA and HLCA-LCA. Therefore, HMAF, which is benefited from hybrid architecture, can adapt the power grid systems in any scale. The design of Local Scheduler Agent and Local Coordinator Agent will be elaborated in the following subsection, and the detailed interaction protocol will be depicted in the next section.

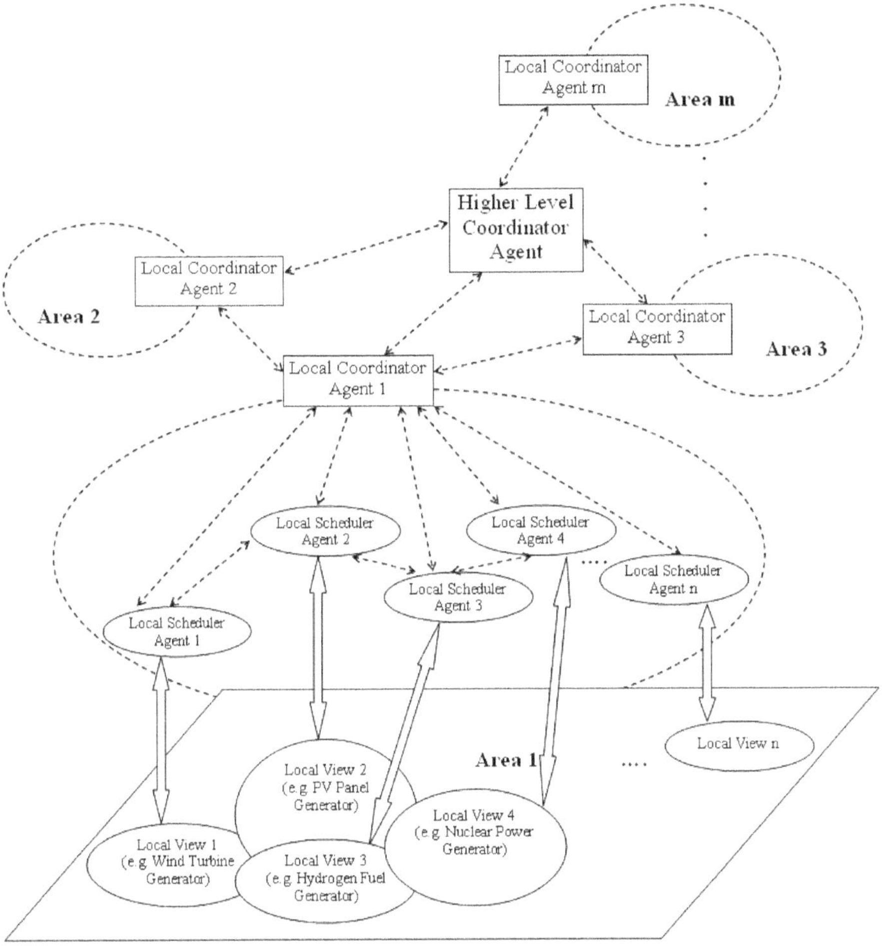

Fig. 1 Hybrid Multi-Agent Framework (HMAF)

3.4 Agents in HMAF

As introduced in Subsection 3.3, HMAF has a hybrid architecture. In the lowest two layers, we include two types of agents, which are Local Scheduler Agent and Local Coordinator Agent. The roles and functionalities of each agent will be described in the rest of this section in detail.

3.4.1 Local Scheduler Agent

A Local Scheduler Agents (LSA) is fixed on each node to monitor all related information of the node, e.g. voltage and loss. A Local Scheduler Agent collects and preprocesses relevant data from sensors, and analyses the information which has been preprocessed in order to detect abnormities in the associated local view. Moreover, a Local Scheduler Agent periodically reports the preprocessed information to the up level Local Coordinator Agent which is in the same area with it. A Local Scheduler Agent then makes decisions on whether there are suspected failures happening in the local view and what actions should be taken. As a Local Scheduler Agent only gleans relevant information in a particular location, the limited information sometime is not sufficient for a precise decision. In this situation, a Local Scheduler Agent will request related information from neighbour nodes in the power system. However, if the information from neighbor nodes is still not enough for making a decision, a Local Scheduler Agent will ask the Local Coordinator Agent to provide relevant information from other indirect linked nodes, e.g. when a Local Scheduler Agent facing a big decision making with regard to cutting its load in order to avoid the damage of the generator.

Fig. 2 shows the design of Local Scheduler Agent. The belief of a Local Scheduler Agent consists of $LSA(AgentID, Status, Capacity, Load, NumNeigbors, NeigIDs)$, where $AgentID$ is the agent identification number, $Status$ indicates the status of the monitored local view which has three states, i.e. $Normal$, $Attention$ and $Emergency$, $Capacity$ is the generator maximum capacity which is monitored by the Local Scheduler Agent, $Load$ means the current load of the generator, $NumNeighbors$ demonstrates the number of neighboring Local Scheduler Agents, $NeigIDs$ stores the identification numbers of these neighbors and the up level Local Coordinator Agent.

A Local Scheduler Agent has the following plans.

1. *DataPreprocess*: This plan preprocesses relevant data from sensors to a standard format for further analysis. When the preprocess is done, this plan posts a message with conversation ID "DataReady" and content as the data entry. This message is handled by the *DataAnalysis* plan.
2. *DataAnalysis*: This plan analyses the preprocessed data to discover abnormities. It will send a message with conversation ID "Request" and the type of information which is needed. This message is handled by other neighbor Local Scheduler Agents *DataAnalysis* plan. Similarly, when this plan receives information request messages from other neighbors, it will send back another message with conversation ID "Response" and the information content which is requested. This message is also handled by other neighbor Local Scheduler Agents *DataAnalysis* plan. In addition, when this plan is not satisfied with neighbors information, it will send another message with conversation ID "MoreNode" to the Local Coordinator Agent. This message is handled by the Local Coordinator Agent *InfoCollection* plan.
3. *InfoProvision*: This plan exchanges information with the Local Coordinator Agent. It periodically sends a message to the Local Coordinator Agent with conversation ID "Provision". The content of the message is the current state of the Local Scheduler Agent. This message is handled by the Local Coordinator Agent *InfoCollection* plan.

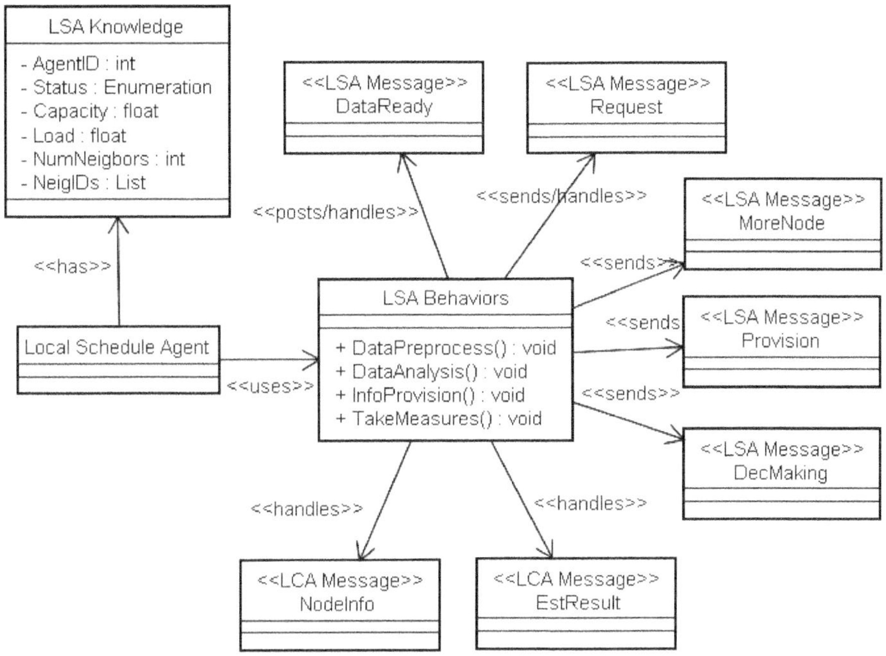

Fig. 2 Local Scheduler Agent

4. *TakeMeasures*: This plan reasons current state and makes decisions about taking measures against faults. When this plan makes a decision, it will send a message with conversation ID "DecMaking" to the Local Coordinator Agent. This message is handled by the Local Coordinator Agent *DecEstimation* plan. The countermeasures include transformer tap change, switch open/close and Strategic Load Shedding, which were listed in [7].

3.4.2 Local Coordinator Agent

A Local Coordinator Agent (LCA) is responsible for periodically collecting information from Local Scheduler Agents which reside at local nodes in the same area with the Local Coordinator Agent. A Local Coordinator Agent should also supply the collected information to its lower level Local Scheduler Agents if needed. In addition, another role of a Local Coordinator Agent is to communicate with other Local Coordinator Agents which are in different areas, and to report the local condition to the Higher Level Coordinator Agent.

Fig. 3 displays the design of Local Coordinator Agent. The belief of a Local Coordinator Agent is composed of $LCA(AgentID, NumLSA, LSA\_IDs,$

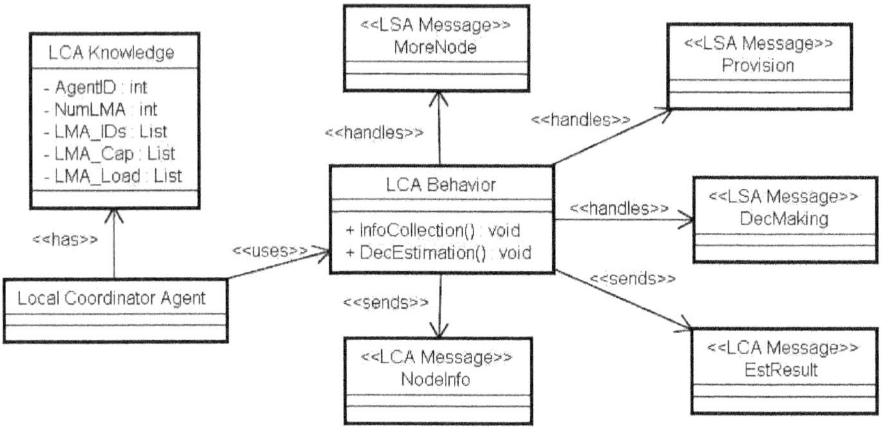

Fig. 3 Local Coordinator Agent

$LSA\_Cap, LSA\_Load)$, where $AgentID$ is the agent identification number, $NumLSA$ is the number of Local Scheduler Agents in the same area with the Local Coordinator Agent, LSA_IDs stores the identification numbers of these Local Scheduler Agents, and LSA_Cap and LSA_Load demonstrate the capacity and current load of each Local Scheduler Agent respectively.

A Local Coordinator Agent has two plans.

1. *InfoCollection*: This plan collects and distributes information from/to the Local Scheduler Agents. When this plan receives a "MoreNode" message, it sends a message with conversation ID "NodeInfo" and content as states of the nodes back to the requesting Local Scheduler Agent. The "NodeInfo" message is handled by the Local Scheduler Agent *DataAnalysis* plan.
2. *DecEstimation*: This plan estimates whether decisions, made by LSAs, are reasonable. The estimation is based on local balance in the associated area. When this plan receives a "DecMaking" message, it responds a message with conversation ID "EstResult" and content as the result of estimation and its suggestion, if applicable, to the requesting Local Scheduler Agent. The "EstResult" message is handled by the Local Scheduler Agent *Take-Measures* plan.

People might argue that, in our framework, if the Local Coordinator Agent is out of order, the associated area could not operate properly. However, the Local Coordinator Agent in our framework is just an information provider rather than a manager, which is different from the Facilitator Agent in [10]. Hence, if Local Coordinator Agents are out of order, this framework only degrades to a general decentralized framework, which can still work as other decentralized architectures.

4 Operation of HMAF

To demonstrate the operation of HMAF, an example is introduced in Fig. 4, where *Gen 1*, 2, 3 and 4 denote *Generator 1*, 2, 3 and 4, respectively, and the arrows indicate loads. As displayed in Fig. 4, *Local View 1* consists of *Generator 1* and *Load 1*, 2 and 4; *Local View 2* is composed of *Generator 2*, *Load 2* and *Load 5*; *Local View 3* is formed with *Generator 3*, *Load 7* and *Load 8*; *Local View 4* contains *Generator 4*, *Load 6* and *Load 9*.[1] Based on HMAF, the power grid system shown in Fig. 4 can be mapped into a multi-agent framework which is displayed in Fig. 5. It is assumed that there is a fault on *Generator* 1 and the *Local Scheduler Agent* 1 (*LSA* 1) has detected this fault.[2] Then, the *LSA* 1 will attempt to request other *LSA*s for help to restore its local power.

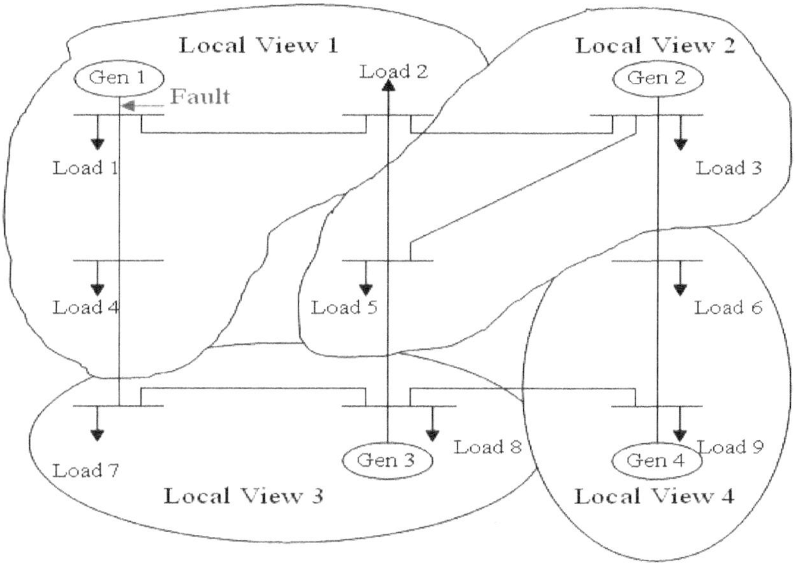

Fig. 4 An Example of a Power Grid System with Fault

As in Fig. 1, the dashed arrows in Fig. 5 indicate the communication links, while the solid arrows in Fig. 5 denote the power flow directions. The current load and capacity of each generator are listed in Table 1.

LSA 1 first requests its neighbors for help with conversation ID "Request" and the request content is "Need Power: 60kw" (60kw=*Load 1*+*Load 2*+*Load 4*). The neighbors of *LSA* 1 are *LSA* 2 and *LSA* 3, which reply *LSA* 1 with

[1] Please note that each *Local View* could contain more than one generator. For simplicity, there is only one generator in each *Local View* in this example.

[2] The detection algorithm for discovering a fault in power grid systems is one of our future works.

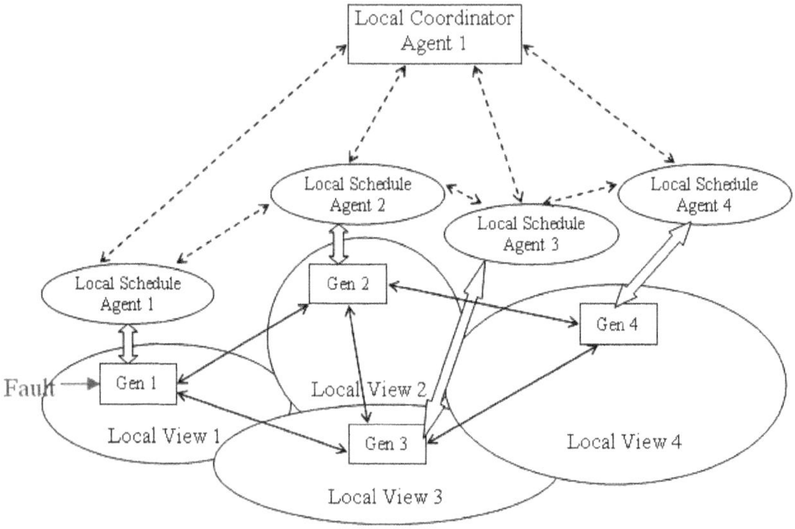

Fig. 5 Mapping of the Power Grid System to Our Framework

Table 1 Current Load and Capacity of Each Generator

Generator	Current Load(kw)	Capacity(kw)
1	*Load 1*=30, *Load 2*=20, *Load 4*=10	80
2	*Load 3*=25, *Load 5*=30	70
3	*Load 7*=15, *Load 8*=20	60
4	*Load 6*=5, *Load 9*=35	60

conversation ID "Response" and the response content is "Available Power: 15kw" (Available Power=Capacity-Current Load) of *LSA* 2 and "Available Power: 25kw" of *LSA* 3. *LSA* 1, then, sends two request messages with conversation ID "Request" and content "Power Supply Request: 15kw" and "Power Supply Request: 25kw" to *LSA* 2 and *LSA* 3 separately. Then, *LSA* 2 controls *Generator* 2 to supply power 15kw to the area which was supplied by *Generator* 1, and *LSA* 1 arranges this power to *Load 1* that is supposed to be vital load in *Local View 1*. Meanwhile, *LSA* 3 designates *Generator* 3 to supply power 25kw to *Generator* 1 area, and, similarly, *LSA* 1 allocates this power to both *Load 1* and *Load 4* (*Load 1* gets 15kw while *Load 4* obtains 10kw). Since the current load of *Generator* 1 is 60kw which is higher than the power *Generator* 2 and *Generator* 3 could supple, *LSA* 1 will ask Local Coordinator Agent 1 (*LCA* 1) for more information in the system. *LSA* 1 sends a "MoreNode" message to *LCA* 1. *LCA* 1 then responds a message with conversation ID "NodeInfo", while the content includes the *AgentIDs* of other nodes in the system and their current load and capacity. When *LSA* 1 receives this message, it reasons that which agents can supply its power. In

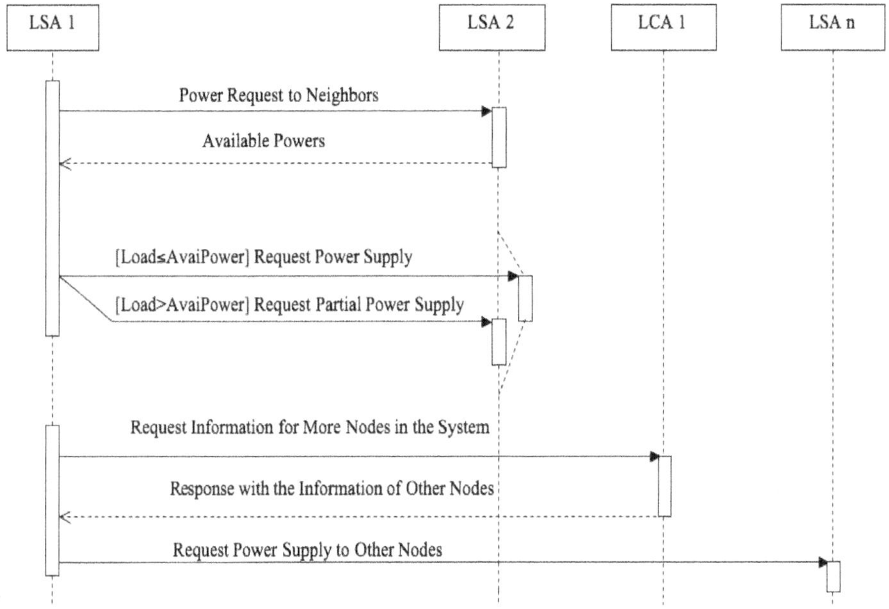

Fig. 6 Interaction Process between LSAs and LCA

this example, the reasoning result is that *Generator* 4 can supply the power 20kw. Thereby, *LSA* 1 sends a request message to the selected agent, namely *LSA* 4 in this example, according to its *AgentID* with the conversation ID "Request" and content "Power Supply Request: 20kw". As long as *LSA* 4 receives the request messages, it supplies the designated power to the area which was supplied by *Generator* 1, and *LSA* 1 assigns this power to *Load 2*. The interaction process is displayed in Fig. 6. In this example, *Generators* 2, 3 and 4 can supply sufficient power for *Generator* 1. However, in some cases, if other generators could not provide enough power for *Generator* 1, *Generator* 1 would have to make a decision about discarding current load, and asks LCA to estimate whether this decision is reasonable. The detailed estimation process is one of our future works.

5 Conclusion

In this paper, a hybrid multi-agent framework for power grid systems load management was proposed. The contribution of this framework is that it combines centralized and decentralized architecture together. Compared with centralized architectures, this framework can avoid single point of failure in some extent. In contrast to complete decentralized architectures, this framework can provide nodes sufficient information for making accurate decisions and quick response when a fault occurs in the system.

In the future, a formal detection model for quick fault discovery in power grid systems will be developed. In addition, an efficient resource allocation algorithm will also be devised to efficiently distribute power to the local view in which the generator is out of order. Finally, this framework will be implemented and tested in some real cases.

Acknowledgements. This research is supported by Australian Research Council Linkage Projects (LP0991428) and a URC Research Partnerships Grants Scheme, from the University of Wollongong with the collaboration of TransGrid, Australia.

References

1. Buse, D.P., Sun, P., Wu, Q.H., Fitch, J.: Agent-based substation automation. IEEE Power Energy 1(2), 50–55 (2003)
2. Coury, D.V., Thorp, J.S., Hopkinson, K.M., Birman, K.P.: Agent technology applied to adaptive relay setting for multi-terminal lines. In: Proc. IEEE Power Eng. Soc. Summer Meeting, pp. 1196–1201 (2000)
3. Green, S., Hurst, L., Nangle, B., Cunningham, P., Somers, F., Evans, R.: Software Agents: A Review. Tech. Rep. TCD-CS-1997-06, Trinity Collega, Dublin, Ireland (1997)
4. Huang, K., Srivastava, S.K., Cartes, D.A., Sloderbeck, M.: Intelligent agents applied to reconfiguration of mesh structured power systems. In: Proc. International Conference on Intelligent Systems Applications to Power Systems, Toki Messe, Niigata, pp. 1–7 (2007)
5. Jung, J., Liu, C.: Multi-agent technology for vulnerability assessment and control. In: Proc. IEEE Power Eng. Soc. Summer Meeting, Vancouver, BC, Canada, vol. 2, pp. 1287–1292 (2001)
6. Kezunovic, M., Xu, X., Wong, D.: Improving circuit breaker maintenance management tasks by applying mobile agent software technology. In: Proc. IEEE Power Eng. Soc. Asia Pacific Transm. Distrib. Conf., pp. 782–787 (2002)
7. Lachs, W., Sutanto, D.: Voltage instability in interconnected power systems: A simulation approach. IEEE Transactions on Power Systems 7(2), 753–761 (1992)
8. Momoh, J., Diouf, O.: Optimal reconfiguration of the navy ship power system using agents. In: Proceedings of IEEE PES Transmission and Distribution Conference and Exhibition, pp. 562–567 (2006)
9. Nagata, T., Fujita, H., Sasaki, H.: Decentralized approach to normal operations for power system network. In: Proceedings of the 13th International Conference on Intelligent Systems Application on Power Systems, pp. 407–412 (2005)
10. Nagata, T., Sasaki, H.: A multi-agent approach to power system restoration. IEEE Transactions on Power Systems 17(2), 457–462 (2002)
11. Nordman, M.M., Lehtonen, M.: An agent concept for managing electrical distribution networks. IEEE Transactions on Power Delivery 20(2), 696–703 (2005)
12. Nordman, M.M., Lehtonen, M.: Distributed agent-based state estimation for electrical distribution networks. IEEE Transactions on Power Systems 20(2), 652–658 (2005)

13. Poutakidis, D., Padgham, L., Winikoff, M.: Debugging Multi-Agent Systems Using Design Artefacts: The Case of Interaction Protocols. In: Proceedings of the First International Workshop on Challenges in Open Agent Systems at AAMAS 2002, Bologna, Italy, pp. 960–967 (2002)
14. Rao, A., Georgeff, M.: An Abstract Architecture for Rational Agents. In: Proceedings of the Third International Conference on Principles of Knowledge Representation and Reasoning, San Mateo, USA, pp. 439–449 (1992)
15. Solanki, J.M., Khushalani, S., Schulz, N.N.: A multi-agent solution to distribution systems restoration. IEEE Transactions on Power Systems 22(3), 1026–1034 (2007)
16. Srivastava, S.K., Xiao, H., Butler-Purry, K.L.: Multi-agent system for automated service restoration of shipboard power systems. In: Proceedings of the 15th International Conference on Computer Applications in Industry and Engineering, San Diego, CA, USA, pp. 119–123 (2002)
17. Wooldridge, M.: An Introduction to Multiagent Systems. John Wiley & Sons Ltd., Chichester (2002)
18. Wooldridge, M., Jennings, N.: Intelligent Agents: Theory and Practice. The Knowledge Engineering Review 10(2), 115–152 (1995)

Part II

Agent-Based Simulation for Complex Systems: Application to Economics, Finance and Social Sciences

Agent-based simulation is now a well-established research domain and has become more and more used in various areas of sciences and industry. One of the major contributions of multi-agent systems is perhaps their capacity to model (quite intuitively) and simulate systems having a high degree of complexity (high number of agents or parameters, complex interactions between them...), such as the ones needed in ecology, epidemiology or sociology. The complexity of the systems to be modeled also induces a highly complex modeling process: for example, it involves the participation of many stakeholders with various skills (and thus various languages and habits...).

Application to Finance and Economics

An agent-based computational economics approach has been active as applications of agent-based technologies to financial and economic systems. There are two kinds of application to economics of an agent simulation; the verification of an existing economic theory and the proposal and estimation of a certain economic behavior. The following two papers use agent-based models to test economic theories. Cao-Alvira builds a macro economic model with financial frictions and studies the impact of serially correlated monetary shocks on the variability of velocity. Matsuda, Kaihara, and Fujii construct an agent model of a goods market with agents negotiating the resource allocations under demand constraints of consumers in the market. The following papers are researches for detailed economic activity. Goyal designs a novel auction strategy with the fuzzy logic technique and conducts an experimental evaluation on the bases of the historical auction records. Yamashita, Takahashi, and Terano analyze learning methods for pension investment management that consider liability using the business game technique.

One of the best results of the application of agent-based simulations to finance is an artificial market. From theoretical to empirical studies, various results are obtained by artificial markets in these ten years. Toriumi, Izumi, and Matsui propose an estimation method based on inverse simulation to estimate the combinations of traders who participate in an artificial market. Artificial markets also use for a market design. Yagi, Mizuta, and Izumi discuss the effectiveness of short-selling regulation using their artificial markets. Studies on artificial markets have attained some successes in market analysis in recent years.

Application to Social Sciences

Social Science domain is certainly one of the research domains that contributes the most to the advance of Multi-Agent Systems. A lot of agent-based models and simulators have been developed to study real social systems (*e.g.* a town) with various points of view such as demography, transportation, social organisation and so one. Following papers illustrate the use of agent paradigm in various application fields of social sciences. El Saadi, Bah and

Belarbi construct an agent-based model from the analytical Todaro Model in order to simulate and analysis the mechanism of labor migration and urban unemployment. Genre-Grandpierre and Banos investigate a new metric of road networks that aims at favoring the efficacy of the short range travels. Its calibration and its impact on the traffic are evaluated thanks to agent-based simulation running on the S3 platform. Finally El-Gemayel, Sibertin-Blanc and Chapron aim at simulating social organizations formalized in the theory of the Sociology of the Organized Actions, theory that focuses upon the actual behaviors of the organization members. Among parameters influencing members' behaviors, this paper is focused on the Tenacity.

Takao Terano
Tohgoroh Matsui
Kiyoshi Izumi
MASFE'09 Workshop Chairs

Alexis Drogoul
Benoit Gaudou
Nicolas Marilleau
AAECS'09 Workshop Chairs

Financial Frictions and Money-Driven Variability in Velocity

José J. Cao-Alvira

Abstract. Frictions are introduced in the financial structure of a cash-in-advance dynamic stochastic general equilibrium model with the interest of studying their impact on the variability of velocity due to serially correlated monetary shocks. Possessing no analytical solution this dynamic environment, a projection method which parameterizes expectations and employs finite elements in the approximation of the system's policy functions is executed on the approximation a solution for the equilibrium of the economy and is able to efficiently handle the occasionally binding cash-in-advance constraint on transactions. This last characteristic permits a robust analysis on the impact of frictions on the variability of velocity. It is concluded that frictions on the financial structure of the economy accentuate a precautionary demand for money balances, increasing the incidence of adjustments on the velocity of transactions as an answer to money growth rate shocks.

1 Introduction

Money-driven general equilibrium models with great difficulty are capable to explain the variability on velocity observed in the data, due to the insufficient propagation mechanisms associated with monetary shocks. Alas, a proper assessment on the variability of consumption velocity of money is fundamental to understanding the role of money in the business cycle and bestows insight on the microfoundations of monetary policy.

Attempting to answer for a significant allotment of its variability, financial frictions are introduced in the information structure of a cash-in-advance (CIA) dynamic stochastic general equilibrium model with the interest of studying the impact on velocity of serially correlated monetary shocks. On

José J. Cao-Alvira
Universidad de Puerto Rico, Rio Piedras PR 00931, USA
e-mail: JoseJulian.Cao@UPR.edu

Q. Bai and N. Fukuta (Eds.): Advances in Practical Multi-Agent Systems, SCI 325, pp. 149–166.
springerlink.com

the proposed modeling environment, at each period the agent is required to make its choice on real money holdings prior to the realization of a monetary shock. Uncertainty regarding the current period's realization of the money growth rate incentives the agent to carry additional units of cash than those that would be chosen if instead the portfolio is to be formed ex-post the realization of the shock. Frictions on the information structure accentuate a precautionary demand for money balances, causing an increase of the variability in velocity when compared to a frictionless environment. On the model, velocity variability will occur more succinctly at low rates of money growth for consumption smoothing purposes, on these cases the agent chooses to hold money for the next period because of expectations of future low realizations of the growth rate.

Given that this particular dynamic environment posses no analytical solution, a projection method which parameterizes expectations and employs finite elements in its approximations of the policy functions is employed to solve for the equilibrium of the economy. The solution method is a more complex application of the methodologies proposed in Cao-Alvira (2006) and Cao-Alvira (2010b), in [4] and [3], which solve for frictionless CIA environments using finite elements and a parameterization of the agent's expectations. The approximation of the model's functional equations using finite elements is an advantage over other commonly used perturbation or projection methods, such as a parameterization using a Chebyshev polynomial or the linearization of the system around its steady state, because it allows for the fit of numerous low-order polynomials over nonintersecting subdomains of the state space, rather than high-order polynomials over the complete domain. McGrattan (1998), in [12], stresses that the fractionation of the space results in an improvement on the precision of the approximation of the policy functions near regions of the state space that are of higher order or highly nonlinear. Aruoba et. al (2006) in [1] concurs with this result, and finds that finite element approximations proved being the most accurate, stable and of fastest convergence from a considered wide range of projection and perturbation methods. Cao-Alvira (2010a), in [2], studies a version of the employed algorithm applied to an optimal growth model with leisure constrained by an irreversible investment requirement, and is able to assess the algorithm's significant level of accuracy and speed; as well as the algorithm's efficient handling of the occasionally binding slackness multiplier. Cao-Alvira (2010b) reports the algorithm's convergence speed and accuracy for a CIA model similar to the one studied in this paper without considering the presence of frictions.

Following a procedure first introduced by den Haan and Marcet (1990), in [7], the algorithm is able to efficiently handle the CIA constraint on transactions by parameterizing the optimal choice rule on the expectations of future real money holdings. The procedure allows for the inequality constraint on the planner's problem to selectively and occasionally bind, enabling an analysis on the variability of money velocity.

The next section contains a detailed description of the modeling environment and the functional forms of its equilibrium conditions. Section 3 discusses the proposed solution methodology and the implementation of its algorithm. Section 4 shows the business cycle properties of the economy for a benchmark calibration and the implications on velocity of information and financial frictions. Section 5 concludes.

2 Cash-in-Advance Model Economy

The economic environment is modeled after the information structure originally introduced by Svensson (1985), in [14], where a cash-in-advance constrained agent is required to choose money holdings prior to the realization of a stochastic shock to the growth rate of money supply, and further developed by Christiano and Eichenbaum (1992), in [5], where credit goods are incorporated in an ex-ante choice of portfolio holdings. Except for the specification of the information structure, the economic environment is modeled similar to the Cooley and Hansen (1989) CIA model, in [6]. This section presents the planner's problem, its equilibrium conditions, its analytical steady states, and the functional forms for which the solution procedure, discussed in Section 3, involving finite elements and a parameterized expectations algorithm, is implemented.

2.1 Discussion of the Economy

In an economy exhibiting information frictions in its future realization of a monetary shock and money is used for transaction purposes, a benevolent social planner maximizes the lifetime utility function of an infinitely lived representative agent by making choices over consumption, labor supply, next period capital, and real money holdings. The consumption good is subject to a CIA constraint, and leisure and capital investment are considered to be credit goods. The timing digression is as follows, the agent begins period t with money M_{t-1} and, at this point, determines its holdings on real money, then receives the lump-sum real transfer T_t, current period's prices are determined and the goods market opens.

Given k_0, the social planner chooses infinite sequences of consumption $\{c_t\}_{t=0}^{\infty}$, labor supply $\{n_t\}_{t=0}^{\infty}$, and real money balances $\{M_{t-1}/p_{t-1}\}_{t=0}^{\infty}$ for an infinitely lived agent, and an infinite sequence of capital stock $\{k_{t+1}\}_{t=0}^{\infty}$ to solve:

$$\max_{\left\{\frac{M_{t-1}}{p_{t-1}}\right\}} E_{t-1}\left\{\max_{\{c_t,n_t,k_{t+1}\}} E_t\left\{\sum_{t=0}^{\infty}\beta^t\left\{\frac{c_t^{1-\tau}-1}{1-\tau}-\gamma_n\frac{n_t^{1+\gamma}}{1+\gamma}\right\}\right\}\right\} \tag{1}$$

subject to the budget constraint:

$$c_t + k_{t+1} + \frac{M_t}{p_{t-1}} = Y_t + (1-\delta)k_t + \frac{M_{t-1}}{p_{t-1}} + T_t \tag{2}$$

and a cash-in-advance constraint:

$$c_t \leq \frac{M_{t-1}}{p_{t-1}} + T_t. \tag{3}$$

Where $Y_t = Ak_t^{\alpha}n_t^{1-\alpha}$, $0 < \alpha < 1$, A is a production technology parameter, $\beta \in (0,1)$ is the time discount factor, and $\delta \in [0,1]$ is the capital depreciation rate. Investment is defined as the next period's capital stock minus the current period's undepreciated level of capital, i.e. $i_t = k_{t+1} - (1-\delta)k_t$. Changes in the money supply are realized through a lump-sum real transfer T_t to the agent, i.e. $T_t = (M_t - M_{t-1})/p_{t-1}$.

The gross growth rate of money supply θ_t, i.e. $\theta_t = M_t/M_{t-1}$, evolves according to a Markovian process with transitional probabilities Π. Q is the number of possible states of nature of θ, $\sum_{z=1}^{Q} \Pi_{wz} = 1$. For each $w = \{1,...,Q\}$ & $r = \{1,...,Q\}$, typical element Π_{wr} is the probability of being on state r on time $t+1$ given the realization of state w in time t:

$$\Pi_{wr} = \Pr\left[\theta_{t+1} = \theta(r) | \theta_t = \theta(w)\right]. \tag{4}$$

The budget constraint in eq. (2) implies that current period's consumption, real money holdings and next period capital stock are financed by actual production, undepreciated capital stock, and the real money balances held from the previous period plus a lump-sum transfer. The CIA constraint requires the agent to hold sufficient real money balances in order to purchase the consumption good.

Defining λ_t as the Lagrangian multiplier associated with the budget constraint and μ_t as the multiplier for the CIA constraint, the planner's problem is identified by the first order conditions in eqs. (5) – (8):

$$\lambda_t = c_t^{-\tau} - \mu_t \tag{5}$$

$$\gamma_n n_t^{\gamma} = \lambda_t (1-\alpha) \frac{Y_t}{n_t} \tag{6}$$

$$\lambda_t = \beta E_t \left\{ \lambda_{t+1} \left(\alpha \frac{Y_{t+1}}{k_{t+1}} + (1-\delta) \right) \right\} \tag{7}$$

$$\frac{E_t\{\lambda_t\}}{p_{t-1}} = \beta E_t \left\{ \frac{c_{t+1}^{-\tau}}{p_t} \right\} \tag{8}$$

the Kuhn-Tucker condition in eq. (9):

$$c_t \leq \left(\frac{M_{t-1}}{p_{t-1}} + T_t \right) \quad \text{and} \quad \mu_t \left[c_t - \left(\frac{M_{t-1}}{p_{t-1}} + T_t \right) \right] = 0 \tag{9}$$

Table 1 Expressions of some endogenous variables

Variable	Expression
Consumption Velocity	$V_t = \frac{C_t}{M_t/p_t}$
Inflation	$\pi_t = \frac{p_t}{p_{t-1}}$
Nominal Interest	$I_t = r_t \cdot E_t\left\{\pi_{t+1}\right\}$
Real Interest	$r_t = E_t\left\{\lambda_{t+1}\left(\alpha\frac{Y_{t+1}}{k_{t+1}} + (1-\delta)\right)\right\}$

and market clearing conditions for the goods and the money markets, in eqs. (10) – (11), respectively:

$$c_t + k_{t+1} = Y_t + (1-\delta)k_t \tag{10}$$

$$M_t = M_{t-1} + p_{t-1}T_t. \tag{11}$$

Eq. (5) shows the marginal benefit of consumption to be equal to the marginal utility of wealth plus the value of liquidity services needed to finance the transaction. A binding CIA constraint works as a necessary transaction cost which increases the marginal benefit of consumption at the equilibrium allocation. Eq. (6) is the condition for labor market equilibrium, where the marginal cost of labor supplied equals the utility value of its marginal productivity. Eq. (7) is the Euler equation, portraying the relationship between current and expected future consumption decisions; if wealth was to be slightly reduced at the current period and carried over to the next period, the loss of current utility must equal the future discounted value of its real gross return. Equilibrium conditions in eqs. (6) & (7) show how a binding CIA constraint creates a distortion away from consumption (cash goods) towards leisure and capital investment (credit goods). By raising the price of consumption above its production cost, the CIA constraint acts as a tax on consumption, diverting the planner towards acquiring goods which are not subject to this constraint. Eq. (8) indicates that real money holdings are chosen such that ex-ante the realization of the monetary shock, the expected value of holding a unit of money in terms of utility, i.e. $E_t\left\{\lambda_t\right\}/p_{t-1}$, equates the next period's expected real marginal benefit of consumption. The Kuhn-Tucker condition in eq. (9) tells that if the constraint is binding, the slackness multiplier is positive, $\mu_t > 0$, and consumption must equal real money holdings, $c_t = M_t/p_{t-1}$; otherwise, if the constraint does not bind, the slackness multiplier is zero, $\mu_t = 0$, and real money holdings may be held for next period, $c_t \leq M_t/p_{t-1}$, in the form of precautionary demand for money. Table 1 indicates the formulae for calculating the consumption velocity of money, inflation, nominal interest, and real interest at time period t. Given that there is no nominal bond in the economy, nominal interest is derived from the Fisher equation.

2.2 Analytical Steady States

Steady states for the CIA model economy are defined by eqs. (12) – (16):

$$\frac{k^{ss}}{y^{ss}} = \frac{\alpha\beta}{1-\beta(1-\delta)} \tag{12}$$

$$\frac{i^{ss}}{y^{ss}} = \delta\frac{k^{ss}}{y^{ss}} \tag{13}$$

$$\frac{c^{ss}}{y^{ss}} = 1 - \frac{i^{ss}}{y^{ss}} \tag{14}$$

$$n^{ss} = \left[\frac{\beta}{\theta^{ss}}\frac{(1-\alpha)}{\gamma_n}\left(\frac{c^{ss}}{k^{ss}}\right)^{-\tau}\left(\frac{y^{ss}}{k^{ss}}\right)^{\frac{\tau-\alpha}{1-\alpha}}\right]^{\frac{1}{\gamma+\tau}} \tag{15}$$

$$\mu^{ss} = \lambda^{ss}\left(\frac{\theta^{ss}}{\beta} - 1\right). \tag{16}$$

The steady state of the CIA constraint is derived from the equilibrium condition of real money balances, eq. (8). Notice the constraint is binding in the steady state for a steady state growth of money supply greater than the time discount factor.

2.3 Equilibrium of the Economy

The equilibrium for the CIA economy is denoted by the sequence of real variables $\{\digamma_t\}_{t=0}^{\infty} = \{c_t, n_t, k_{t+1}, M_t/p_{t-1}\}_{t=0}^{\infty}$, given a sequence of monetary growth gross rates $\{\theta_t\}_{t=0}^{\infty}$ evolving according to the transition matrix Π_{wr} in (4), a sequence of money supply and real lump-sum transfers $\{M_t, T_t\}_{t=0}^{\infty}$, and an initial stock of capital k_0, which satisfy the first order conditions (5), (6), (7), & (8), the CIA constraint (9), and the market clearing conditions (10) & (11).

2.4 State Space and Functional Forms of the Economy

Θ is the state space of the economy, which can be sub-divided in two subsets; one, Ω, containing the continuous variables of the state space, and a second, Λ, containing the discrete state variables. At time t, the partial state space Ω_t is composed of the possible realizations of the capital stock at time t and the money supply at time $t-2$. Ω has a well defined compact support, i.e. $\Omega = [\underline{\mathrm{k}}, \bar{\mathrm{k}}] \times [\underline{\mathrm{M}}, \bar{\mathrm{M}}]$. Λ_t is composed of the possible realizations of the money growth rate parameters at time t and $t-1$. Λ also has a well defined compact support, i.e. $\Lambda = [\theta(1), ..., \theta(Q)] \times [\theta(1), ..., \theta(Q)]$.

The solution methodology employs the use of time invariant policy functions Υ_{uw} & P_u for all $u = [1, ..., Q]$ and all $w = [1, ..., Q]$, to express the equilibrium conditions of the CIA model economy. The policy function Υ_{uw} is dependent on the past and present realizations of the money growth parameters, while P_u is dependant only on the past realization of the parameter. Conditional on $\theta_{t-1} = \theta(u)$ and $\theta_t = \theta(w)$, $\Upsilon_{uw}(\Omega_t)$ is defined to map the current state of capital and money stock into the control of the conditional expectation function of the Euler equation, i.e. $\Upsilon_{uw}(\Omega_t) \equiv \Upsilon_{\Theta_t} = \Upsilon_w(k_t, M_{t-2}|\theta_{t-1} = \theta(u), \theta_t = \theta(w))$ where $\Upsilon_{\Theta_t} = \beta E_t\{\lambda_{t+1}[\alpha \frac{Y_{t+1}}{k_{uw,t+1}} + 1 - \delta]\}$. Conditional on $\theta_{t-1} = \theta(u)$, $P_u(\Omega_t)$ is defined to map the current state of capital and money stock into the control of the price function, i.e. $P_u(\Omega_t) \equiv p_{\Theta_t} = P_u(k_t, M_{t-2}|\theta_{t-1} = \theta(u))$.

Define $\boldsymbol{\Upsilon}$ as a $[Q \times Q]$ matrix, where any given row $r \in \{1, ..., Q\}$ contains a vector $\bar{\Upsilon}_r$ of policy functions Υ_{rw}, such that $\boldsymbol{\Upsilon} = [\bar{\Upsilon}_1, ..., \bar{\Upsilon}_Q]'$. The rth row vector can be represented as $\bar{\Upsilon}_r = [\Upsilon_{r1}, ..., \Upsilon_{rQ}]$. Let $\bar{P}$ be a column vector containing the policy functions P_u for all $u \in \{1, ..., Q\}$, i.e. $\bar{P} = [P_1, ..., P_Q]'$. Using the functional form of the policy functions $\Upsilon_{uw}(k_t, M_{t-2})$ and $P_u(k_t, M_{t-2})$, for all u & w, and the Markovian nature of the exogenous parameters, the residuals of the Euler equation and the equilibrium condition for real money holdings can be written as in eqs. (17) & (18), respectively:

$$R^K_{uw}(k, M; \boldsymbol{\Upsilon}, \bar{P}) = \Upsilon_{uw}(k, M) - \beta \sum_{z=1}^{Q} \Pi_{wz}\{\Upsilon_{wz}(\tilde{k}_{uw}, \tilde{M})[\alpha \frac{\tilde{y}_{wz}}{\tilde{k}_{uw}} + 1 - \delta]\} \quad (17)$$

$$R^M_u(k, M; \boldsymbol{\Upsilon}, \bar{P}) = \sum_{w=1}^{Q} \Pi_{uw} \left\{ \frac{\Upsilon_{uw}(k, M)}{P_u(k, M)} - \sum_{z=1}^{Q} \Pi_{wz} \frac{\tilde{c}_{wz}^{-\tau}}{P_w(\tilde{k}_{uw}, \tilde{M})} \right\}, \quad (18)$$

for all $u = \{1, ..., Q\}$ & $w = \{1, ..., Q\}$. The real variables are defined by

$$y_{uw} = Ak^{\alpha} n_{uw}^{1-\alpha} \quad (19)$$

$$\tilde{k}_{uw} = Ak^{\alpha} n_{uw}^{1-\alpha} + (1 - \delta) k - c_{uw} \quad (20)$$

$$n_{uw} = \left[\frac{(1 - \alpha)}{\gamma_n} \Upsilon_{uw}(k, M) Ak^{\alpha} \right]^{\frac{1-\alpha}{\gamma+1}} \quad (21)$$

$$c_{uw} = \begin{cases} \Upsilon_{uw}(k, M)^{-\tau} & \text{if} \quad \mu_{uw} = 0 \\ M_{+1}/P_u(k, \tilde{M}) & \text{if} \quad \mu_{uw} > 0 \end{cases}. \quad (22)$$

The money supplies ex-ante and ex-post the realization of the growth rate shocks are defined by $\tilde{M}$ and M_{+1}, respectively

$$\tilde{M} = \theta(u) M \quad (23)$$

$$M_{+1} = \theta(w)\tilde{M}. \tag{24}$$

Policy functions in $\boldsymbol{\Upsilon}$ & $\bar{P}$ are the solutions to $R^K_{uw}(k, M; \boldsymbol{\Upsilon}, \bar{P}) = 0$ for all u & w and $R^M_u(k, M; \boldsymbol{\Upsilon}, \bar{P}) = 0$ for all u, and in combination with y_{uw}, $\tilde{k}_{uw}$, n_{uw}, c_{uw}, $\tilde{M}$ & M_{+1} from eqs. (19), (20), (21), (22), (23) & (24), and the sequence of money growth rates θ, evolving according the transition matrix Π in (4), generate the sequences $\{c_t,\ n_t,\ k_{t+1},\ M_t/p_{t-1}\}_{t=0}^{\infty}$, that solve for the equilibrium of this economy along the functional state space Θ.

3 Solution Methodology

The solution procedure involves the usage of finite elements in its approximation of the policy functions and a parameterized expectations algorithm to minimize the weighted absolute value of residual functions $R^K_{uw}(k, M; \boldsymbol{\Upsilon}, \bar{P})$ and $R^M_u(k, M; \boldsymbol{\Upsilon}, \bar{P})$, for all u & w; where the true decision rules $\Upsilon_{uw}(k, M)$ & $P_u(k, M)$ are replaced by the parametric approximations $\upsilon^h_{uw}(k, M)$ & $p^h_u(k, M)$. υ^h_{uw} & p^h_u are approximated using an implementation of the finite element method, that follows that advocated in McGrattan (1996), see [13].

To create the approximate time invariant functions υ^h_{uw} & p^h_u, the space $\Omega = [\underline{\mathrm{k}}, \bar{\mathrm{k}}] \times [\underline{\mathrm{M}}, \bar{\mathrm{M}}]$ is divided in n_e nonoverlapping rectangular subdomains called "elements". Parameterization of the policy functions for each element, at each realization of u & w, are constructed using linear combinations of low order polynomials or "basis functions". This procedure creates local approximations for each function. Given the discrete nature of Λ, this state space need not to be divided.

The parameterized functions $\upsilon^h_{uw}(k, M)$ & $p^h_u(k, M)$ are built as follows:

$$\upsilon^h_{uw}(k, M) = \sum_{ij}^{IJ} \upsilon^{uw}_{ij} W_{ij}(k, M) \tag{25}$$

$$p^h_u(k, M) = \sum_{ij}^{IJ} p^u_{ij} W_{ij}(k, M). \tag{26}$$

Where $i = \{1, ..., I\}$ indicate capital nodes, $j = \{1, ..., J\}$ indicate money supply nodes, $W_{ij}(k, M)$ is a set of linear basis functions dependent on the element $[k_i, k_{i+1}] \times [M_j, M_{j+1}]$, for all i, j, over which the local approximations are performed. υ^{uw}_{ij} & p^u_{ij} are vectors of coefficients to be determined. The parameterized value of the conditional expectation function and the price function over the full state space are obtained by piecing together all the local approximations. The approximate solutions for $\upsilon^h_{uw}(k, M)$ & $p^h_u(k, M)$ on Θ are then "piecewise linear functions".

$W_{ij}(k, M)$ are the basis functions that the finite element method employs. These are constructed such that they take a value of zero for most of the space Ω, except for a small interval where they take a simple linear form. The basis functions adopted for these approximations are set such that $W_{ij}(k, M) = \Psi_i(k)\Phi_j(M)$, where

$$\Psi_i(k) = \begin{cases} \frac{k-k_{i-1}}{k_i-k_{i-1}} & \text{if } k \in [k_{i-1}, k_i] \\ \frac{k_{i+1}-k}{k_{i+1}-k_i} & \text{if } k \in [k_i, k_{i+1}] \\ 0 & \text{elsewhere} \end{cases} \quad (27)$$

$$\Phi_j(M) = \begin{cases} \frac{M_{-1}-M_{j-1}}{M_j-M_{j-1}} & \text{if } M \in [M_{j-1}, M_j] \\ \frac{M_{j+1}-M}{M_{j+1}-M_j} & \text{if } M \in [M_j, M_{j+1}] \\ 0 & \text{elsewhere} \end{cases}, \quad (28)$$

for all i, j. $\Psi_i(k)$ & $\Phi_j(M)$ have the shape of a continuous pyramid which peaks at nodal points $k = k_i$ & $M = M_j$, respectively, and are only non-zero on the surrounding elements of these nodes.

The approximations $v_{uw}^h(k, M)$ & $p_u^h(k, M)$ are chosen to simultaneously satisfy the equations:

$$\int_{\underline{\mathrm{M}}}^{\bar{\mathrm{M}}} \int_{\underline{\mathrm{k}}}^{\bar{\mathrm{k}}} \omega(k, M) R_{uw}^K\left(k, M; \boldsymbol{v}^h, \bar{p}^h\right) dk dM = 0 \quad (29)$$

$$\int_{\underline{\mathrm{M}}}^{\bar{\mathrm{M}}} \int_{\underline{\mathrm{k}}}^{\bar{\mathrm{k}}} \omega(k, M) R_u^M\left(k, M; \boldsymbol{v}^h, \bar{p}^h\right) dk dM = 0, \quad (30)$$

for all $u = \{1, ..., Q\}$ & $w = \{1, ..., Q\}$. $\omega(k, M)$ is a weighting function, and $R_{uw}^K(k, M; \boldsymbol{v}^h, \bar{p}^h)$ & $R_u^M(k, M; \boldsymbol{v}^h, \bar{p}^h)$ are the residuals from the Euler equation and the equilibrium condition for real money holdings, where the true policy functions $\boldsymbol{\Upsilon}$ & $\bar{P}$ are replaced by the vectors of parametric approximations $\boldsymbol{v}^h$ & $\bar{p}^h$. A Galerkin scheme is employed to find the vectors of coefficients v_{ij}^{uw} & p_{ij}^u, for all i, j and all u & w, which solves for the weighted residual equations (29) & (30) over the complete space Θ. The Galerkin scheme employs the basis functions $W_{ij}(k, M)$ as weights for $R_{uw}^K(k, M; \boldsymbol{v}^h, \bar{p}^h)$ & $R_u^M(k, M; \boldsymbol{v}^h, \bar{p}^h)$:

$$\int_{\underline{\mathrm{M}}}^{\bar{\mathrm{M}}} \int_{\underline{\mathrm{k}}}^{\bar{\mathrm{k}}} W_{ij}(k, M) R_{uw}^K\left(k, M; \boldsymbol{v}^h, \bar{p}^h\right) dk dM = 0 \quad (31)$$

$$\int_{\underline{\mathrm{M}}}^{\bar{\mathrm{M}}} \int_{\underline{\mathrm{k}}}^{\bar{\mathrm{k}}} W_{ij}(k, M) R_u^M\left(k, M; \boldsymbol{v}^h, \bar{p}^h\right) dk dM = 0 \quad (32)$$

for all i, j and all states u & w. Since the basis functions are only nonzero surrounding their nodes, eqs. (31) & (32) can be rewritten in terms of the individual elements:

$$\sum_{e=1}^{n_e} \int_{\Omega_e} W_{ij}\left(k, M\right) R_{uw}^{K}\left(k, M; \boldsymbol{v}^h, \bar{p}^h\right) dkdM = 0 \tag{33}$$

$$\sum_{e=1}^{n_e} \int_{\Omega_e} W_{ij}\left(k, M\right) R_{u}^{M}\left(k, M; \boldsymbol{v}^h, \bar{p}^h\right) dkdM = 0, \tag{34}$$

for all i, j and all states u & w. n_e is the total number of elements and Ω_e is the capital and money stock domain covered by the element e.

A Newton algorithm is used to find the coefficients for $[\boldsymbol{v}_s, \bar{p}_s]$ which solve for the nonlinear system of equations H:

$$H\left([\boldsymbol{v}_s, \bar{p}_s]\right) = 0. \tag{35}$$

Where $H\left([\boldsymbol{v}_s, \bar{p}_s]\right)$ is denoted by eqs. (33) & (34). The first step is to choose initial vectors of coefficients $[\boldsymbol{v}_{s_0}, \bar{p}_{s_0}]$, and iterate as follows:

$$\left[\boldsymbol{v}_{s_{l+1}}, \bar{p}_{s_{l+1}}\right] = \left[\boldsymbol{v}_{s_l}, \bar{p}_{s_l}\right] - J\left(\left[\boldsymbol{v}_{s_l}, \bar{p}_{s_l}\right]\right)^{-1} H\left(\left[\boldsymbol{v}_{s_l}, \bar{p}_{u_l}\right]\right). \tag{36}$$

J is the Jacobian matrix of H, and $[\boldsymbol{v}_{s_l}, \bar{p}_{s_l}]$ is the lth iteration of $[\boldsymbol{v}_s, \bar{p}_s]$. The algorithm solves for the parameterized version of the conditional expectation function of the Euler equation $v_{uw}^h(k, M)$ and for the price function $p_u^h(k, M)$, for $\{\theta(u), \theta(w)\} \subset \Lambda$ and $\{k, M\} \subset \Omega$. Convergence is assumed to have occurred once $\left\|\left[\boldsymbol{v}_{s_{l+1}}, \bar{p}_{s_{l+1}}\right] - \left[\boldsymbol{v}_{s_l}, \bar{p}_{s_l}\right]\right\| < 10^{-7}$.

3.1 Algorithm Chronology

The following steps summarize the chronology of the parameterized expectations algorithm, which employs finite elements in the approximation of the policy functions used to solve the stochastic model:

1. Choose the location and quantity I & J of nodes along the capital and money supply domain (i.e. k_i & M_j for $i = \{1, ..., I\}$, $j = \{1, ..., J\}$), which will delimit the n_e nonoverlapping elements in Ω, and use a Gaussian-Legendre quadrature rule to identify abscissas and weights of capital and money stock on each element.
2. Identify the Q states of the Markovian monetary shock and the transition matrix Π_{wr} of the process.
3. For each realization $\theta_{t-1} = \theta(u)$ approximate P_u with p_u^h using (26), and for each $\theta_t = \theta(w)$ approximate Υ_{uw} with v_{uw}^h using (25), for all Gauss-Legendre capital and money stock abscissas on each element along the state space Ω_t.
4. Initiate a recursive solution procedure creating a conjecture that the CIA constraint for this economy, in eq. (9), does not bind, i.e. $\mu_{uw} = 0$. Consumption is automatically solved from eq. (22): $c_{uw} = \left[v_{uw}^h(\Omega_t)\right]^{-\tau}$.

5. Compute the values of employment n_{uw}, actual output y_{uw}, next period capital $\tilde{k}_{uw}$, using eqs. (21), (19) and (20) respectively, and of real money holdings $M_{+1}/p_u^h(\Omega_t)$. Where $M_{+1} = \theta(w) \cdot \tilde{M}$, and $\tilde{M} = \theta(u) \cdot M$.
6. Check whether the initial conjecture was correct. If "yes", go to next step. If not, the constraint binds: $\mu_{uw} > 0$ & $c_{uw} = M_{+1}/p_u^h(\Omega_t)$. The value of the CIA multiplier becomes $\mu_{uw} = [c_{uw}^{-\tau} - v_{uw}^h(\Omega_t)]$.
7. For each realization $\theta_t = \theta(w)$ approximate P_w with p_w^h, and for each $\theta_{t+1} = \theta(z)$ approximate Υ_{wz} with v_{wz}^h, for all Gauss-Legendre capital and money stock abscissas of each element along the state space Ω_{t+1}.
8. Create a conjecture that the CIA constraint does not bind at time $t+1$, i.e. $\mu_{wz} = 0$. Consumption becomes $c_{wz} = \left[v_{wz}^h(\Omega_{t+1})\right]^{-\tau}$.
9. Compute the value of each possible n_{wz}, y_{wz}, and $\tilde{M}_{+1}/p_w^h(\Omega_{t+1})$. Where $\tilde{M}_{+1} = \theta(z) \cdot M_{+1}$.
10. Check whether the conjecture in Step 8 was correct. If "yes", go to next step. If not, the constraint binds: $\mu_{wz} > 0$, & $c_{wz} = \tilde{M}_{+1}/p_w^h(\Omega_{t+1})$. The value of the CIA multiplier becomes $\mu_{wz} = [c_{wz}^{-\tau} - v_{wz}^h(\Omega_{t+1})]$.
11. Construct the residual functions $R_{uw}^K(k, M; \boldsymbol{v}^h, \bar{p}^h)$ & $R_u^M(k, M; \boldsymbol{v}^h, \bar{p}^h)$ using eqs. (17) & (18) for each state u & w.
12. If the weighted approximations of the residual functions for each element, in eqs. (33) & (34), are sufficiently close to zero then "stop", else update the vectors of coefficients v_{ij}^{uw} & p_{ij}^u, for all u & w, as in (36) and go to Step 3.

4 Business Cycle Properties

4.1 Calibration

When calibrating the model, the time interval is consistent with that of a quarter of a year. The time discount factor is set to 0.99, depreciation rate at 0.019, the capital elasticity of output to 1/3, and the inverse of labor supply elasticity is set to 0, denoting Hansen's indivisible labor (see [9]), and the steady state labor supply is set to 0.31. Steady state inflation rate is set to the U.S. quarterly data mean value from 1959:I to 1998:III, that is 1.727%. This calibration of parameters is comparable to that of Walsh (1998), in [16]. For the benchmark calibration, following Kocherlakota (1996) in [11], the intertemporal elasticity of substitution in consumption is set to equal 1/3, yielding $\tau = 3.00$. These parameters are summarized in Table 2.

Table 2 Calibration parameters of the benchmark economy

A	α	β	δ	γ	τ	θ^{ss}
1.00	0.33	0.99	0.02	0.00	3.00	1.01727

The growth rate of the money supply θ_t is assumed to evolve according to the process in eq. (37) :

$$\theta_{t+1} = (1-\rho)\theta^{ss} + \rho\theta_t + \varepsilon_{t+1} \; where \; \varepsilon \sim N\left(0, \sigma^2\right). \tag{37}$$

A first order autoregressive estimation of M2, using quarterly data from 1959:I to 1998:III, yields $\rho = 0.727$ and $\sigma_\varepsilon = 0.01$. The money growth AR(1) process is approximated by a discrete Markov chain using the methodology advanced by Tauchen and Hussey (1991), see [15]. Also, given the high persistence of the process, an adjustment to the weighting function recommended by Flodén (2008), in [8], is performed. Five states for the money supply growth rate are considered; these return the state vector $\theta = [0.9846, 1.0017, 1.0173, 1.0328, 1.0500]$, and the state transition matrix in (38).

$$\Pi = \begin{bmatrix} 0.51136 & 0.45235 & 0.03601 & 0.00028 & 0.00000 \\ 0.07677 & 0.58264 & 0.32302 & 0.01751 & 0.00005 \\ 0.00361 & 0.19100 & 0.61076 & 0.19100 & 0.00361 \\ 0.00005 & 0.01751 & 0.32302 & 0.58264 & 0.07677 \\ 0.00000 & 0.00028 & 0.03601 & 0.45235 & 0.51136 \end{bmatrix}. \tag{38}$$

4.2 Model Economy Steady States

Finding the steady state values of the endogenous variables of the model requires the calculation of $v_{uw}^h(k, M)$ and $p_u^h(k, M)$ for a determinate environment, and finding k such that $k = \tilde{k}_{uw}$, and the realization of the steady state ratios contained in eqs. (12) – (16). These objectives are achieved by minimizing the geometric distance between the vector $\bar{x}^{ss}$, containing the values of the analytical steady state and the vector $\bar{x}_{uw}^h$, containing the steady state approximations. Money growth is fixed at $\theta(u) = \theta(w) = \theta(z) = \theta^{ss}$. Vectors $\bar{x}^{ss}$ & $\bar{x}_{uw}^h$ are defined in eqs. (39) & (40), respectively, and the geometric distance is defined in eq. (41). Table 3 presents the steady state values for the endogenous variables of the economy at the benchmark calibration; as expected, $I = \pi/\beta$ and the CIA constraint is binding at the steady state.

$$\bar{x}^{ss} = \left[k^{ss}, \frac{k^{ss}}{y^{ss}}, \frac{i^{ss}}{y^{ss}}, \frac{c^{ss}}{y^{ss}}, n^{ss}, \mu^{ss}\right]' \tag{39}$$

$$\bar{x}_{uw}^h = \left[\tilde{k}_{uw}, \frac{k}{y_{uw}}, \frac{\tilde{k}_{uw} - (1-\delta)\,k}{y_{uw}}, \frac{c_{uw}}{y_{uw}}, n_{uw}, c_{uw}^{-\tau} - \Upsilon_{uw}(k, M)\right]' \tag{40}$$

$$\left\|\bar{x}^{ss} - \bar{x}_{uw}^h\right\| = \sqrt{\sum_{i=1}^{6}\left(x_i^{ss} - x_{uw,i}^h\right)^2}. \tag{41}$$

Table 3 Steady states of the benchmark economy

c^{ss}	k^{ss}	n^{ss}	Y^{ss}	μ^{ss}	π^{ss}	r^{ss}	I^{ss}	V^{ss}
0.82	11.94	0.31	1.04	0.05	1.017	1.01	1.03	1.00

4.3 Velocity at Serially Correlated Money Growth Rates

Velocity differs from one when the agent's real money holdings on a period are greater than the expenditures on consumption, and will stay constant at one when the choice of investing in an interest bearing venture, such as capital, is preferred to holding additional units of cash. The latter occurs when real interest rate is sufficiently high and the variation of expected marginal utility of consumption is relatively small, see Hodrick R. et. al. (1991) in [10].

Fig. 1 contains two comparable cases that illustrate the effects of financial frictions on the variability of velocity. The left panel of Fig. 1 considers the scenario where frictions are not present in the information structure of the benchmark economy and the right panel the scenario where they are. Each panel contains a mesh depicting the level values of μ_t over the state space Ω_t for each possible realization of θ_t, given that $\theta_{t-1} = \theta^{ss}$. μ_t increases with θ_t. Considering first the frictionless environment, μ_t is positive and velocity is equal to one on almost the totality of the state space Θ_t. The CIA constraint binds at the steady state level of capital stock on all but the lowest realization of the money growth parameter. Serially correlated money growth shocks, on a frictionless environment, can only yield variable velocity at sufficiently low money growth rates combined with sufficiently low real interest rates. It is

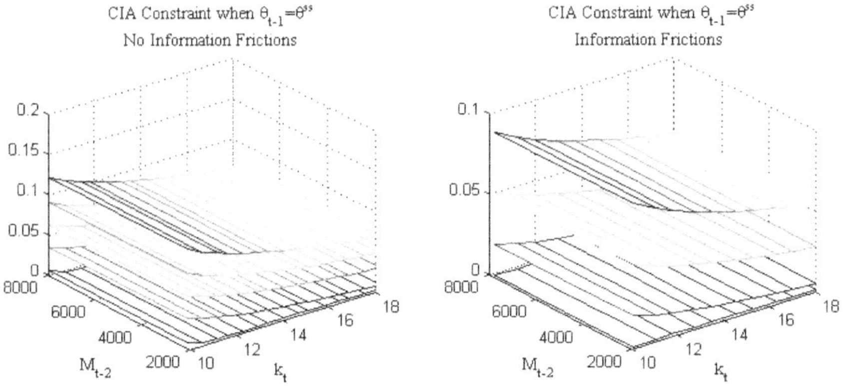

Fig. 1 The left and right panels illustrate the values assumed by μ_t on the benchmark economy when frictions are absent and present on the information structure, respectively. On the environment with frictions, the CIA-constraint binds at the steady state money growth rate but it does not at lesser growth rates.

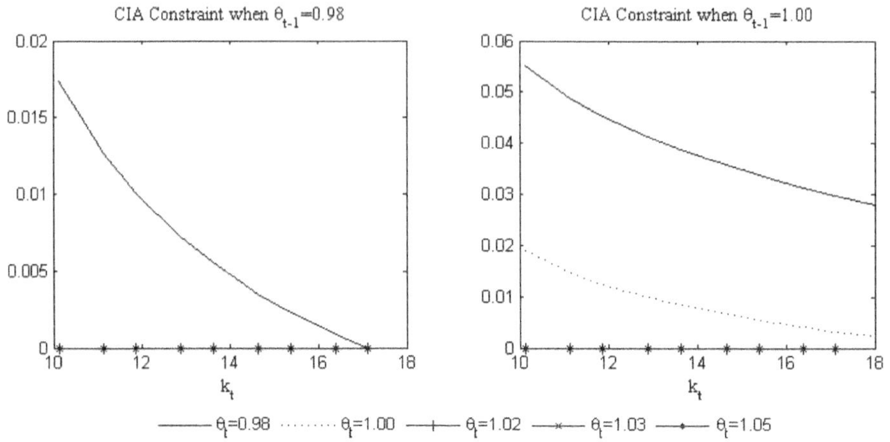

Fig. 2 The two panels depict those cases where the previously realized growth parameter is greater than the steady state value. The lesser the realization of the previous period's money growth parameter, the more possible cases will occur that the agent will decide to hold on cash for next period and the CIA-constraint will not bind.

possible to notice that below a critical level of capital stock, the real interest is sufficiently high to dissuade the agent from holding any money for precautionary reasons, and instead acquire interest bearing physical capital in its portfolio. Cao-Alvira (2010b) presents a thorough description of a projection methodology that solves for a similar cash-in-advance model economy with no frictions in its information structure, as well as a robustness analysis and cutoff points for the existence of variability on velocity. The considered model economy in Cao-Alvira (2010b) contains a more comprehensive state space partition and, as a consequence, includes more possible realizations of the money growth rate where the slackness multiplier is zero.

Observing the case depicted in panel 2 of Fig. 1, the environment where frictions are present, it is possible to notice two key features. First, μ_t is positive, or the CIA constraint binds, at the steady state of the benchmark economy. This result is consistent with those findings reported in Table 3 and, as expected, information or financial frictions do not affect the steady state specification of the model. And second, μ_t equals zero, on all considered values of Ω_t, for those parameters $\theta_t < \theta^{ss}$. At the considered state space partition, serially correlated shocks on monetary growth, combined with frictions on the information structure, pushes the agent to accumulate excessive real money holdings on states of nature that exhibit a lower than average money growth rate. On these states, cash is being transferred from one period to the next in order to smooth consumption and decrease its expected marginal utility variability.

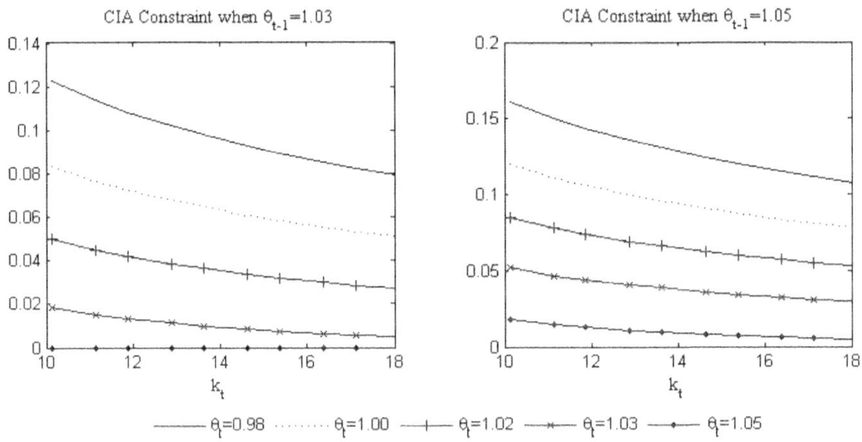

Fig. 3 The two panels depict those cases where the previously realized growth parameter is greater than the steady state value. Only on the scenario where previously the highest possible growth parameter is realized, the right panel, the agent will decide to never carry excessive cash holdings and the CIA constraint will always bind.

The results depicted on Fig. 2 and Fig. 3 better assist on illustrating this feature of the environment containing frictions. Each figure contains two panels where the value of μ_t is depicted for all considered values of the capital stock and the money growth parameters, given a fixed value of M_{t-2} and a previous realization of the growth parameter. The panels on Fig. 2 depict those cases where the previously realized growth parameter is lesser than the steady state value, and the panels in Fig. 3 depict the cases where these parameters are greater than the steady state value. As can be observed, the lesser the previous period's monetary growth, the more possible scenarios will occur that the agent decides to hold on cash for the future. Examining both panels on Fig. 2, it is worth noticing that agents will decide to carry money on to the next period even when the current realization of monetary growth is at the steady state level, if the previous value of the parameter is less than the steady state. Only on the scenario where the highest possible growth parameter is realized in the previous period, the right panel of Fig. 3, the agent will decide to never carry excessive cash holdings.

Table 4 documents the simulated moments of the friction and frictionless version of the benchmark economy and the sample moments of the US economy for the time period 1959:I to 1998:III of three economic indicators mostly used in the literature to measure the variability of velocity. These are the coefficient of variation of velocity, measured as $cv(V) = (\sigma_V/\mu_V) \cdot 100$, the correlation between velocity and gross consumption growth, where consumption growth is defined as the ratio between current and past consumption, and the correlation of inflation and real interest rate. While the first index is

Table 4 Simulated moments of three key indexes for the friction and frictionless version of the benchmark economy and those for the US economy for the time period 1959:I to 1998:III. Numbers in parenthesis are standard deviations over 500 simulations. US economy values as reported by Wang W., Shi S. (2006).

Indexes	Benchmark	Information Friction	Data
$cv\,(V)$	0.0306	0.4600	1.7632
	(0.0127)	(0.0480)	
$corr\left(V, \frac{C_t}{C_{t-1}}\right)$	-0.1534	-0.4072	-0.2834
	(0.0663)	(0.073)	
$corr\,(\pi, I)$	0.7300	0.0102	0.5046
	(0.0413)	(0.0268)	

a straightforward measure of variability, the last two are good indicators on how monetary variables affect the real economy.

Comparing the fit of both simulated modeling environments, that with the richest information structure represents a serious improvement in performance. On the sample state space partition, a setting which allows for frictions, money shocks are able to explain over a fourth of the observed variability in velocity, while a frictionless environment can explain only 1.74% of it. The inclusion of frictions in this partition of the state space, which particularly considers the money growth rate realizations close to the state, increases by fifteen times the explained observed variability in the date. When considering the correlation between velocity and consumption growth, the model containing frictions is able to explain its full comovement, while the frictionless model is able to explain only half of it. The US economy sample value for this estimate has a negative sign, which both simulated models can replicate. The reasoning for this result is that on high realizations of the money growth parameter, due to inflation, the purchasing power of money decreases, causing the need to increase the velocity of balances, and households choose to decrease consumption in order to smooth it with that of previous periods. The inability of a standard cash-in-advance environment to explain the amount of correlation between velocity and consumption growth is well documented in the literature and, as in this paper, some researchers on velocity have parted from the original formulation of the model to be able to replicate it; some with better luck than others. A nonexhaustive alphabetically ordered list includes Hodrick R. et. al (1991), in. [10], Svensson L. (1985), in [14], and Wang W., Shi S. (2006), in [17].

As can be observed in the simulated values for the correlation of inflation and real interest rate, the cost of improving the fit on velocity by considering frictions is a degraded specification of inflation on the calibration. The frictions in the information structure over smooths inflation decreasing its variability, compared to the frictionless case.

5 Conclusion

This paper introduces a modeling environment where frictions are incorporated on the selection of assets on an agent's portfolio with the interest of studying the impact of serially correlated monetary shocks on the variability of velocity. The solution technique proposed and developed in finding the equilibrium of the system employs the usage of finite elements in the approximation of the policy functions, and a parameterization of the agent's expectations over the choice of future real money balances. The method converges with rapid speed and is efficient approximating the studied highly nonlinear environment. Because the algorithm allows for the cash-in-advance constraint on consumption expenditures to selectively bind, instead of exogenously assuming it always does as other more conventional methods require, it is possible for the equilibrium conditions to achieve variability in the velocity of money holdings. By considering frictions on the financial structure of the economy, a solely money-driven monetary benchmark economy is able to significantly increase the explanation of observed variability on the data, when judged against a comparable frictionless environments.

Acknowledgements. The author wishes to recognize the comments on the direction of the paper and tutelage received by Yi Wen, Tao Zhu and Karl Shell, and Ellen McGrattan and Fabrice Collard for their great guidelines on minimum weighted residual methods. Of course, all mistakes are my own. Previous versions of this paper benefited from comments of attendees to the 15th International Conference of the Society for Computational Economics on Computing in Economics and Finance, at University of Technology Sydney. The author benefited from financial support from Decanato de Estudios Graduados y de Investigación at UPR.

References

1. Aruoba, S.B., Rubio, J., Fernandez-Villaverde, J.: Comparing Solution Methods for Dynamic Equilibrium Economies. J. Econ. Dyn. Control 30, 2477–2508 (1998)
2. Cao-Alvira, J.: Finite elements in the presence of occasionally binding constraints. Comput. Econ. 35(4), 355–370 (2010)
3. Cao-Alvira, J.: Velocity variability on cash-in-advance economies due to precautionary demand. EGAE UPR-RP Working paper (2010)
4. Cao-Alvira, J.: Non-Linear Solution Methods, Occasionally Binding Constraints, Information Frictions, Inventories, Cash-In-Advance Economies & Velocity of Money. Economics Department, Cornell University. Ph.D. Dissertation (2006)
5. Christiano, L., Eichenbaum, M.: Liquidity Effects and the Monetary Transmission Mechanism. Am. Econ. Rev. 82(2), 346–353 (1992)
6. Cooley, T.F., Hansen, G.: The inflation tax in a real business cycle model. Am. Econ. Rev. 79, 733–748 (1989)
7. den Haan, W., Marcet, A.: Solving the stochastic growth model by parameterizing expectations J. Bus. Econ. Stat. 8, 31–34 (1990)

8. Flòden, M.: A Note on the Accuracy of Markov-Chain Approximations to Highly Persistent AR(1)-Processes. Econ. Lett. 99, 516–520 (2008)
9. Hansen, G.: Indivisible labor and the business cycle. J. Monetary. Econ. 16, 309–325 (1985)
10. Hodrick, R., Kocherlakota, N., Lucas, D.: The Variability of Velocity in Cash-in-Advance Models. J. Polit. Econ. 99, 358–384 (1991)
11. Kocherlakota, N.: The Equity Premium: It's Still a Puzzle. J. Econ. Lit. 34, 42–71 (1996)
12. McGrattan, E.: Application of Weighted Residual Methods to Dynamic Economic Models. In: Marimon, R., Scott, A. (eds.) Computational Methods for the Study of Dynamic Economies. Oxford University Press, Oxford (1999)
13. McGrattan, E.: Solving the Stochastic Growth Model With a Finite Element Method. J. Econ. Dyn. Control 20, 19–42 (1996)
14. Svensson, L.: Money and Asset Prices in a Cash in Advance Economy. J. Polit. Econ. 93, 919–944 (1985)
15. Tauchen, G., Hussey, R.: Quadrature-based methods for obtaining approximate solutions to nonlinear pricing models. Econometrica 59, 371–396 (1991)
16. Walsh, C.E.: Monetary Theory and Policy, 2nd edn. The MIT Press, Cambridge (1998)
17. Wang, W., Shi, S.: The variability of velocity of money in a seach model. J. Monetary. Econ. 53, 537–571 (2006)

Automated Fuzzy Bidding Strategy Using Agent's Attitude and Market Competition

Madhu Goyal, Saroj Kaushik, and Preetinder Kaur

Abstract. This paper designs a novel fuzzy competition and attitude based bidding strategy (FCA-Bid), in which the final best bid is calculated on the basis of the attitude of the bidders and the competition for the goods in the market. The estimation of attitude is based on the bidding item's attribute assessment, which adapts the fuzzy sets technique to handle uncertainty of the bidding process as well it uses heuristic rules to determine attitude of bidding agents. The bidding strategy also uses and determines competition in the market (based on the two factors i.e. no. of the bidders participating and the total time elapsed for an auction) using Mamdani's Direct Method. Then the final price of the best bid will be determined based on the assessed attitude and the competition in the market using fuzzy reasoning technique.

1 Introduction

Online auctions have become increasingly important area of research with its popularity, because it provides the traders the flexibility of time and geographical location for trading. Software agent technology is one of the most popular mechanisms used in online auctions for buying and selling the goods. Software agent is a software component that can execute autonomously, communicates with other agents or the user and monitors the state of its

Madhu Goyal
Faculty of Engineering and Information Technology, UTS, Australia
e-mail: madhu@it.uts.edu.au

Saroj Kaushik
Indian Institute of Technology, Delhi, India
e-mail: saroj@cse.iitd.ernet.in

Preetinder Kaur
Faculty of Engineering and Information Technology, UTS, Australia
e-mail: preetinder.kaur@student.uts.edu.au

Q. Bai and N. Fukuta (Eds.): Advances in Practical Multi-Agent Systems, SCI 325, pp. 167–180.
springerlink.com

execution environment effectively [2] [4] [7]. The agents can use different auction mechanisms (e.g. English, Dutch, Vickery etc.) for procurement of goods or reaching agreement between agents. The agent makes decisions on behalf of consumer and endeavors to guarantee the delivery of item according to the buyer's preferences. In these auctions buyers are faced with difficult task of deciding amount to bid in order to get the desired item matching their preferences. The bidding strategies for the software agents can be static or it may be dynamic [13]. The static agents may not be appropriate for the negotiating market situations like extent of competition may vary as traders leave or enter into the market, deadlines and new opportunities may increase the pressure. The dynamic or we can say flexible negotiation capabilities for software agents in the online auctions have become a central concern [11]. Agents need to be able to prepare bids and evaluate offers on behalf of the users they represent with the aim of obtaining the maximum benefit [8] for their users according to the changing market situation.

Much research has already been done by the researchers to formulate different bidding strategies according to the changing market situations [9] [10] [1] [15] [14]. Strategies based on flexible negotiation agents perform better as compared to the strategies based on fixed negotiation agents [11] [5]. Faratin et al in [5] developed strategies based on time, attitude, resources, but many more factors such as competition, trading alternatives are not considered. In this paper we focus on the design of a novel bidding strategy based on the above mentioned factors to be used by the software agent in online auction. A fuzzy competition and attitude based bidding strategy(FCA-Bid) is designed, in which the final best bid is calculated on the basis of the attitude of the bidders as well as the competition for the goods in the market.

2 Fuzzy Competition and Attitude Based Bidding Strategy (FCA-Bid)

The agent's decision making about bidding involves various internal and external environmental factors. The internal factors include good or item's attributes, attitude of the agents on the assessment of attributes, and current available number of the goods and the external environmental factors may include like competition for the goods in the market, nature of the market supply (demand), other opportunities available in the market and many more.

In fuzzy competition and attitude based bidding strategy (FCA-Bid) (Fig. 1), the factors which are focused are attitude of the agents with respect to the goods' attributes and competition for the goods in the market. For estimation of the price for a bid for winning an auction, the agent must have a balanced behavior between these factors i.e. the attitude (eagerness) to win the auction based on the attributes of the goods and finding the competition for the goods in the market. The attitude towards bidding the quality goods is more as compared to the less quality goods. The bidding price also affects

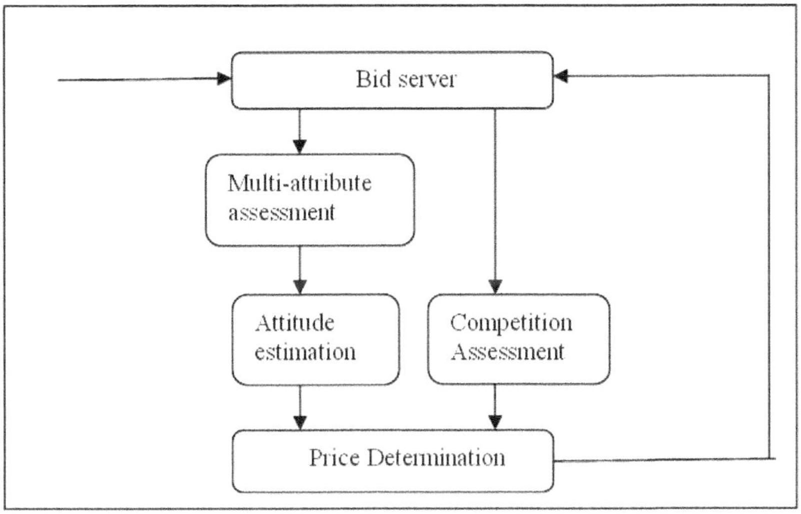

Fig. 1 A Fuzzy Bidding Strategy (FCA-Bid) Model.

the attitude of the agents. The higher bid price dampens the attitude of the agents towards the goods. Also the increasing competition for the goods in the market increases the attitude for that good. The competition in turn depends on the number of bidders and the time elapsed for the auction. As the number of bidders increases, the competition among them also increases, resulting in a higher price. In the beginning of the auction the competition is less and it increases as time elapses and it is at the peak when time approaches approximately in the middle of the auction period. At the end of the auction period the competition among the bidders decreases. The steps of the design of fuzzy competition and attitude based bidding strategy (FCA-Bid) are as follows:

- first, each attribute is evaluated and then the assessment of all these attributes will be aggregated
- then attitude of the agent will be found based on these assessments,
- next the level of competition as the function of no. of bidders and time elapsed for the auction will be found
- Finally the best bid is calculated on the basis of the above attitude of the agents and the competition for the goods in the market.

In this paper we have used fuzzy set methods to deal with the uncertainty, which exists during the determination of overall assessment of the goods for their attributes, the attitude of the agent based on the assessment of goods and the level of competition in the market. First of all, this paper uses a satisfactory degree measure as the common universe of assessment, i.e., an assessment is treated as a fuzzy set on the satisfactory degree. Secondly, an attitude is expressed as a fuzzy set on the set of assessments, i.e., the assessment set is the universe of attitude (eagerness). Thirdly, competition is expressed as a fuzzy set on the fuzzy sets of the no. of bidders and the time elapsed of the auction.

2.1 Attribute Evaluation

The attribute evaluation is done in two parts [6]. First the expression for the assessment of the attributes is found then these assessments will be aggregated to find the overall assessment of the attributes of the goods. Let $C = \{c_0, c_1, ..., c_K\}$ be the set of $K + 1$ attributes and $W = \{w_0, w_1, ..., w_K\}$ is the set of weights for attributes in C.

Attribute Assessment
The assessment of the attributes is expressed in terms of a fuzzy set. Let $A = \{a_1, a_2, ..., a_n\}$ be a set of assessment terms on the universe i.e. the satisfactory degree [0,1]. This is the satisfactory degree of the agent to a particular attribute. All the fuzzy sets have same universe which is convenient for the aggregation of various assessments. Let $g_k(k = 0, 1, ..., K)$ is the satisfactory degree measure for attribute c_k. Then an agent's opinion on the goods in terms of attribute c_k is denoted by $g_k(u)$ where $u(\in Uk)$ is the real attribute value of attribute c_k and U_k is the real universe for attribute c_k. For instance, departing time is an attribute for a flight ticket. The possible departing time in a day is from 0:00 to 23:59. For any time slot u, a client may present a satisfactory degree such as departing at 7:30 is with satisfactory degree 0.9 and departing at 3:00 is with 0.3. In the following, $A = \{a_1, ..., a_n\}$ be the set of used assessment terms which are fuzzy sets on satisfactory degree [0,1]. Then a numeric satisfactory degree is transformed to a linguistic term. In the above example [11] a_7 is with the biggest the membership degree for 0.9, the assessment for departing at 7:30 is a_6 by the maximum membership degree principle. Similarly, the assessment for 0.3 is a_2.

Aggregation of Assessments
All the goods have a number of different attribute. So to find the overall estimation on the good, the assessment of these all attributes will be aggregated together. Take booking a flight ticket for example, an assessment is made on a ticket usually based on the airlines, flight departure and arrival time, flight type, aircraft types, seat positions, as well as price. The change of an attribute's value may leads to the alternation of an assessment. Instinct natures of different attributes increase the difficulty and uncertainty for obtaining an overall assessment. Notice that an agent's preference on an individual attribute can be expressed through the agent's satisfactory degree on that attribute. This paper uses a satisfactory degree measure as the common universe of assessment. Based on assessment on each individual attribute, an overall assessment can be obtained as follows. Suppose the individual assessments of all attributes are $v_0, v_1, ..., v_K$ and the weights of

them are $w_0, w_1, ..., w_k$ respectively. Then an overall assessment is obtained by taking the difference between ã and $a_i \in A$ [11], where ã is a fuzzy set on [0,1] as follows

$$d(\tilde{a}, a_i) = \mid \tilde{a} - a_i \mid d\lambda \tag{1}$$

Finally, we select the nearest term(s) a to ã as the overall assessment.

2.2 Attitude Estimation

Attitude is a learned predisposition to respond in a consistently favorable or unfavorable manner with respect to a given object [6] [11]. In other words, the attitude is a preparation in advance of the actual response, constitutes an important determinant of the ensuing behavior. In AI, the fundamental notions to generate the desirable behaviors of the agents often include goals, beliefs, intentions, and commitments. The exhibited behavior is based on a number of factors which depends on the nature of the dynamic world. Once an agent chose to adopt an attitude, it strives to maintain this attitude, until it reaches a situation where the agent may choose to drop its current attitude towards the object and adopt a new attitude towards the same object. Thus, an agent's attitude towards an object refers its persistent degree of commitment towards achieving one or several goals associated with the object, which give rise to an overall favorable or unfavorable behavior with regard to that object. In online auctions the attitude of an agent towards the goods is the eagerness that measures agent's interest in negotiating and coming to a deal [14]. The level of interest may be categorized as: must deal, desirable, nice to have, optional, unessential, and absolutely unessential [12]. Attitude is related to the overall assessment on the given goods. It is expected to change as per the changes in internal and external environmental conditions. Like the attitude for the goods having better assessment have more positive attitude of bidding those goods and also the attitude towards more competitive goods is stronger. In this paper we will estimate the attitude on the bases of the assessment on the goods and will consider competition as an independent factor in calculating the final bid. After conducting new assessment on the goods according to current price p_c, estimation of agent's attitude is implemented. In order to do so, the relationship between attitude and assessments is required. As said earlier, the better the assessment on the given goods is, the stronger the attitude of bidding for those goods will be.

Suppose $E = \{e_1, ..., e_m\}$ is the set of attitude expressions, $A = \{a1, ..., a_n\}$ is the set of assessments, and $T = \{t_1, ..., t_L\}$ is the agent's transaction records such that $t_i = 1$ if the client won the transaction t_i, otherwise $t_i = 0$. Because in each transaction, the agent's assessment and attitude occur simultaneously, a set of formal rule, denoted by R, thus can be extracted from T such that any $r \in R$ is of form

$$r : (a_i \rightarrow e_j, \alpha_{ij}) \tag{2}$$

where $a_i \in A$, $e_j \in E$, and α_{ij} is the reliability degree obtained by

$$\alpha_{ij} = \frac{|\{t \in T | \exists a_i, e_j \in t \wedge t = 1\}|}{|\{t \in T | \exists a_i \in t \wedge t = 1\}|} \quad (3)$$

Such rule depicts the approximate degree of agent's attitude e_j to which the agent can win the bid under the assumption that the overall assessment is a_i [6]. Furthermore, these rules can be treated as a set of fuzzy sets on A such that the membership degree in a fuzzy set f_j corresponding to eagerness e_j is α_{ij}. Obviously, f_j is an integration of rules $(a_i \Rightarrow e_j, \alpha_{ij})(i = 1, ..., n)$, which is able to be treated as an alias of e_j . Hence, the fuzzy set f_j is also called attitude in the following without other specification. Based on the rules in R, an agent can estimate the possible attitude [6] of the agent when it learns the current overall assessment. Set of fuzzy sets is obtained through the following way: suppose the overall assessment is a_c, then the attitude at the moment is determined by the maximum membership degree principle

$$e_c \in E(a_c) = \{e_j \in E | f_j(a_c) \geq f_i(a_c) \;\; if \;\; i \neq j\} \quad (4)$$

Notice that such determined e_c may not necessarily be unique. In the following, we call $E(a_c)$ the candidate attitude set under a_c.

2.3 Competition Assessment

The level of competition in an auction may be captured by the number of bidders and the time elapsed. Competition among bidders plays an integral role in price formation [16]. As the number of bidders increases, the competition among them also increases (Fig. 2), resulting in a higher price. Bapna, Jank and Shmueli [3] found the number of bidders to be positively associated with the current price of the item. Furthermore, it is observed that, typically, the middle of the auction experiences a smaller amount of bidder participation as compared to the early and later stages of the auction. Bidders generally utilize this time to scrutinize the auctioned item or just simply wait to see how other bidders behave. Therefore, it would be interesting to see how this competition characteristic affects the on-line auction's price formation. We anticipate that the number of bidders has a significant positive relationship with price levels. In the beginning of the auction the competition is less and it increases as time elapses and it is at the peak when time approaches approximately in the middle of the auction period. At the end of the auction period the competition among the bidders decreases (Fig. 3). Here we will describe the competition factor in terms of no. of bidders (b) and the total time elapsed (t) for the auction of items. We will consider the competition as a set fuzzy set of values $c_1, c_2, ..., c_n$, no. of bidders B as a fuzzy set of values $y_1, y_2, ..., y_n$. And the time elapsed as another fuzzy set T of values $x_1, x_2, ..., x_n$.

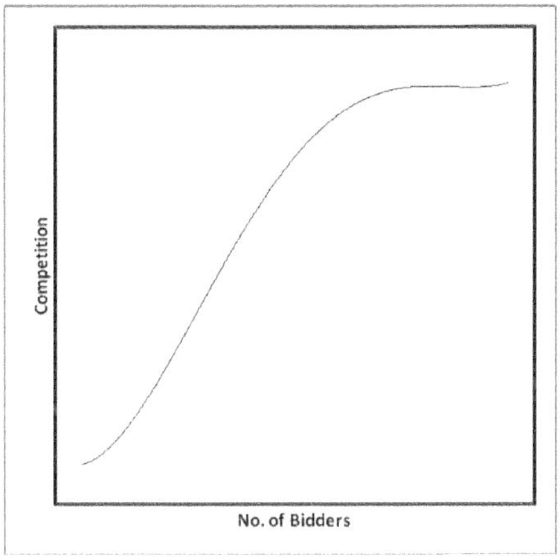

Fig. 2 Competition versus No. of Bidders.

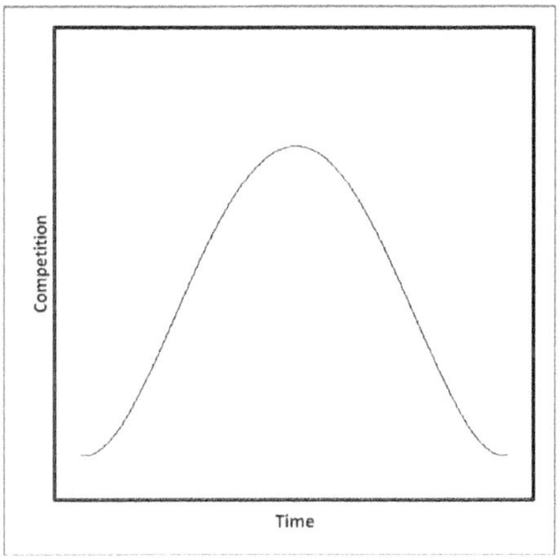

Fig. 3 Competition versus Time Elapsed.

According to Mamdani's Direct Method [17] we can find adaptability n no. of rules $w_1, w_2, ..., w_n$ as follows

$$\begin{aligned} w_1 &= \mu x_1(T) \vee \mu y_1(B) \\ w_2 &= \mu x_2(T) \vee \mu y_2(B) \\ &\dots. \\ w_n &= \mu x_n(T) \vee \mu y_n(B) \end{aligned}$$

Then the conclusion of each rule can be found as follows

$$\begin{aligned} \mu c_1^{'}(C) &= W_1(T) \vee \mu c_1 \\ \mu c_2^{'}(C) &= W_2(T) \vee \mu c_2 \\ &\dots\dots \\ \mu c_n^{'}(C) &= W_n(T) \vee \mu c_n \end{aligned}$$

These conclusions can be aggregated to find the final conclusion

$$\mu c(C) = \mu c_1^{'}(C) \wedge \mu c_2^{'}(C) \wedge ... \wedge \mu c_n^{'}(C)$$

To find the definite value for the conclusion, here center of gravity of the fuzzy set has been applied as follows

$$c = \frac{\int \mu z(c) c dc}{\int \mu z(c) dc} \quad (5)$$

2.4 Agent Price Determination

Price of the goods depends on the attitude towards that good and the competition in the market for that good. If the attitude for the goods is positive and also competition for that product in the market is high then the price of the item is high. If the attitude is negative and competition is also low then the price for that item is obviously be low. If the attitude is positive and the competition is low then the price is going to be medium and so on.

We can calculate the price of the good based on the assessed attitude and the competition determined which is based on the no. of bidders and time elapsed for the auction by applying Mamdani's Method for fuzzy relations and compositional rule of inference [17]. Here we will describe the price of goods in terms of attitude of agent towards the good and competition in the market for that good. We will consider bid displacement factor $\triangle P$ as a fuzzy set of values $p_1, p_2, ..., p_n$, attitudes E as a fuzzy set of values $e_1, e_2, ..., e_n$ and competition C as a fuzzy set of values $c_1, c_2, ..., c_n$. According to Mamdani's Method for fuzzy relations and compositional rule of inference the rule e_i and $c_j \rightarrow p_k$ can described by

$$\mu R(E, C, \triangle P) = \mu e_i(E) \wedge \mu c_j(C) \wedge \mu p_k(\triangle P) \quad (6)$$

For n no. of rules, the compiled fuzzy relation R is given as

$$R = R_1 \cup R_2 ... \cup R_n$$

For the input of fuzzy set E' on E and fuzzy set C' on C, the output fuzzy set $\triangle P'$ on $\triangle P$ can be obtained as follows

$$\triangle P' = (E' \; and \; C') \circ R = E' \circ (C' \circ R) = C' \circ (E' \circ R) \quad (7)$$

and then the final price for the bid will be

$$Final \; bid = Current \; bid + \triangle P'$$

3 Experimental Evaluations

In this section, an experiment implements the fuzzy bidding strategy in a scenario in which an agent intends to book flight tickets. Six factors (as shown in Table 2) are concerned in this situation, i.e. ticket price (c_0), depart time (c_1), arrival time (c_2), number of stops (c_3), seat positions (c_4), and travel season (c_5). The flight ticket bid for is a return ticket to destination D with the following properties:

- price: \$800 - \$2000;
- depart time: 18:00 PM, Wednesday;
- return arrival time: 10:00 AM, Friday;

- number of stops: 1;
- seat position: window;
- travel season: April (off-peak season).

Suppose the identified perspective of an agent is summarized as below:

Table 1 Concerned Attributes of a Flight Ticket.

Attributes	**Symbol**	**Values range**
price	c_0	$[800-2000]
depart time	c_1	Sun. 0:00 - Sat. 24:00
arrival time	c_2	Sun. 0:00 - Sat. 24:00
stops	c_3	0, 1, 2, 3
seat position	c_4	window, aisle, middle
flight season	c_5	Jan. 01 - Dec. 31

- The agent prefers to a cheaper ticket and agrees to that the cheaper the better.
- The agent prefers to travel at the weekend rather than at working day.
- The agent prefers to no stop travel.
- The agent prefers to aisle seat then window seat.
- The agent prefers to travel during off-peak season rather than peak season.
- The agent thinks the flight price is the most important factor, secondly the travel season, and other factors are of same importance.

Based on the agent's perspective, the agent evaluates the attitudes using seven terms, very bad (a_1), bad (a_2), slightly bad (a_3), acceptable (a_4), fairly good (a_5), good (a_6), and very good (a_7).The seven terms are expressed by fuzzy sets on the satisfactory degree [0,1] as below

$$f_{ai} = e^{-162(x-(i-1)\frac{1}{6})\wedge 2} \tag{8}$$

The assessment on each individual factor is in illustrated in Table 2.

Now, a fuzzy set $\tilde{a}(u)$ is obtained, the most nearest assessment to $\tilde{a}$ is a_6. So the new overall assessment for the ticket is a_6 [6]. Then the agent needs to estimate the agent's attitude according to this assessment. Suppose

Table 2 Attribute Assessment.

Attribute	Assessment
c_0	(no assessment)
c_1	good (a_6)
c_2	fairly good (a_5)
c_3	slightly bad (a_3)
c_4	acceptable (a_4)
c_5	good (a_6)

Table 3 Rule set for attitude estimation

	Attitude				
Ass.	e_1	e_2	e_3	e_4	e_5
a_1	0.17	0.23	0.2	0.27	0.13
a_2	0.1	0.28	0.22	0.26	0.13
a_3	0.1	0.26	0.18	0.32	0.13
a_4	0.17	0.26	0.23	0.23	0.12
a_5	0.12	0.25	0.27	0.21	0.16
a_6	0.12	0.26	0.26	0.23	0.13
a_7	0.12	0.24	0.31	0.24	0.1

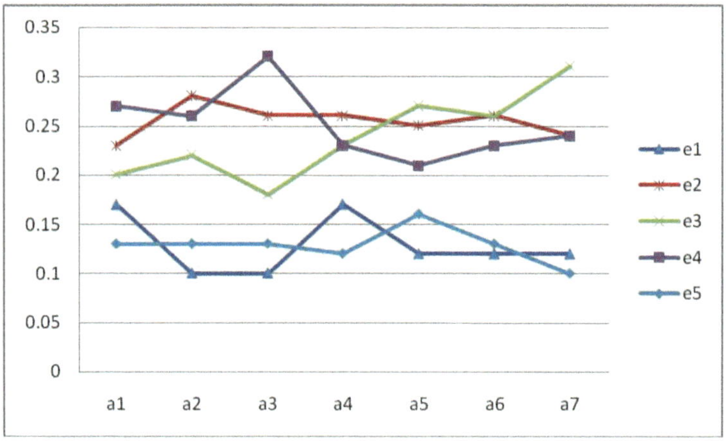

Fig. 4 Illustration for Rule Set.

the agent uses five terms to distinguish the attitude, i.e., none (e_1), slightly (e_2), medium(e_3), strong (e_4), and very strong (e_5). In order to estimate the agent's attitude, a set of rules are extracted from a historical auction records, which are illustrated in Table 3 and Fig. 4.

By Fig. 4, the agent's attitudes at this moment are e_2 and e_3 because they have the highest reliability. Because e_3 is stronger than e_2, the agent first searches possible bids under the attitude e_3.

For finding the price of the good, we will apply fuzzy logic by considering two factors attitude and competition as described in the section 2.4. Let us consider the following set of rules for the logic using various fuzzy sets.

Rule 1: IF attitude of agent for buying the goods is E_1
AND competition in the market for that product is C_1
THEN price for that item will be displaced by P_1

Rule 2: IF attitude of agent for buying the goods is E_1
AND competition in the market for that product is C_2
THEN price for that item will be displaced by P_2
Rule 3: IF attitude of agent for buying the goods is E_2
AND competition in the market for that product is C_1
THEN price for that item will be displaced by P_2
Rule 4: IF attitude of agent for buying the goods is E_2
AND competition in the market for that product is C_2
THEN price for that item will be displaced by P_3

These fuzzy sets represents the linguistic variables as follows: attitudes low as E_1 and high as E_2, Competition less as C_1 and more as C_2 and Negative displacement as P_1, no displacement as P_2 and positive displacement as P_3. We assume that the set of attitudes for buying any item as $E = \{e_1, e_2, e_3\} = \{0, 0.5, 1\}$ and set of competition for the good in the market as $C = \{c_1, c_2, c_3\} = \{0, 0.5, 1\}$. Also, the bid displacement as $\triangle P = \{p_1, p_2, p_3\} = \{-100, 0, +100\}$. The fuzzy sets used in the preceding four rules can be quantized as shown in the Fig. 5:

$E_1 = [1.0, 0.5, 0]$ $C_1 = [1.0, 0.5, 0]$ $P_1 = [1.0, 0, 0]$
$E_2 = [0, 0.5, 1.0]$ $C_2 = [0, 0.5, 1.0]$ $P_2 = [0, 1.0, 0]$
$P3 = [0, 0, 1.0]$

Note that the number of elements in E, C and P are three and the fuzzy sets are also quantized into three elements. Now let us construct fuzzy relations by Mamdani's Method for fuzzy relations

$$\mu R(E, C, \triangle P) = \mu e_i(E) \wedge \mu c_j(C) \wedge \mu p_k(\triangle P), \quad i, j, k = 1, 2, 3$$

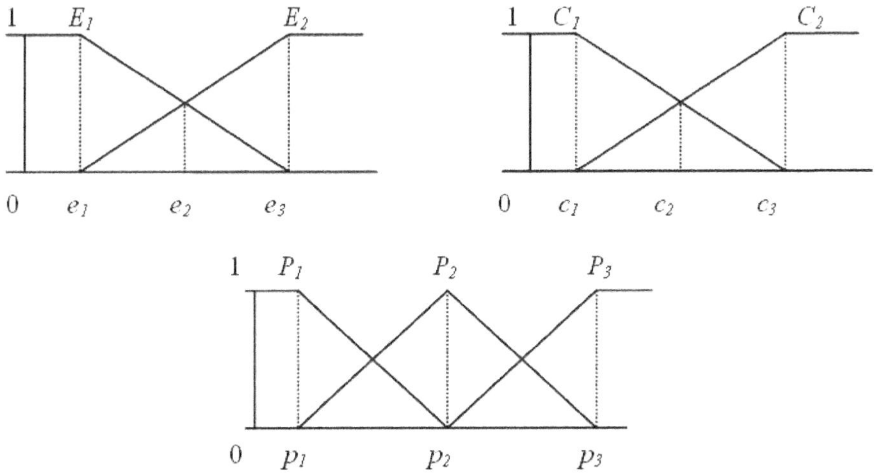

Fig. 5 Fuzzy Sets for Bidding Logic.

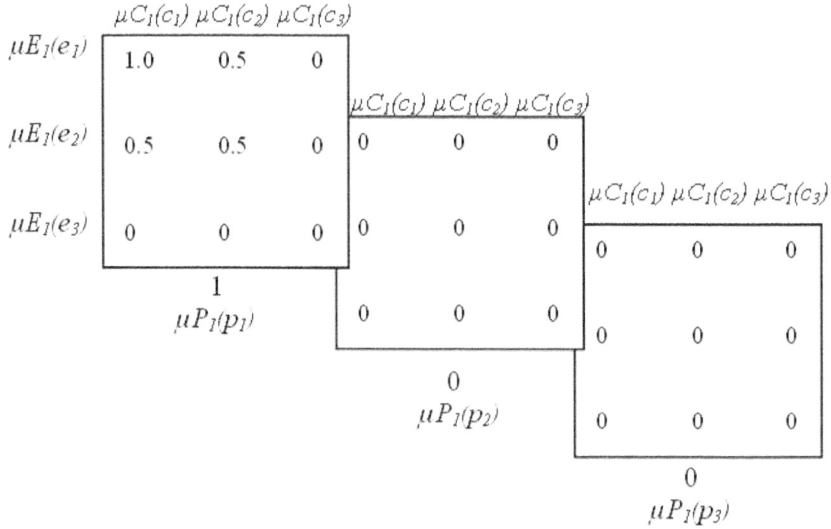

Fig. 6 Fuzzy Relation R_1.

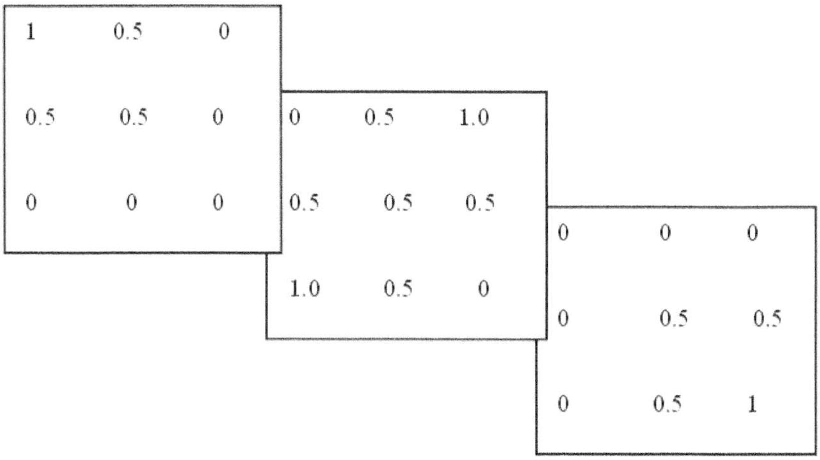

Fig. 7 Fuzzy Relation R.

By the preceding conversion formula we get the fuzzy relation R_1 from the first rule as in Fig. 6.

Similarly we can convert Rules 2, 3 and 4 into fuzzy relations R_2, R_3 and R_4 accordingly. The total fuzzy relation R is given using Mamdani's Method for compilation of fuzzy relations as shown in Fig.7 by

$$R = R_1 \cup R_2 \cup R_3 \cup R_4$$

Let attitude of agent for buying the goods is high i.e. 1 and the competition in the market for that product is more i.e. 1. Such a situation can be described by fuzzy sets E' and C' as $E' = [0, 0, 1]$ $C' = [0, 0, 1]$. Now the conclusion of the reasoning can be calculated by applying Mamdani's compositional rule of Inference [3] as follows $P' = C' \circ (E' \circ R)$ where $\circ$ is the composition process. After implementing this we will get $P' = [0, 0, 1]$. Defuzzification of *P'* by taking center of gravity with the weighted mean we get definite value for the bid displacement factor $\triangle P$ as +100. So the final bid price will be $P = P + \triangle P$.

4 Conclusions

In this paper we have designed a fuzzy competition and attitude based bidding strategy (FCA-Bid), which uses a soft computing method i.e. fuzzy logic technique to compute the final bid price based on the attitude of the agent and the competition in the market. Another unique idea presented in this paper is that to deal quantitatively the imprecision or uncertainty of multiple attributes of items to acquire in auctions, fuzzy set technique is used. The bidding strategy also allows for flexible heuristics both for the overall gain and for individual attribute evaluations. Specifically, the bidding strategy is adaptive to the environment as the agent can change the bid amount based its assessment of the various attributes of item, eagerness of agent as well as competition in the auction . The attitude of the agents is found with respect to the goods' attributes and the competition is calculated based on the number of bidders and the time elapsed for the auction. It was noticed that the strategies in which agent's behavior depends on attitudes and competition, are easily adaptable to the dynamic situations of the market [11] [16]. An experimental evaluation is also conducted to find the final bid price on the bases of the historical auction records and by considering some set of rules for the attitude and the competition factors. In future we will investigate about the development of the bidding strategies for multiple auctions. We will also compare our bidding techniques with the other strategies to find out the relative strengths and the weaknesses.

References

1. Anthony, P., Jennings, N.R.: Developing a Bidding Agent for Multiple Heterogeneous Auctions. ACM transactions on Internet Technology 3(3), 185–217 (2003)
2. Anthony, P., Jennings, N.R.: Evolving Bidding Strategies for Multiple Auctions. In: Proceedings of the 15th European Conference on Artificial Intelligence, pp. 187–192 (2002)
3. Bapna, R., Jank, W., Shmueli, G.: Price Formation and its Dynamics in Online Auctions. Decision Support Systems 44(3), 641–656 (2008)

4. Byde, A., Priest, C., Jennings, N.R.: Decision Procedures for Multiple Auctions. In: Proceedings of the First International Joint Conference on Autonomous agents and Multiagent Systems, pp. 613–620 (2002)
5. Faratin, P., Sierra, C., Jennings, N.R.: Negotiations Decision Functions for Autonomous Agents. International Journal of Robotics and Autonomous Systems 24(3), 159–182 (1998)
6. Goyal, M.L., Ma, J.: Using Agents Attitudes and Assessments in Automated Fuzzy Bidding Strategy. In: Proceedings of the 1st International Conference on Agents and Artificial Intelligence (2009)
7. Greenwald, A., Stone, P.: Autonomous Bidding Agents in the Trading Agent Competition. IEEE Internet Computing 5(2), 52–60 (2001)
8. He, M., Leung, H.f., Jennings, N.R.: A Fuzzy-Logic Based Bidding Strategy for Autonomous Agents in Continuous Double Auctions. IEEE Transaction on Knowledge and Data Engineering 15(6), 1345–1363 (2003)
9. Kowalczyk, R., Bui, V.: On Fuzzy e-Negotiation Agents: Autonomous Negotiation with Incomplete and Imprecise Information. In: Proceedings of the 11th International Workshop on Database and Expert Systems Applications, p. 1034 (2000)
10. Luo, X., Jennings, N.R., Shadbolt, N., Leung, H.F., Lee, J.H.M.: A Fuzzy Constraint based Model for Bilateral, Multi-issue Negotiations in Semi-competitive Environments. Artifical Intelligence 148(1-2), 53–102 (2003)
11. Ma, H., Leung, H.F.: An Adaptive Attitude Bidding Strategy for Agents in Continuous Double Auctions. Electronic Commerce Research and Applications 6(4), 383–398 (2007)
12. Ma, H., Leung, H.f.: Bidding Strategies in Agent-Based Continuous Double Auctions. Birkhäuser, Basel (2008)
13. Murugesan, S.: Negotiation by Software Agents in Electronic Marketplace. In: Proceedings of the 2000 IEEE Region 10 Conference on Intelligent Systems and Technologies for the new Millenium, pp. 286–290 (2000)
14. Sim, K.M., Wang, S.Y.: Flexible Negotiation Agent With Relaxed Decision Rules. IEEE Transactions on Systems, Man and Cybernetics- Part B: Cybernetics 34(3), 1602–1608 (2004)
15. Stone, P., Littman, M.L., Singh, S., Kearns, M.: ATTac-2000: an Adaptive Autonomous Bidding Agent. Journal of Artificial Intelligence Research 15, 189–206 (2001)
16. Tagra, H.: Generic Architecture for Agents in E-Commerce (Auction), Ph. D dissertation. Indian Institute of Technology New Delhi (2006)
17. Tanaka, K.: An Introduction to Fuzzy Logic for Practical Applications. Springer, Heidelberg (1991)

Resource Allocation Analysis in Perfectly Competitive Virtual Market with Demand Constraints of Consumers

Tetsuya Matsuda, Toshiya Kaihara, and Nobutada Fujii

Abstract. Virtual market mechanism solves resource allocation problems by distributing the scheduled resources based on software agent interactions in the market. We formulate agent behaviours negotiating the resource allocations under demand constraints of consumers in the market, and demonstrate the applicability of the virtual market concept to this framework. In this paper we demonstrate the proposed virtual market successfully calculates Pareto optimal solutions in resource allocation problem under the demand constraints of consumers.

1 Introduction

Recently, it is difficult to manage and control artificial systems with conventional approaches because these systems become large-scale and complex. We focus on the concept of market in social science approach as a technique to achieve the adaptability of artificial systems as well as the optimality [1]. In perfectly competitive market defined by the neo-classical economics, it is guaranteed that the equilibrium resource allocation is Pareto optimal [2]. Market-Oriented programming (MOP) is the technique that obtains the Pareto optimal solution by the construction of a virtual perfectly competitive market (virtual market) in the computer [3] [4]. MOP has been also extended into oligopolistic virtual market in our previous study [5].

Tetsuya Matsuda · Toshiya Kaihara · Nobutada Fujii
Kobe University, 1-1, Rokkodai, Nada, Kobe 657-8501, Japan
e-mail: matsuda@kaede.cs.kobe-u.ac.jp, kaihara@kobe-u.ac.jp, nfujii@phoenix.kobe-u.ac.jp

Q. Bai and N. Fukuta (Eds.): Advances in Practical Multi-Agent Systems, SCI 325, pp. 181–200.
springerlink.com

Even in the perfectly competitive market, it is sometimes required to assume additional constraints, such as consumers' demand or producers' supply due to their capacity. For example, it is quite common that producers have bottleneck, such as buffer or transport facilities, to supply their products in real cases [6]. Consumers also have normally several constraints, such as stock yard or storage, to keep the supplied goods for their future consumption. It is difficult for conventional MOP to handle such cases due to the strong assumption of pure competitive market [7], so it is attractive to enhance its target into real market situation with several constraints.

In this paper we focus on consumers problem in economic system with reality, and assume the consumers constraint additionally to MOP. We newly formulate the Walrasian virtual market with the constraints and try to analyse the proposed virtual market successfully calculates Pareto optimal solutions in resource allocation problem under the demand constraints of consumers.

2 Virtual Market Structure

There are two types of entities in the virtual market: agents and goods. Agents sell, buy, consume and produce goods in the economy. The price of goods represents the exchange ratio, and there exists an auctioneer for goods that receive bids from agents and changes the price of the corresponding goods according to its current aggregate demand.

A consumer is an agent that can buy, sell and consume goods, and the consumer's preference for consuming various combinations or bundles of goods is specified by its utility function. Before exchanging goods in the economy, consumers may also start with an initial allocation of several goods, termed their endowment. In our market model, consumers don't only consume but also supply goods for input goods. It means that people supply like labour forces or capital to companies.

A producer is an agent that can transform some sorts of goods into some others, subject to its technology. The technology specifies the combinations of inputs and outputs feasible for the producer. Different from consumers, producers do not have any endowment. Producers can profit from buying inputs and selling outputs. The production function of a producer maps a bundle of input goods into the bundle of output goods that can be produced by transforming input. To simplify the discussion, we assume that every producer can produce only one kind of goods, so a producer agent can have only a single production function in this case.

We show the notations for the agent formulations in this paper as follows:

p_i F price of good i
G_{num} F number of kind of goods in the market
S_{num} F number of kind of producers in the market
C_{num} F number of kind of consumers in the market
I^{C_k} F class of goods in the market
$e_i^{C_k}$ F amount of consumer C_k's initial good i
$x_i^{C_k}$ F demand to good i of consumer C_k
$y_j^{C_k}$ F supply to good j of consumer C_k
B^{C_k} F budget of consumer C_k
u^{C_k} F utility function of consumer C_k
$\bar{x}_l^{C_k}$ F demand limit to good l of consumer C_k
γ^{S_k} F profit of producer S_k
c^{S_k} F cost of producer S_k
r^{S_k} F income of producer S_k
$x_i^{S_k}$ F demand to good i of producer S_k
$y_j^{S_k}$ F supply to good i of producer S_k
f^{S_k} F production function of producer S_k

3 Agent Formulations

3.1 Consumer Agents

3.1.1 Utility Function

The utility function is defined as a scalar function of a bundle of consuming goods. In this model, the Cobb-Douglass utility function u^{C_k}, that is basic function in Microeconomics, for consumer C_k is represented as follows:

$$u^{C_k} = A^{C_k} \prod_i (x_i^{C_k})^{\alpha_i^{C_k}} \qquad (0 < A^{C_k}, \sum_{i \in I^{C_k}} \alpha_i^{C_k} = 1) \tag{1}$$

3.1.2 Constraints

Budget constraint

Consumer C_k's budget B^{C_k} is defined by initial endowments and current prices $\boldsymbol{p}$, and the budget is represented as (2) . The variable γ^{C_k} is producers' profits returned to consumer C_k.

$$B^{C_k} = \boldsymbol{p} \cdot \boldsymbol{e}^{C_k} + \gamma^{C_k} \tag{2}$$

Therefore the constraint of budget for consumer C_k is represented as the following equation:

$$\sum_{i \in I^{C_k}} p_i x_i \leq B^{C_k} \tag{3}$$

We assume that consumers buy goods as many as possible, because consumers can get more utility as they attain more goods under their utility function in their perfect competitive market.

Demand constraint

It is sometimes observed for consumers to have their capacity of allocated goods in real market. The capacity constraint normally causes to reduce the resource allocation to the consumers. For instance, the stock capacity of consumer agents must impose upper limit on their demanded goods. In this model, we assume that the consumers have constraints on the amount of their demand. Consumer C_k can be allocated goods i below the capacity $\bar{x}_i{}^{C_k}$. It means that assuming a real allocation problem, we cannot allocate resource too much by constraint of size.

Therefore the constraint of demand for consumer C_k is represented as the next equation:

$$x_i^{C_k} \leq \bar{x}_i{}^{C_k} \qquad (i \in I^{C_k}) \tag{4}$$

3.1.3 Demand / Supply Functions

Consumer C_k is to choose a feasible bundle of goods, x^{C_k}, so as to maximize the consumer's utility function. The allocated bundle of goods is feasible for consumer C_k if the equations (3) , (4) are satisfied. The consumer's choice can be expressed at the following optimization problem:

$$\max \qquad u^{C_k} = A^{C_k} \prod_{i \in I^{C_k}} (x_i^{C_k})^{\alpha_i^{C_k}} \tag{5}$$

$$\text{s.t.} \qquad \sum_{i \in I^{C_k}} p_i x_i = B^{C_k} \tag{6}$$

$$x_i^{C_k} \leq \bar{x}_i^{C_k} \quad (i \in I^{C_k}) \tag{7}$$

Here consumer C_k's demand function of goods i; $x_i^{C_k}(p_i)$ is introduced from equations (5), (6), (7), and it is represented as follows:

$$x_i^{C_k} = \bar{x}_i^{C_k} \qquad (i \in \theta^{C_k}) \tag{8}$$

$$\text{or} x_i^{C_k}(p_i) = \frac{\alpha_i^{C_k}(B^{C_k} - \sum_{j \in \theta^{C_k}} p_j x_j^{Ck})}{p_n(1 - \sum_{j \in \theta^{C_k}} \alpha_j^{C_k})} \qquad (i \notin \theta^{C_k}) \tag{9}$$

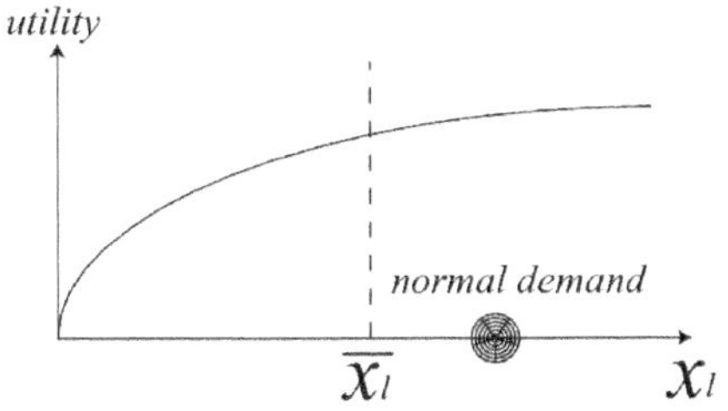

Fig. 1 Utility function (i)

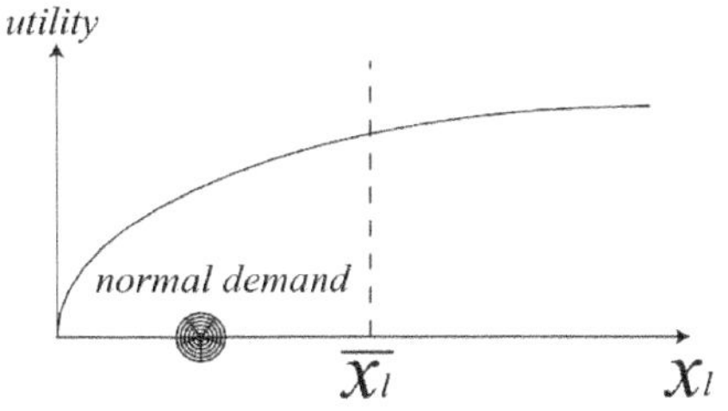

Fig. 2 Utility function (ii)

The set θ^{C_k} includes the exceeded goods over $\bar{x}_i{}^{C_k}$. At initial conditions we assume $\theta^{C_k} = \{\phi\}$. The demand of goods i in $x_i^{C_k} \geq \bar{x}_i^{C_k}$ (Fig.1) is defined as equation (8) and add goods i to θ^{C_k}. On the other hand, the demand of goods i in $x_i^{C_k} < \bar{x}_i^{C_k}$ (Fig.2) is defined as equation (9).

In this model, we assume each consumer supplies all its endowments into the market to try to attain maximum utility. So the supply function $y_j^{C_k}$ is represented as follows:

$$y_j^{C_k} = e_j^{C_k} \tag{10}$$

3.2 Producer Agents

3.2.1 Production Function

In this model, we assume producer s_k also has the basic Cobb-Douglass production function $y_j^{s_k}$ represented as the following equation:

$$y_j^{S_k} = A^{S_k}(x_i^{S_k})^{\alpha^{S_k}} \qquad (0 < A^{S_k}, 0 < \alpha^{S_k} < 1) \tag{11}$$

In this equation $x_i^{S_k}$ and $y_j^{S_k}$ represent input goods i by S_k and output goods j, respectively. The power variable α^{S_k} must be below 1 to satisfy the fundamental theorem of welfare economics defined by micro economics, and that means production function must follow the law of diminishing return [2].

3.2.2 Profit Function

Producer S_k is assumed to transform goods i to goods j. Then producer S_k is required to pay purchasing cost represented as equation (12), and the amount of proceeds after clearing market is represented as equation (13).

$$c^{S_k} = p_i x_i^{S_k} \tag{12}$$

$$r^{S_k} = p_j y_j^{S_k} \tag{13}$$

Then producer S_k can profit γ^{S_k} represented as follows:

$$\gamma^{S_k} = r^{S_k} - c^{S_k} \tag{14}$$

3.2.3 Demand / Supply Functions

The objective of producer S_k is to choose a production plan that maximises profit (14) subject to its technology and the current price of both its output and input goods. The producer's choice can be represented as the following constrained optimization problem:

$$max \;\; \gamma^{S_k} = p_j y_j^{P_k} - p_i x_i^{S_k} \tag{15}$$

$$s.t. \;\; y_j^{S_k} = A^{S_k}(x_i^{S_k})^{\alpha^{S_k}} \tag{16}$$

Then producer S_k's demand function of goods i, $x_i^{S_k}(p_i)$, is conducted from equations (15) and (16) to solve the NLP problem, and represented as the next equation:

$$x_i^{S_k}(p_i) = (\frac{p_i}{A^{S_k}\alpha^{S_k}p_j})^{\frac{1}{\alpha^{S_k}-1}} \tag{17}$$

Producer S_k's supply function of goods j, $y_j^{S_k}(p_j)$, is also conducted from equations (15), (16), and represented as follows:

$$y_j^{S_k}(p_j) = (\frac{(p_i)^{\alpha^{S_k}-1}}{A^{S_k}(\alpha^{S_k})^{\alpha^{S_k}}(p_j)^{\alpha^{S_k}}})^{\frac{1}{\alpha^{S_k}-1}} \tag{18}$$

3.3 Pricing Mechanism

We modify the pricing mechanism of MOP with considering the demand constraints shown in equations (9), (8). The algorithm is described as follows:

STEP1
: The auctioneer of goods sets an initial price of the good. (In this model, initial price of every goods is 1.0).

STEP2
: Consumers bid demand functions based on equations (9), (8).

STEP3
: Consumers bid supply functions.

STEP4
: Producers bid demand functions.

STEP5
Producers bid supply functions.
STEP6
The auctioneer of goods set up the price corresponded with the supply and demand.
STEP7
We define the previous price of goods i is p_i^{before}, and updated current price is p_i^{after}. If all goods satisfied the equation (19), then p_i^{after} is equilibrium price and go STEP8(σ is very small number. In this model, $\sigma = 10^{-13}$). If we cannot set up equilibrium price, return to STEP2.

$$|p_i^{after} - p_i^{before}| < \sigma \tag{19}$$

STEP8
Prices of all the goods are cleared, and all agents start trading.

The demand function of consumers (in STEP 2) followed by equations (9), (8) is finally represented as Fig.3. The maximum demand value faces the ceiling at $\bar{x}_i^{C_k}$ shown in this figure.

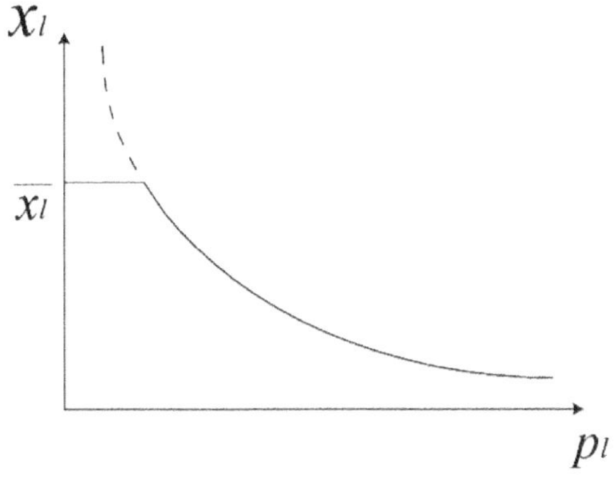

Fig. 3 Demand curve

4 Computer Simulation and Experimental Results

We construct two types of virtual markets, such as exchange market and economic market, and try to verify demand constraint effect to resource allocation. Pareto optimality is also investigated as the basic analysis in the exchange market, and the effect of the number of agents is analysed additionally in the economic market.

4.1 Exchange Market Analysis

4.1.1 Effect of Demand Constraint

Only consumer agents exist in the exchange market because no goods are produced in the market. The experimental conditions are described as follows:
$G_{num} = 3,\ S_{num} = 0,\ C_{num} = 2$
We unintentionally choose parameters.

[Experimental results]
4 scenarios are experimented in terms of the demand constraint. There is no constraint in Case A, whereas agent C_1 has strong constraints in both goods 1 and 2 in Case D. The sets of input and resuts at Case A, B, C, D are shown in Table 1, 2, 3, 4, respectively. Results in each table shows equilibrium price of goods 1, 2 and 3 $(\hat{p_1}, \hat{p_2}, \hat{p_3})$ and allocations.

Table 1 Input and result in Case A

(Input)

Agent	Utility function (u^{C_k})	$(e_1^{C_k}, e_2^{C_k}, e_3^{C_k})$	$(\bar{x}_1^{C_k}, \bar{x}_2^{C_k}, \bar{x}_3^{C_k})$	Utility value
C_1	$(x_1^{C_1})^{0.7}(x_2^{C_1})^{0.2}(x_3^{C_1})^{0.1}$	(5.0, 8.0, 10.0)	(∞, ∞, ∞)	5.8870
C_2	$(x_1^{C_2})^{0.1}(x_2^{C_2})^{0.1}(x_3^{C_2})^{0.8}$	(8.0,10.0,5.0)	(∞, ∞, ∞)	5.6168

(Results)

Price	Allocation		Utility value	
$(\hat{p_1}, \hat{p_2}, \hat{p_3})$	$(x_1^{C_1}, x_2^{C_1}, x_3^{C_1})$	$(x_1^{C_2}, x_2^{C_2}, x_3^{C_2})$	C_1	C_2
(1.4519,0.38773,1.3430)	(11.470,12.272,1.7715)	(1.5296,5.7277,13.229)	9.6452	9.7249

Table 2 Input and result in Case B

(Input)

Agent	Utility function (u^{C_k})	$(e_1^{C_k}, e_2^{C_k}, e_3^{C_k})$	$(\bar{x}_1^{C_k}, \bar{x}_2^{C_k}, \bar{x}_3^{C_k})$	Utility value
C_1	$(x_1^{C_1})^{0.7}(x_2^{C_1})^{0.2}(x_3^{C_1})^{0.1}$	(5.0, 8.0, 10.0)	$(9, \infty, \infty)$	5.8870
C_2	$(x_1^{C_2})^{0.1}(x_2^{C_2})^{0.1}(x_3^{C_2})^{0.8}$	(8.0,10.0,5.0)	(∞, ∞, ∞)	5.6168

(Results)

Price	Allocation		Utility value	
$(\hat{p_1}, \hat{p_2}, \hat{p_3})$	$(x_1^{C_1}, x_2^{C_1}, x_3^{C_1})$	$(x_1^{C_2}, x_2^{C_2}, x_3^{C_2})$	C_1	C_2
(0.51865,0.87771,1.5639)	(9.0000,15.636,4.3878)	(4.0000,2.3636,10.612)	9.3545	8.4702

The comparison between Table 1 and 2 conducts basic characteristics on the demand constraint. Consumer agent C_1 has constraint in goods 1 $(\bar{x}_1^{C_1})$ in Case B.

Consumer agent C_1 bids into goods 1 by the constraint limit $\bar{x}_1^{C_1}$ at maximum, so the allocation of $x_1^{C_1}$ in Table 2 (Results) is 9.0000 and less than in Table 1 (Results). Consequently, the surplus budget in agent C_1 was transferred into the bids for both goods B and C. As a result, $x_2^{C_1}$, $x_3^{C_1}$ in Table 2

Table 3 Input and result in Case C

(Input)

Agent	Utility function (u^{C_k})	$(e_1^{C_k}, e_2^{C_k}, e_3^{C_k})$	$(\bar{x}_1^{C_k}, \bar{x}_2^{C_k}, \bar{x}_3^{C_k})$	Utility value
C_1	$(x_1^{C_1})^{0.7}(x_2^{C_1})^{0.2}(x_3^{C_1})^{0.1}$	(5.0, 8.0, 10.0)	$(9, 13, \infty)$	5.8870
C_2	$(x_1^{C_2})^{0.1}(x_2^{C_2})^{0.1}(x_3^{C_2})^{0.8}$	(8.0,10.0,5.0)	(∞, ∞, ∞)	5.6168

(Results)

Price	Allocation		Utility value	
$(\hat{p_1}, \hat{p_2}, \hat{p_3})$	$(x_1^{C_1}, x_2^{C_1}, x_3^{C_1})$	$(x_1^{C_2}, x_2^{C_2}, x_3^{C_2})$	C_1	C_2
(0.46278,0.37022,2.2213)	(9.0000,13.000,8.3333)	(4.0000,5.0000,6.6667)	9.6126	6.5093

Table 4 Input and result in Case D

(Input)

Agent	Utility function (u^{C_k})	$(e_1^{C_k}, e_2^{C_k}, e_3^{C_k})$	$(\bar{x}_1^{C_k}, \bar{x}_2^{C_k}, \bar{x}_3^{C_k})$	Utility value
C_1	$(x_1^{C_1})^{0.7}(x_2^{C_1})^{0.2}(x_3^{C_1})^{0.1}$	(5.0, 8.0, 10.0)	$(9, 10, \infty)$	5.8870
C_2	$(x_1^{C_2})^{0.1}(x_2^{C_2})^{0.1}(x_3^{C_2})^{0.8}$	(8.0,10.0,5.0)	(∞, ∞, ∞)	5.6168

(Results)

Price	Allocation		Utility value	
$(\hat{p_1}, \hat{p_2}, \hat{p_3})$	$(x_1^{C_1}, x_2^{C_1}, x_3^{C_1})$	$(x_1^{C_2}, x_2^{C_2}, x_3^{C_2})$	C_1	C_2
(0.44660,0.22330,2.4117)	(9.0000,10.000,9.0741)	(4.0000,8.0000,5.9259)	9.1992	6.3670

(Results) is more than in Table 1 (Results). The demand constraint of consumer agent C_1 influences the resource allocation in consumer agent C_2 at the same time. The allocation of goods 1 into consumer agent C_2 ($x_1^{C_2}$) in Table 2 (Results) is more than in Table 1 (Results) due to the decreased allocation into consumer agent C_1 in Table 2.

Price of goods 1 decreases in Table 2, because market demand function of goods 1 in case B is less than in case A due to the demand constraint of consumer agent C_1 in case B shown in Fig.4. We can observe the same tendencies amongst Case A, B, C and D.

4.1.2 Verification of Pareto Optimality

Computer simulation shows the price clearing dynamism during their trade in the exchange market. Now we try to verify the Pareto optimality of attained equilibrium prices by comparing our proposed method with conventional analytical method, named fixed point algorithm.

4.1.3 Fixed Point Algorithm

Fundamental theorem of welfare economics ensure that equilibrium price is Pareto optimality in perfect competitive market. In neo classical economics, equilibrium price is set up by fixed point; fixed point algorithm [9]. Fixed

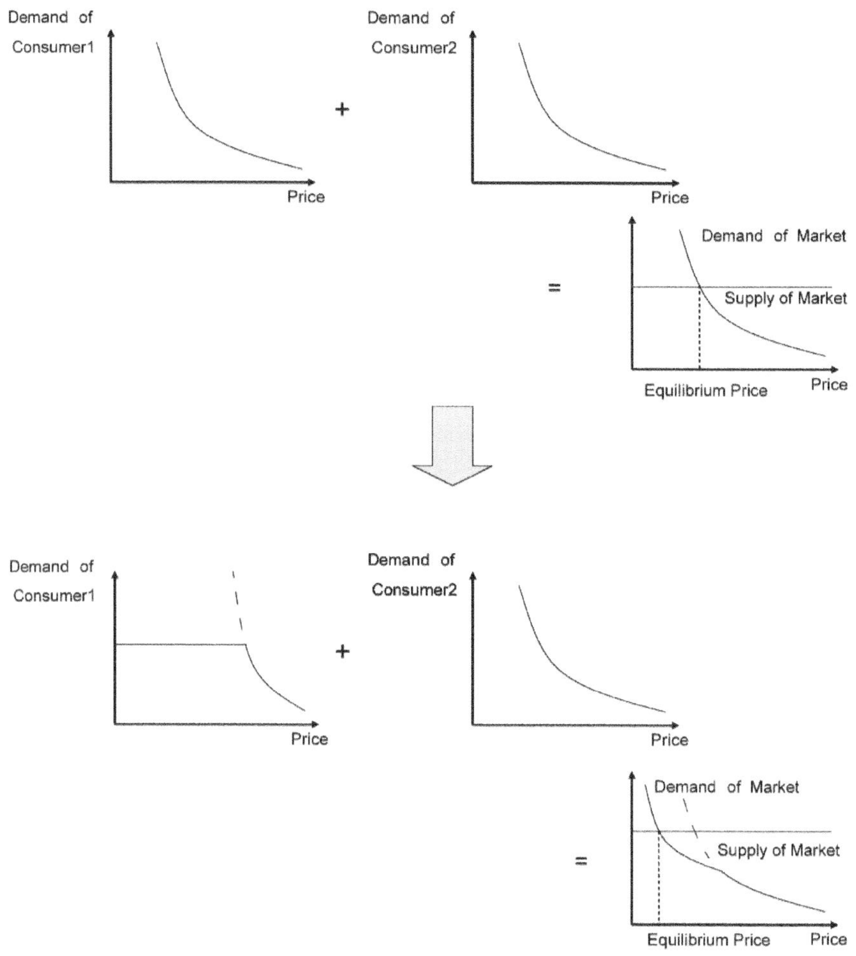

Fig. 4 Price change mechanism

point algorithm set up one-on-one Pareto solution with initial endowments. And we can use fixed point algorithm only in perfect competitive market. In this research, we use fixed algorithm called Scarf algorithm and compare with equilibrium price cleared by our approach.

Excess demand function

The difference between demand $\boldsymbol{x}$ and supply $\boldsymbol{y}$ is called excess demand, and excess demand function $\boldsymbol{E}(\boldsymbol{p})$ is shown in equation (20).

$$\boldsymbol{E}(\boldsymbol{p}) = \boldsymbol{x} - \boldsymbol{y} \tag{20}$$

It has the following characteristics.

1. It is continuum function for $\boldsymbol{p}$.
2. It is zero-homogeneous; $\boldsymbol{E}(\lambda \boldsymbol{p}) = \boldsymbol{E}(\boldsymbol{p})$ $(\forall \lambda > 0)$.
3. It is satisfied Walras rule; $\boldsymbol{p} \cdot \boldsymbol{E} = 0$.

Equilibrium price and fixed point algorithm

X is set, f is function from X to X. If $f(\boldsymbol{x}^*) = \boldsymbol{x}^*$ is true for any $\boldsymbol{x}^* \in X$, $\boldsymbol{x}^*$ is defined as fixed point of f. Brouwer's fixed-point theorem tells that there is a fixed point in continuum function which domain and codomain are in same bounded closed set. And there are equilibrium prices $\boldsymbol{p}^*$ [9].

In economic science, fixed algorithm is major to set up equilibrium prices because fixed poin and equilibrium price is the same.

Scarf algorithm

Scarf algorithm is one of the most basic fixed point algorithm. First, normalize the number i which is kinds of goods and make price set T ((I-1)dimensional basic elementary substanceS_I). Next, we divide each side of S_I equally and make small elementary substance which has I vertex. The number D is dividing number and called grid size. If D is large, solution is high-precision.

In Scarf algorithm, we search solution from the small elementary substance which has the vertex of S_I to next small elementary substance.

4.1.4 Preto Optimiality in Equilibrium Solution

Table 5 shows the normalized equilibrium prices attained by the proposed approach. On the other hand, the equilibrium prices attained by Scarf algorithm are shown in Table 6.

Differences of the equilibrium prices between Scarf algorithm and our approach are 1.0×10^{-8} and 1.0×10^{-5} at maximum in D=1.0 10^8 and D=1.0 10^5, respectively. The difference was occurred by the calculation error in Scarf algorithm due to the grid size. We observed the price vector in equilibrium attained by our approach is a Pareto optimal solution by comparing the solution in Scarf algorithm.

The calculation time to get the solution by our approach is much shorter than Scarf algorithm. Scarf algorithm needs higher computer power, because simulation time increases remarkably as the number of grids are increased.

Table 5 Experimental results(MOP)

	(p_1, p_2, p_3)	Time
(a)	(0.45619492,0.12182478,0.42198030)	0.007
(b)	(0.17520216,0.29649596,0.52830189)	0.029
(c)	(0.15151515,0.12121212,0.72727273)	0.037
(d)	(0.14492754,0.07246377,0.78260870)	0.031

Table 6 Experimental results(Scarf)

	D=1.0 10^8		D=1.0 10^5	
	(p_1, p_2, p_3)	Time	(p_1, p_2, p_3)	Time
(a)	(0.45619491,0.12182478,0.42198031)	96.859	(0.45620,0.12182,0.42198)	0.094
(b)	(0.17520215,0.29649596,0.52830189)	151.328	(0.17520,0.29649,0.52831)	0.16
(c)	(0.15151516,0.12121213,0.72727271)	157.31	(0.15152,0.12122,0.72726)	0.14
(d)	(0.14492754,0.07246377,0.78260869)	158.219	(0.14492,0.07247,0.78261)	0.16

4.2 Economic Market Analysis

4.2.1 Effect of Demand Constraint

There are consumers and producers in the market. The experimental conditions are described as follows:

$G_{num} = 2,\ S_{num} = 1,\ C_{num} = 1$

Initial parameters of agents are shown in Table 7 and 8.

Table 7 Initial states of consumer

Utility function (u^{C_1})	Endowment $(e_1^{C_1}, e_2^{C_1})$
$(x_1^{C_1})^{0.1}(x_2^{C_9})^{0.9}$	(10, 1)

Table 8 Initial states of producer

Input	Output	Production function
goods1	goods2	$y_2^{S_1} = 2.0(x_1^{S_1})^{0.4}$

[Experimental results]

Fig. 5 show allocations, utility of consumer, profit of producer and equilibrium prices of goods 1 and goods 2 $(\hat{p_1}, \hat{p_2})$. Transverse is $\bar{x_2}^{C_1}$.

Goods 2 is transformed from goods 1 to maximize utility and profit. Because the consumer agent prefers goods 2 to goods 1, the producer agent transforms from goods 1 to goods 2. Fig.5 (a) describes that the amount of allocation of goods 2 increases as the demand constraint is larger, and goods 1 that is input goods decreases because consumer agent can be allocated goods 2 less or equal of demand constraint. The amount of the allocation doesn't change even if the demand constraint is increased more than 5 in Fig.5 (a). That is because the consumer agent can get more utility from goods 2 than from goods 1 in this case due to the law of diminishing marginal utility.

Fig.5 (b) tells us that the price of goods 1(/goods 2) decreases(/increases) as the demand constraint increases. That is because the demand of goods 2 of consumer agent is high in case the demand constraint is large. Demand of goods 1 of consumer agent is decreased in case the demand constraint is

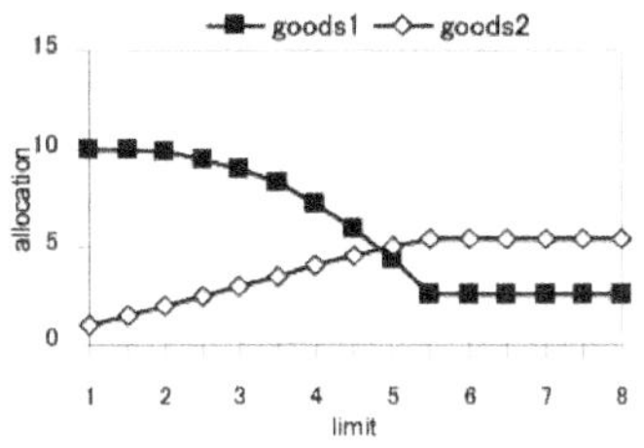

(a) Consumer's allocation

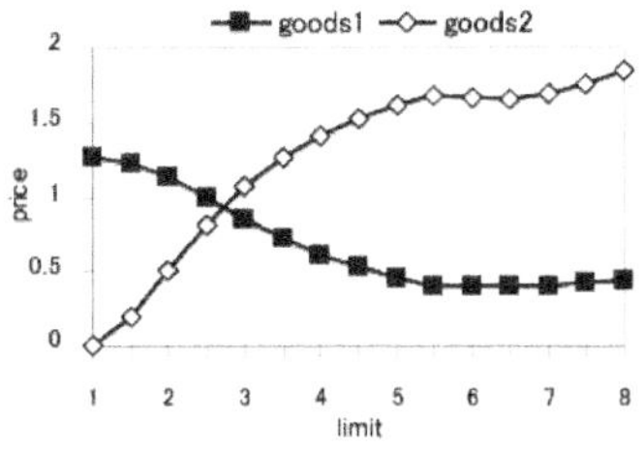

(b)Price of goods

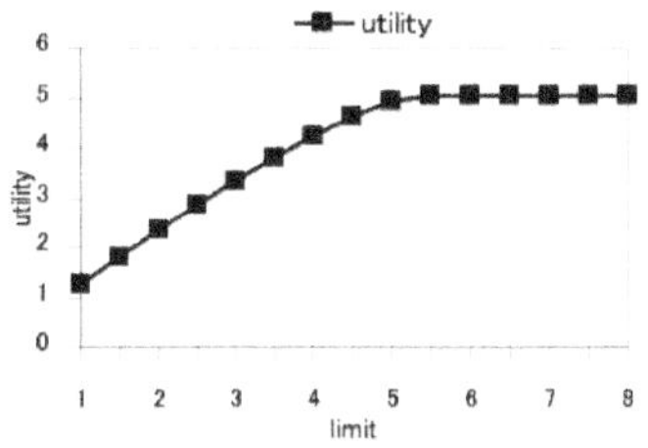

(c)Utility

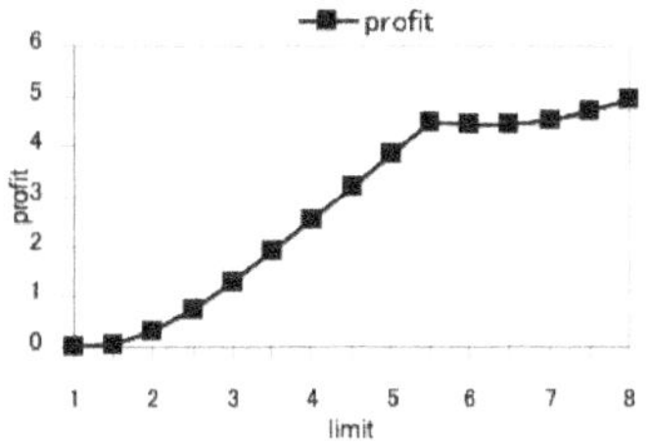

(d)Profit

Fig. 5 Experimental results

large, instead. Fig.5 (c) tells us that the utility of consumer agent increases because the number of allocated goods 2 is getting higher due to the increased production. Fig.5 (d) describes that the profit of producer agent is increased because the producer agent can produce more volume of goods 2.

4.2.2 Effect of Preference

In this section, we verify the effect of preference of consumers. The experimental conditions are described as follows:

$G_{num} = 2$, $S_{num} = 2$, $C_{num} = 1$ Table 9 is preference parameters of consumer agent. Initial endowments is $(e_1^{C_1}, e_2^{C_1}) = (10, 10)$, and initial parameters of producer agents are Table 10.

Table 9 Initial degree of consumers's preference

Degree of preference$(\alpha_1^{C_1}, \alpha_2^{C_1})$	Experimental result
(0.8,0.2)	(a)
(0.6,0.4)	(b)
(0.5,0.5)	(c)
(0.3,0.7)	(d)
(0.1,0.9)	(e)

Table 10 Initial states of producer

Agent	Input	Output	Production function
S_1	goods1	goods2	$y_2^{S_1} = 2.0(x_1^{S_1})^{0.4}$
S_2	goods2	goods1	$y_1^{S_2} = 2.0(x_2^{S_2})^{0.4}$

[Experimental results]

Fig. 6 shows the allocations of each goods, and Fig. 7 shows the equilibrium prices $(\hat{p_1}, \hat{p_2})$. Fig. 8 and Fig. 9 show the utility of consumer agents and the profit of producer agents, respectively. Horizontal axis in each figure shows the upper constraint of consumer agent 1 $(\bar{x_1}^{C_1})$.

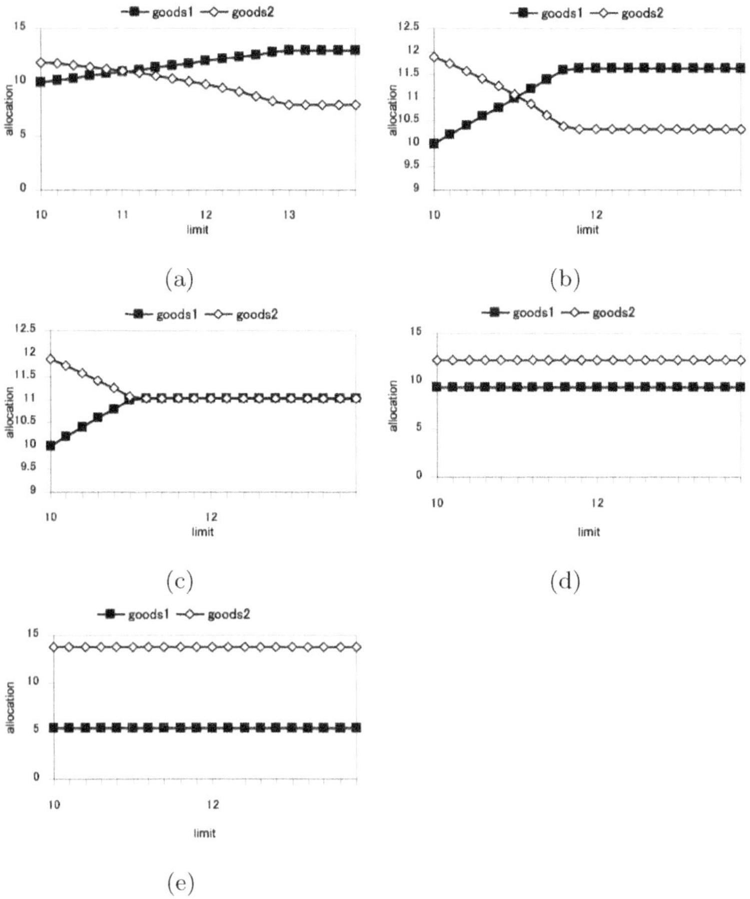

Fig. 6 Allocation

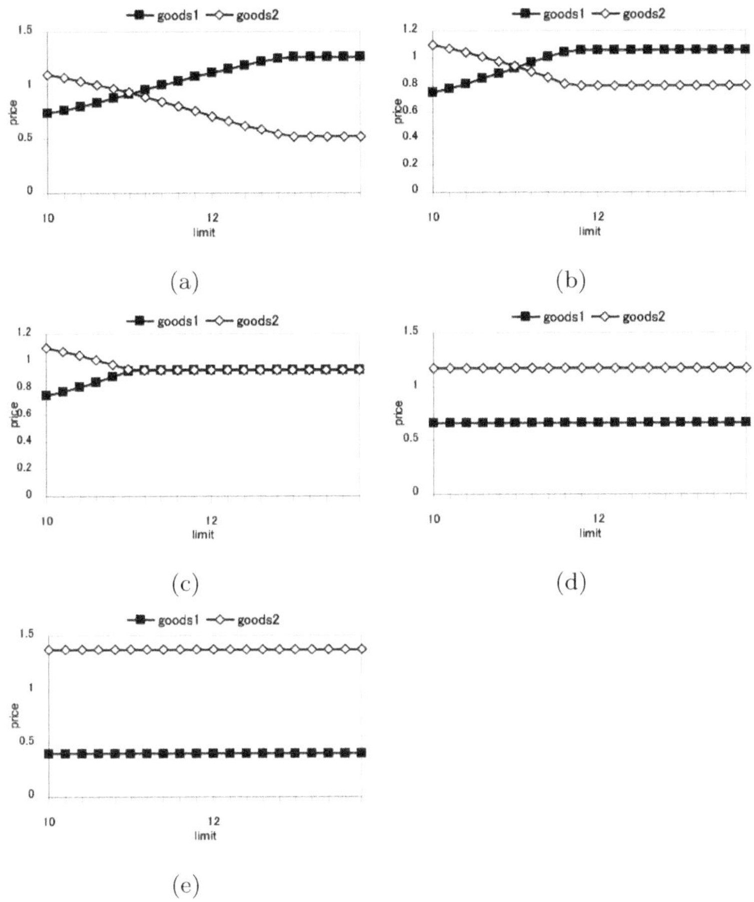

Fig. 7 Price

Fig. 6 shows that if consumer agent prefers goods 1 to goods 2, producer agent produces goods 1 more than goods 2 ((a), (b)). On the other hand, if consumer agent prefers goods 2, then producer agent produces more goods 2 ((d), (e)). But, as shown in case (a) and case (b), if upper constraint $(\bar{x_1}^{C_1})$ is in relatively small number, the allocation of goods 2 is more because of the demand constraint. And, the allocation of goods 1 doesn't increase if upper demand constraint is strong, because the utility of goods 1 is more than the utility of goods 2 (law of diminishing marginal utility).

So, if consumer agents prefer goods 2 to goods 1, the profit of producer agent 2 (transforms goods 2 from goods 1) decreases, and the profit of producer agent 1 (transforms goods 1 form goods 2) increases instead (Fig. 9). Additionally, if the value of $|\alpha_1^{C_1} - \alpha_2^{C_1}|$ decreases, the differences between the allocations of each goods (Fig. 6), the price of each goods (Fig. 7), and the profit of each producder (Fig. 9) are getting smaller, because the utility of

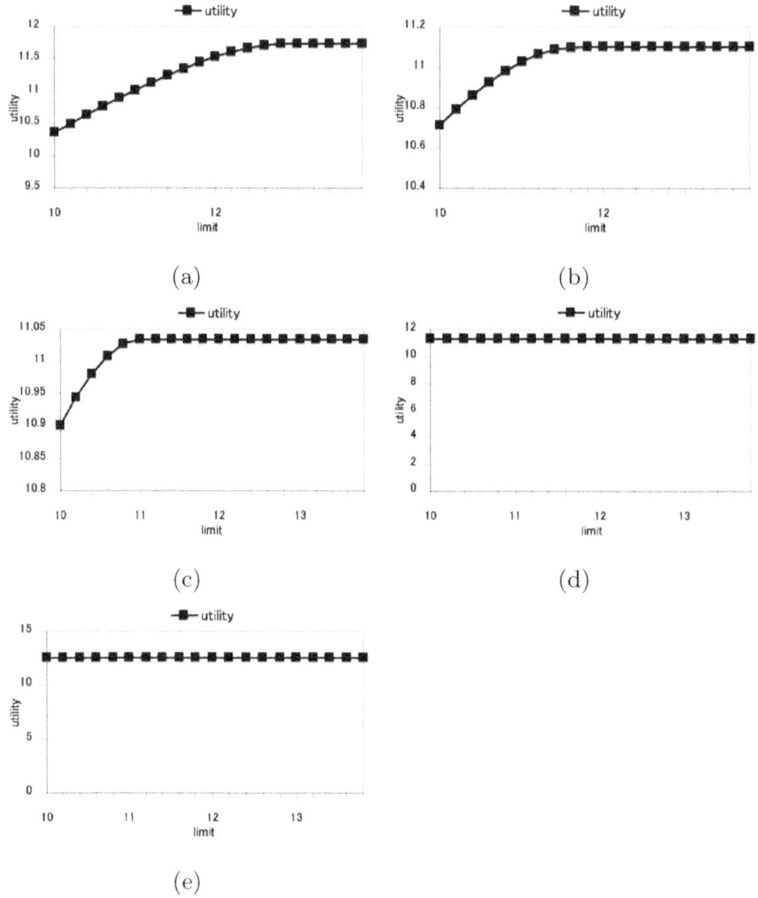

Fig. 8 Utility

each goods is becoming almost equivalent. Especially, if $\alpha_1^{C_1} = \alpha_2^{C_1}$ (in case (c)), the differences between the allocations of each goods, the price of each goods, and the profit of each producder become zero, because the consumer agents utility of each goods is completely the same.

4.2.3 Effect of the Number of Agents

In this section, we verify the effect of the number of agents to calculation time to get Pareto optimal solution. The experimental conditions are described as follows:

$G_{num} = 3, \ S_{num} = 2$

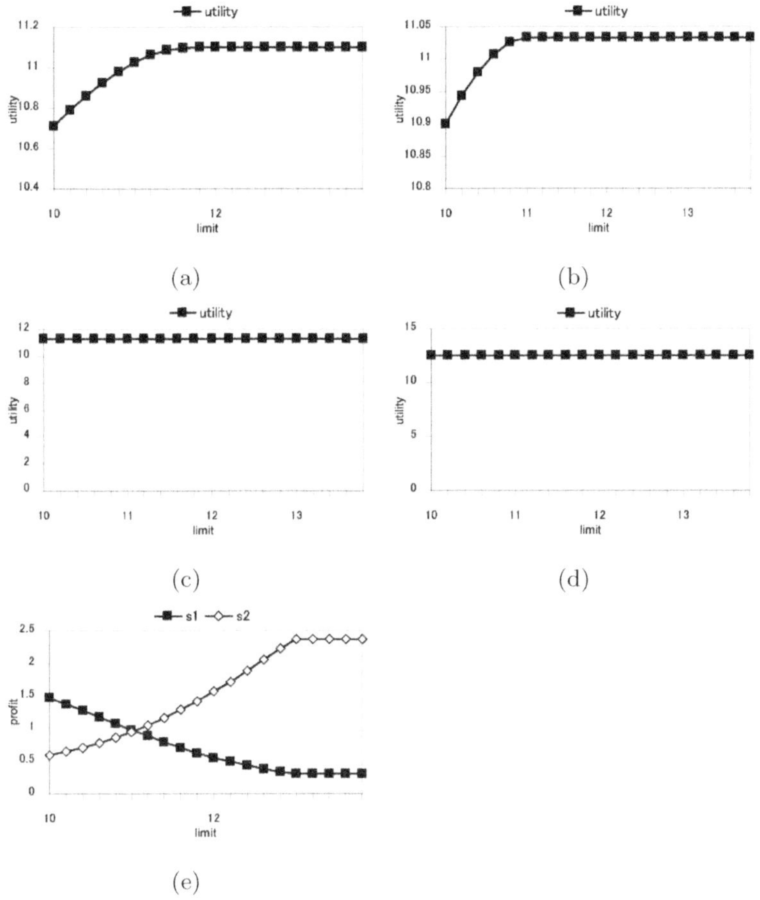

Fig. 9 Profit

Table 11 is the initial sets of parameters in producer agents. Initial endowments, demand constraint and preference of consumers are set followed by uniform random distribution.

Table 11 Initial states of producer

Agent	Input	Output	Production function
S_1	goods1	goods2	$y_2^{S_1} = random(x_1^{S_1})^{random}$
S_2	goods2	goods3	$y_3^{S_2} = random(x_2^{S_2})^{random}$

[Experimental results]

The average of calculation time in 100 trials are shown in Table 12.

Table 12 Computation time (sec)

C_{num}	Time
10	0.04044
50	0.09550
100	12.14

Finally we show one sample in case of $C_{num} = 10$. Table 13 and Table 14 are both in its initial states, and Table 15 and Table 16 describe equilibrium prices acquired after the convergence.

Table 13 Initial states of consumer

Agent	Utility function (u^{C_k})	Endowment ($e_1^{C_k}, e_2^{C_k}, e_3^{C_k}$)	Limit ($\bar{x}_1^{C_k}, \bar{x}_2^{C_k}, \bar{x}_3^{C_k}$)
C_1	$(x_1^{C_1})^{0.37}(x_2^{C_1})^{0.37}(x_3^{C_1})^{0.26}$	(12.0, 3.0, 4.0)	(14.0, 10.0, 5.0)
C_2	$(x_1^{C_2})^{0.58}(x_2^{C_2})^{0.25}(x_3^{C_2})^{0.17}$	(15.0, 1.0, 14.0)	(17.0, 6.0, 16.0)
C_3	$(x_1^{C_3})^{0.09}(x_2^{C_3})^{0.59}(x_3^{C_3})^{0.32}$	(14.0, 10.0, 5.0)	(17.0, 19.0, 12.0)
C_4	$(x_1^{C_4})^{0.07}(x_2^{C_4})^{0.55}(x_3^{C_4})^{0.38}$	(14.0, 5.0, 10.0)	(20.0, 11.0, 12.0)
C_5	$(x_1^{C_5})^{0.17}(x_2^{C_5})^{0.11}(x_3^{C_5})^{0.72}$	(5.0, 13.0, 3.0)	(8.0, 19.0, 3.0)
C_6	$(x_1^{C_6})^{0.48}(x_2^{C_6})^{0.09}(x_3^{C_6})^{0.42}$	(14.0, 9.0, 12.0)	(17.0, 15.0, 25.0)
C_7	$(x_1^{C_7})^{0.56}(x_2^{C_7})^{0.06}(x_3^{C_7})^{0.39}$	(15.0, 11.0, 5.0)	(19.0, 16.0, 13.0)
C_8	$(x_1^{C_8})^{0.24}(x_2^{C_8})^{0.38}(x_3^{C_8})^{0.38}$	(4.0, 5.0, 12.0)	(13.0, 17.0, 24.0)
C_9	$(x_1^{C_9})^{0.03}(x_2^{C_9})^{0.53}(x_3^{C_9})^{0.44}$	(9.0, 6.0, 6.0)	(18.0, 20.0, 8.0)
C_{10}	$(x_1^{C_{10}})^{0.20}(x_2^{C_{10}})^{0.15}(x_3^{C_{10}})^{0.65}$	(9.0, 6.0, 11.0)	(12.0, 14.0, 12.0)

Table 14 Initial states of producer

Agent	Input	Output	Production function
S_1	goods1	goods2	$y_2^{S_1} = 2.0(x_1^{S_1})^{0.4}$
S_2	goods2	goods3	$y_3^{S_2} = 3.0(x_2^{S_2})^{0.5}$

Table 15 Allocation

Agent	$x_1^{C_k}$	$x_2^{C_k}$	$x_3^{C_k}$
C_1	11.75	4.21	3.21
C_2	17.00	6.00	8.22
C_3	5.12	11.88	6.85
C_4	4.02	10.77	8.05
C_5	8.00	12.32	3.00
C_6	17.00	3.51	17.22
C_7	19.00	2.59	13.00
C_8	12.06	6.80	7.33
C_9	1.25	8.02	7.26
C_{10}	12.00	4.40	12.00

Table 16 Equilibrium price

$\hat{p_1}$	$\hat{p_2}$	$\hat{p_3}$
0.49284	1.3711	1.2627

It has been confrimed that all of the consumer's allocation are under their demand constraints in Table 15. Additionally it has been also clarified that the calculation time for Pareto optimal solution increases as the number of consumer agents increases in Table 12. However, the calculation time is within 13 seconds in the biggest model, and that means the calculation time is fully enough for the practical use to attain a Pareto optimal solution.

5 Conclusions

We formulated a perfectly competitive virtual market with demand constraints of consumers, and demonstrated its efficiency on resource allocation problem in this paper. Firstly we mentioned our motivation about the resource allocation problem with the demand constraints, and explained the agent formulation in the virtual market. Then market-oriented programming mechanism with the demand constraints were explained. After a brief explanation of a conventional analytical approach, named Scarf algorithm, several simulation experiments were carried out to analyse the proposed approach. Finally we have obtained the following results:

- It has been confirmed that the proposed perfectly competitive virtual market with demand constraints successfully conducts Pareto optimal solutions.
- After the investigation about the effects of demand constraint, rational resource allocation have been realised by the proposed virtual market.
- The effect of the number of agents have clarified the proposed approach is useful for practical problems both in optimality and calculation time.

The obvious next targets of this research are as follows:

- Application of the proposed approach into a practical problem with demand constraint
- Modelling extension to handle resource allocation with demand constraints both of consumers and producers

References

1. Kaihara, T., Fujii, S.: Towards intelligent artificial systems with adaptive behaviour - Sociological approach, Manufacturing Systems Division Conference 2003, pp. 61–62 (2003)
2. Arrow, K.J., Debreu, G.: Existence of an equilibrium for a competitive economy. Econometrica 22(3), 265–290 (1954)

3. Wellan, M.P.: A market-oriented Programming environment and its application to distributed multicommodity flow problems. Journal of Artificial Intelligence Research 1, 1–23 (1993)
4. Kaihara, T., Fujii, S., Yoshimura, N.: Validation on Pareto optimality of Walrasian virtual market (in Japanese). Journal of the Institute of Systems, Control and Information Engineers 17(10), 444–450 (2004)
5. Kaiahra, T., Fujii, S.: A study on modelling methodology of oligopolistic virtual market and its application into resource allocation problem. Journal of Electrical Engineers in Japan 164(1), 77–85 (2008)
6. Goldratt, E.M.: The Goal (Third edition). North River Press (2004)
7. Daal, J.V., Jolink, A.: Economic of Leon Walras, Bunkashobouhakubunsha (1994)
8. Misaki, K.: Economic Thought of Walras (in Japanese). The University of Nagoya Press (1998)
9. Toyoaki, W.: Climate policy and normal equilibrium (in Japanese), Keisoshobo (2004)

Market Participant Estimation by Using Artificial Market

Fujio Toriumi, Kiyoshi Izumi, and Hiroki Matsui

Abstract. In designing a realistic artificial market, one of the most important points to consider is the combination of agents used. In this study, we propose an estimation method based on inverse simulation to estimate the combinations of traders who participate in the market. The proposed method applies a simulation that estimates market paricipation in different markets. The simulation results indicate that the proposed method is capable of estimating market participants.

1 Introduction

Recently, it has become clear that it is difficult to explain all of the phenomena observed in markets by established market models based only on conventional human psychological reasoning. On the other hand, the study of artificial markets is one of the promising approaches to explaining such phenomena.

An artificial market is an agent-based model of a financial market. Studies on artificial markets have attained some successes in market analysis in recent years. For example, the Santa Fe Artificial Stock Market could simulate financial bubbles with agents using genetic algorithms [1, 10]. Using an artificial market, Lux explained the fat-tailed distribution of financial returns, one of the characteristic styles of financial markets [9]. Artificial markets have also been used to test existing economic theories such as efficient market hypotheses [4]. In a market design using an artificial market, Darley and

Fujio Toriumi
Graduate School of Information Science, Nagoya University
e-mail: tori@is.nagoya-u.ac.jp

Kiyoshi Izumi
Digital HumanResearchCenter, AIST

Hiroki Matsui
CMD Laboratory, Inc

Q. Bai and N. Fukuta (Eds.): Advances in Practical Multi-Agent Systems, SCI 325, pp. 201–215.
springerlink.com

Outkin examined the effect of tick-size reduction in the Nasdaq market [6]. From theoretical to empirical studies, a variety of results have been obtained by artificial markets over the past ten years [2, 3, 5].

One of the most important requirements in designing artificial markets is to create a market that is similar to the real market. If an artificial market has no resemblance to a real market, the results derived from such an artificial market would be unreliable. In particular, the combination of agents determines the reliability of an artificial market. Accordingly, when the agents' behaviors in the artificial market are highly similar to the behavior of traders in the real market, the characteristics of the artificial market become similar to those of the real market.

The combination of agents can thus be considered a key parameter of artificial markets. Therefore, the agent combination problem can be considered a parameter estimation problem. "Inverse Simulation" [8] is one of the parameter estimation methods used for social simulation, and here we propose a market-participant estimation method using the inverse simulation technique.

2 Artificial Market

2.1 Artificial Market Framework

An artificial market (e.g. [5]) is a virtual financial market run on a computer. Agents, which are computer programs that play the role of virtual dealers, participate in artificial markets.

The artificial market consists of the following components (Fig.1).

1. Market mechanism module
2. Order protocol
3. Agent interface
4. Agents

2.2 Market Mechanism Module

The market mechanism module, which is the core of an artificial market service, collects orders to buy or sell from market participants. The gathered orders are stored in its order database. Using these gathered orders, a price is determined continuously whenever a selling price (offer price) and a buying price (bid price) meet. A transaction (payment and exchange) is executed just after a price is determined, and the determined price is stored in a price database. This module can provide market data to market participants in reply to their requests. Market data include price history, trading volume, and market orders.

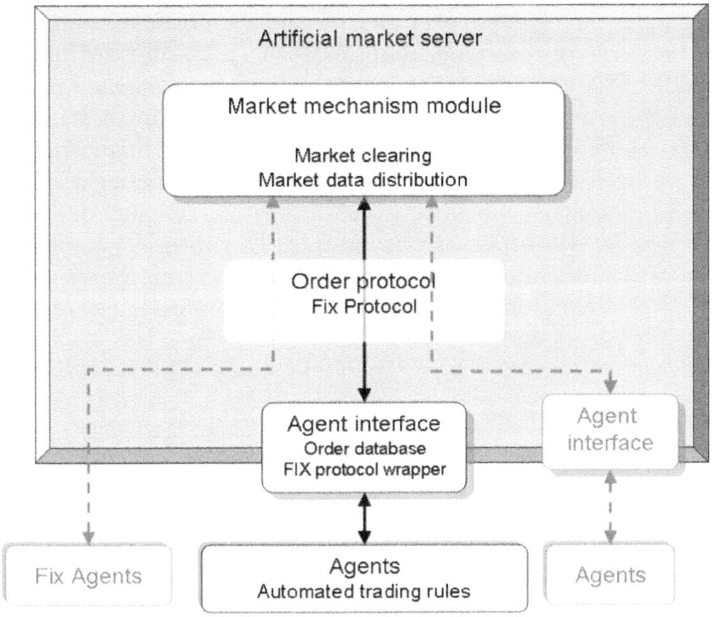

Fig. 1 Artificial market server

There are two types of common trading mechanisms:

- Continuous double auction
- Clearing house

The proposed artificial market applies a continuous double auction as its execution mechanism. Orders are executed using a price-time priority rule that first ranks orders by their price. Orders having the same price are then ranked according to when they were entered.

2.3 FIX Protocol

The leading automated trading tools include TradeStation[1], pylon[2], ZEUS[3], and OmegaChart[4]. Since these trading tools have no compatibility among themselves, it is difficult to test the programs generated by these systems in the same artificial market. To solve this issue, it would be best to use standard order protocols to communicate between trading programs and the market mechanism module.

[1] http://www.tradestation.com/
[2] http://www.pylonsoft.com/
[3] http://www.delight-web.com/zeus/
[4] http://www.omegachart.org/

Accordingly, our system uses the Financial-Information-eXchange(FIX) protocol (e.g. [7]) for communication between agents and the market mechanism module. FIX was originally a protocol for data exchanges in actual electrical trading between securities firms (sell side) and institutional investors (buy side). Many electrical broking systems in actual financial markets use the FIX protocol. The standard of FIX is open to the public and determined by the FIX Committee, which consists of members of many financial institutions and system developers. By using the FIX protocol for communication within an artificial market service, the programs in our system can be easily modified to participate in actual financial markets through electrical broking systems.

2.4 Agents

A virtual dealer that participates in an artificial market is called an "Agent." An agent is the computer program that buys and sells virtual stocks in the artificial market.

Each agent receives such market information as past stock prices, trading volumes, the results of its own orders, and so on. From this information, the agent determines an order based on its trading rules. The order is sent to the artificial market through the agent interface.

2.5 Agent Interface

The agent interface is a wrapper that transforms orders or market information to FIX messages.

An agent using the FIX protocol can directly connect to the artificial market. Consequently, an agent interface is not necessary for such an agent.

On the other hand, most automatic trading agents cannot be connected to the market directly by the FIX protocol. However, agents without the FIX protocol can still communicate order and market data through a FIX wrapper in the agent interface. The FIX wrapper then translates the agent's communication content into a FIX protocol and vice versa. The agent interface stores orders received from agents and sends them to the market mechanism module as a FIX message.

3 Market-Participant Estimation Using Inverse Simulation

3.1 Inverse Simulation

The aim of this simulation is to clarify the parameters (e.g. combination of agents) of an artificial market that is sufficiently similar to a real market.

Since there are so many parameters applicable to simulation, it is difficult to analyze all patterns of parameters comprehensively. To estimate optimal parameters, we adopted Inverse Simulation using the Genetic Algorithm. Inverse simulation is one of the techniques used for parametric estimation that have been adopted in agent-based social simulations.

Estimation by inverse simulation uses the following procedure. First, the artificial society model is designed from the mass parameters representing real society. An evaluation function is established to estimate the similarity of the artificial society to real society. Next, a set of artificial societies having different parameters is generated. Each artificial society is simulated and evaluated for its similarity to the real society. Then, the artificial societies are updated by using a genetic algorithm.

In this simulation, we assume that a market is a society and that agents are parameters. The flow of one step of the estimation simulation proceeds as follows (Figure 2).

1. Select an individual from the population.
2. Generate an artificial market from the individual.
3. Start the simulation.
4. Evaluate the artificial market.
5. Repeat 1-4 until all individuals are simulated.
6. Update the population by selection, crossover and mutation.

3.2 Artificial Market Parameters

Parameters of the artificial market include the type of participating agents. Each length of eight chromosomes defines an agent's parameters (Figure 3).

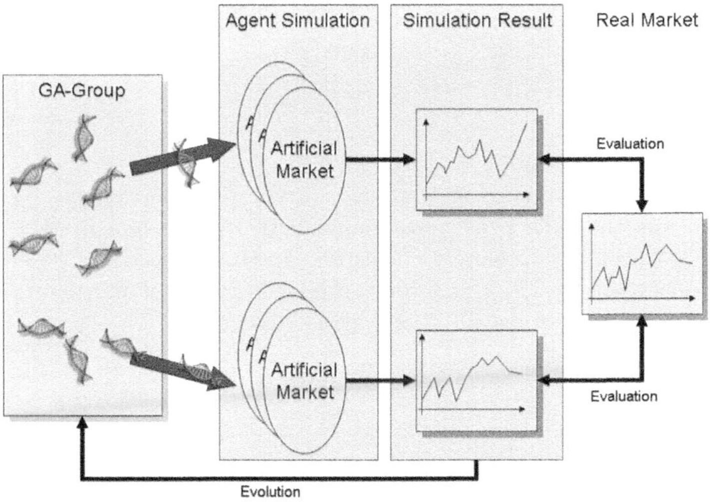

Fig. 2 Flow of Inverse Simulation using GA

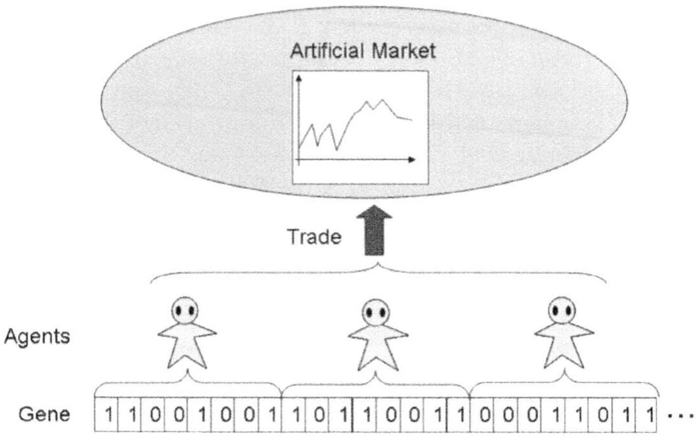

Fig. 3 Generating Agents from Gene

For example, "Active", "Buy", "Fundamentalist", "Trade each minutes", "Order 5 stocks for each trade" agent is represent to genotype "11000000". Thus, the length of a gene is $8 \times N$ (N: number of agents).

The details of the agents are described in Section 4.

3.3 Evaluation Function

To evaluate the similarity between the artificial market and the real market, we use the similarity of price movement between the real prices and the prices generated by the artificial market.

Let the real price be $P = p_1, p_2, \cdots, p_T$, and the price generated by artificial markets be $\hat{P} = \hat{p_1}, \hat{p_2}, \cdots, \hat{p_T}$. The evaluation function of artificial markets is given as follows:

$$e = \frac{\sum_{t=1}^{T}(p_t - \hat{p_t})^2}{T} \tag{1}$$

Evaluation value e shows the similarity of the price movement. When $e = 0$, this means that the price movements between the real market and the artificial market are completely consistent. In this simulation, we attempted trades five times for each parameter and used the median of evaluation values as the artificial market's evaluation value.

4 Trader Agents

In this simulation, we used two types of agents: Fundamentalists and Chartists. Each type of agent has different parameters and trade strategies. The details of each agent are described in this section.

4.1 Fundamentalist Agents

Fundamentalist agents are modeled as institutional investors. Fundamentalists have a decided target price for each stock. This may be a fundamental price or a target price for a VWAP trade. In any case, fundamentalists trade with the aim of matching a target price.

Fundamentalists trade on only one pre-determined side throughout a day. That is, buyer agents only buy and seller agents only sell stocks the whole day. Generally, institutional investors have long-term strategies, and thus they rarely change trading sides in the course of a day. The trade algorithm of fundamentalist agents is shown in Appendix 6.

4.2 Chartist Agents

Chartist agents use only charts to decide on trade matters. Chartists always take account of the daily price trends. There are two types of chartists: the Follower and the Contrarian. The follower agents trade by following the trend. If there is an upward trend, this agent places a buying order, since it forecasts that the trend will continue. The contrarian agents trade against the trend. If there is an upward trend, this agent places a selling order, since it forecasts that the trend will change.

In this simulation, the following four types of agents use different algorithms to judge the trend. Each trend-judgement algorithm is shown in Appendix 6.

1. GoldenCross Strategy
2. HLBand Strategy
3. RSI Strategy
4. BollingerBand Strategy

The trade algorithm of chartists is shown in Appendix 6.

4.3 Agent Parameters

Each agent type has different parameters. This section describes the details of each parameter.

4.3.1 Trade Frequency

Trade frequency t_f is a parameter that defines the step interval of trades. Each agent orders in t_f steps. Fundamentalist agents and Chartist agents have this parameter.

4.3.2 Activeness

Activeness A is a parameter that defines the activeness of the agents. When an *active* agent estimates the upper trend, the agent orders with a higher price limit than the current price.

The parameter *Activeness* takes one of two values, "active" or "passive." An active agent orders at a higher price than the current price for a buying order and at a lower price than the current price for a selling order. On the other hand, a passive agent orders at the current price for both buying and selling orders.

4.3.3 Trade Type

Trade type TT defines the trading side of agents, i.e. buyer or seller. Fundamentalist agents do not changes their side within a day. In other words, they only make either buy or sell orders through one simulation episode.

The parameter trade type takes one of two values, "buy" or "sell." We call agents that have the "buy" trade-type parameter "buy-side," and those that have the "sell" trade-type parameter "sell-side."

4.3.4 Target Quantity

Target quantity Q is a parameter that defines the target number of stocks. Fundamentalists try to buy/sell stocks until the held stocks satisfy target quantity Q as a limit. In the case where the number of held stocks exceeds this limit, the agent stops the trade to avoid holding more than the limit.

4.3.5 Trend Behavior

Generally, there are two types of chartists, "Followers" and "Contrarians." Followers trade by following the trend, and contrarians trade by opposing the trend. Trend behavior TB takes one of two values, "Followers" or "Contrarians."

4.3.6 Measure Span

Measure span m_t is a length of price data used to estimate the trend on a chart for Chartist agents. Each Chartist agent estimates the trend of stock prices from its trend estimation algorithm by using the given length of price data.

Table 1 shows each parameter and the values that each agent can take. Fundamentalist agents use the parameters *Trade Frequency, Belief, Trade Type, Trade Quantity,* and Chartist agents use the parameters *Trade Frequency, Belief, TrendBehavior, Measure Span.*

Table 1 Agent Parameters

	Fundamentalist	Chartists
Trade Frequency	1,5,10,30	
Activeness	Active,Passive	
Trade Type	Buy,Sell	-
Target Quantity	5,30,60,200	-
Trend Behavior	-	Followers/Contrarians
Measure Span	-	10,60

5 Simulation for Estimating Market Participants

5.1 Simulation Settings

5.1.1 Target Market

In this simulation, we use a GA-based inverse simulation to estimate market participants.

The target market is the “Nikkei 225 Futures” of October 2008 to June 2009, for data taken from certain days in every other month. The “Nikkei 225 Futures” is a stock price index for futures trading in the Osaka Securities Exchange. We selected the “Nikkei 225 Futures” for the target market because it has a sufficiently high volume of trades.

The markets in the Osaka Securities Exchange are closed from 11:00 to 12:30. Consequently, a discontinuity occurs in the sequential price data. In order to eliminate the influence of this discontinuous trading, only the data sets of tick data in the early session are used for analysis.

The target days, in the range from October 2008 to June 2009, are just after the the worldwide recession caused by the subprime financial meltdown. Figure 4 shows the price movement of the “Nikkei 225 Futures” during this period.

5.1.2 Genetic Algorithm

In this simulation, the genetic algorithm is used to estimate optimal parameters. The settings of the genetic algorithm are shown in Table 2.

5.2 Results of Simulation

First, we show the change in fitness of each generation in Figure 5. Note that fitness is defined as the distance between simulation results and real market data. Thus, a smaller fitness value means that simulation results are more similar to the real market data.

From the figure, the evaluation values of all markets are small enough to assume that the artificial markets follow the real market.

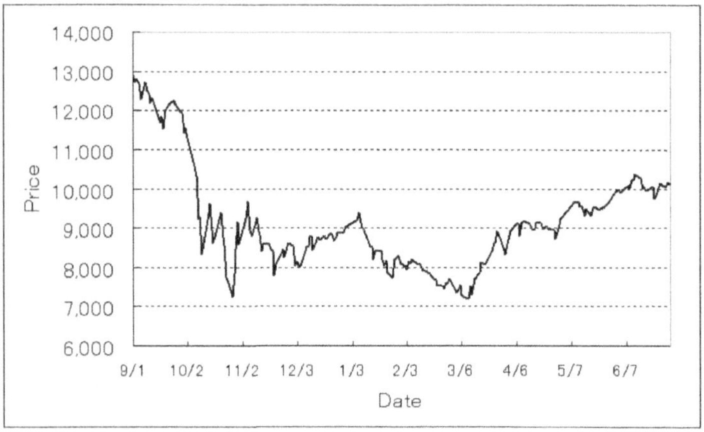

Fig. 4 Prices on Nikkei 225 Futures

Table 2 Settings of genetic algorithm

Population Size	50
Generations	100
Length of gene	400
Selection	Leave the highest-fitness individual Select 5% by roulette selection
Crossover	One-point crossover
Mutation rate	10%

Figure 6 shows the sample of price movement of the artificial market and the real market on April 1, 2009. From the figure, the simulation price rises after once dropping down, which also happened in the real market. As indicated above, the simulation prices follow the rough shape of the real price movement.

5.3 Analysis of Market Participants

First, let's consider the number of fundamentalist agents and chartist agents. Figure 7 shows the estimated rate of buy-side fundamentalist agents, sell-side fundamentalist agents, follower chartist agents, and contrarian chartists agents in each market. From the simulation results, it is estimated that there are few Followers and many Fundamentalists in the October 2008 market, and after the price volatility becomes lower, the rate of followers become larger. In September 2008, market prices collapsed due to the bankruptcy of Lehman Brothers. In the market just after such a special circumstance, it is highly likely that many ordinary investors will stop their trades to avoid the

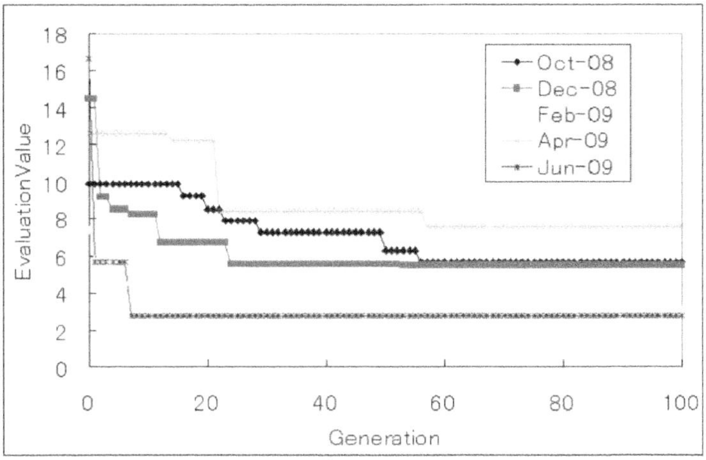

Fig. 5 Convergence of evolved societies

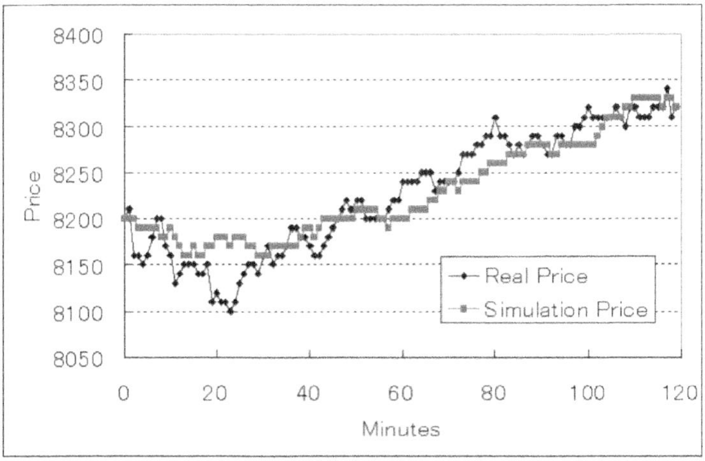

Fig. 6 Price movements in real market and in artificial market(April 2009)

risk of bankruptcy. It is possible that the simulation results express such a situation.

Next, Figure 8 shows the estimated rate of *active* agents and *passive* agents in each market. There are many *passive* agents but few *active* agents in the market in October 2008. In other markets, the rates of *active* agents and *passive* agents are roughly balanced. From these results, this simulation shows the particularity of the market in October 2008.

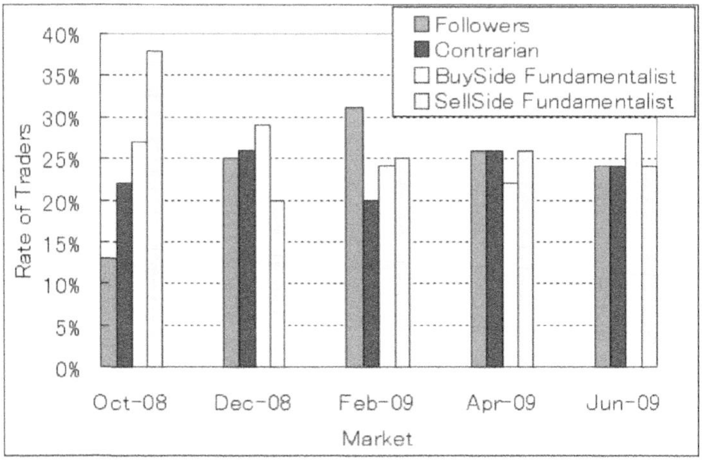

Fig. 7 Rate of Agent Types

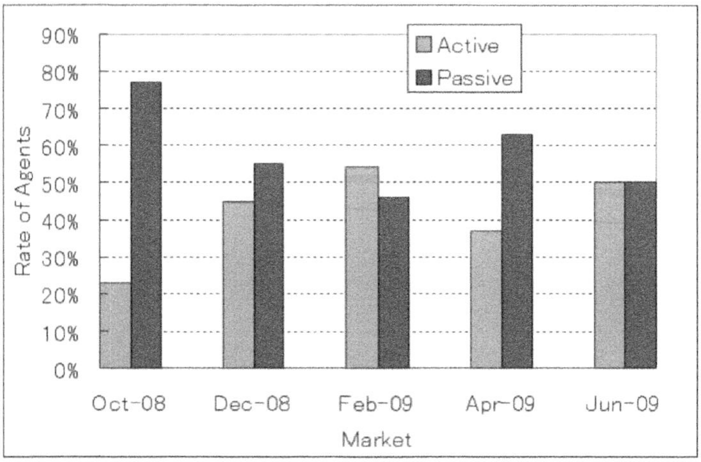

Fig. 8 Rate of Activeness in Agents

Furthermore, the *active* agents make more orders at a price that has a higher chance of trades being executed. Therefore, a low rate of *active* agents causes a smaller amount of trading volume in the market. In fact, there was actually a small amount of trading volume in the real market on the simulated day. This fact also backs up the validity of the simulation results.

From the above results, it is estimated that the types of agents in the artificial market that simulates a market under special circumstances are different from those in other markets. This observation suggests that it is possible to estimate the market participants by using the proposed method.

6 Conclusion

In this paper, we proposed a method to estimate market participants by inverse simulation. We applied the proposed method to markets after the worldwide recession caused by the subprime financial meltdown. The simulation results indicate that the types of agents in the market under such special circumstances are different from those in other markets.

However, there are several points that merit further consideration in this simulation. The first is the sufficiency of the agent types. We used three types of agents in the simulation; however, it has not been verified whether these agent types cover all types of traders. There might be traders who trades using completely different algorithms. To ability to generate enough types of agents to fully represent a real market is one of the future works of this study.

A second point is the adequacy of the evaluation function. In this paper, we use the similarity of price movements as the evaluation function. However, price movement does not reflect all characteristics of a market. For example, the standard deviation, ARCH coefficient, Skewness and Kurtosis have been used to evaluate the price movement of stocks, and a board is also expected to evaluate the market. Developing an evaluation function that can more fully express market characteristics is also a future work of this study.

References

1. Arthur, W., Holland, J., LeBaron, B., Palmer, R., Tayler, P.: Asset pricing under endogenous expectations in an artificial stock market. In: Arthur, B., Durlauf, S., Lane, D. (eds.) The Economy as an Evolving Complex System II, pp. 15–44. Addison-Wesley, Reading (1997)
2. Chen, S.-H., Wang, P.P. (eds.): Computational Intelligence in Economics and Finance. Springer, Heidelberg (2003)
3. Chen, S.-H., Wang, P.P., Kuo, T.-W. (eds.): Computational Intelligence in Economics and Finance: Volume II. Springer, Heidelberg (2007)
4. Chen, S.-H., Yeh, C.-H.: On the emergent properties of artificial stock markets: the efficient market hypothesis and the rational expectations hypothesis. Journal of Economic Behavior & Organization 49(2), 217–239 (2002)
5. Consiglio, A. (ed.): Artificial Markets Modelling: Methods and Applications. Springer, Heidelberg (2007)
6. Darley, V., Outkin, A.: A NASDAQ Market Simulation. World Scientific, Singapore (2007)
7. FIX Protocol Ltd. The FIX Guide: Implementing the FIX Protocol. Xlibris (2005)
8. Kurahashi, S., Minami, U., Terano, T.: Why not multiple solutions: Agent-based social interaction analysis via inverse simulation. In: IEEE International Conference on System, Man, and Cybernetics, page 2048 (1999)
9. Lux, T.: The socio-economic dynamics of speculative markets: Interacting agents, chaos, and the fat tails of return distributions. Journal of Economic Behavior & Organization 33, 143–165 (1998)
10. Palmer, R.G., Arthur, W.B., Holland, J.H., LeBaron, B., Tayler, P.: Artificial economic life: A simple model of a stock market. Physica D 75, 264–274 (1994)

A Fundamental Agents' Trade Algorithm

```
Type := BUYER or SELLER
if Type = BUYER then
  buy()
else if Type = SELLER then
  sell()
end if
```

B Trend Trader Agents' Trade Algorithm

```
Type := Follower or Contrarian
Trend := Upward or Downward
if Type = Follower then
  if Trend = Upward then
    buy()
  else if Trend = Downward then
    sell()
  end if
else if Type = Contrarian then
  if Trend = Upward then
    sell()
  else if Trend = Downward then
    buy()
  end if
end if
```

C Trend Estimation Algorithms

1. **GoldenCross Strategy**

   $n := 6, 30, 60$
   $MA_l(t)$:= Moving Average from $t - n$ to $t - 1$
   $MA_s(t)$:= Moving Average from $t - m/2$ to $t - 1$
   if $MA_s(t-1) < MA_l(t-1)$ and $MA_l(t) > MA_s(t)$ **then**
     Trend := Upward
   else if $MA_s(t-1) > MA_l(t-1)$ and $MA_l(t) < MA_s(t)$ **then**
     Trend := Downward
   end if

2. **HLBand Strategy**

   $n := 6, 30, 60$
   HB := $\max(p(t-1), ..., p(t-n))$
   LB := $\min(p(t-1), ..., p(t-n))$
   if $p(t)$ >HB **then**

```
    Trend := Upward
  else if p(t) <LB then
    Trend := Downward
  end if
```

3. **RSI Strategy**

```
  n := 6, 30, 60
  gain := 0
  loss := 0
  i := 0
  while i < n do
    dp := p(t − i) − p(t − (i + 1))
    if dp > 0 then
      gain := gain + |dp|
    else if dp < 0 then
      loss := loss + |dp|
    end if
    i := i + 1
  end while
  if gain/(gain + loss) > 0.75 then
    Trend := Upward
  else if gain/(gain + loss) < 0.25 then
    Trend := Downward
  end if
```

4. **BollingerBand Strategy**

```
  n := 6, 30, 60
  MA(t) := Moving Average from t − n to t − 1
  V(t) := Standard Deviation from t − n to t − 1
  if p(t) > MA(t) + V(t) then
    Trend := Upward
  else if p(t) < MA(t) − V(t) then
    Trend := Downward
  end if
```

A Study on the Market Impact of Short-Selling Regulation Using Artificial Markets

Isao Yagi, Takanobu Mizuta, and Kiyoshi Izumi

Abstract. Since the subprime mortgage crisis in the United Sates, stock markets around the world have crashed, revealing their instability. To stem the decline in stock prices, short-selling regulations have been implemented in many markets. However, their effectiveness remains unclear. In this paper, we discuss the effectiveness of short-selling regulation using artificial markets. An artificial market is an agent-based model of financial markets. We constructed an artificial market that allows short-selling and an artificial market with short-selling regulation and have observed the stock prices in both of these markets. We found that the market in which short-selling was allowed was more stable than the market with short-selling regulation, and a bubble emerged in the regulated market. We evaluated the values of assets of agents who used three trading strategies, specifically, these agents were fundamentalists, chartists, and noise traders. The fundamentalists had the best performance among the three types of agents. Finally, we observe the price variations when the market price affects the theoretical price.

Isao Yagi
Interdisciplinary Graduate School of Science and Engineering,
Tokyo Institute of Technology, J2-52, 4259, Nagatsuda-Cho, Midori-ku,
Yokohama 226-8502, Japan
e-mail: iyagi2005@gmail.com

Takanobu Mizuta
SPARX Asset Management Co., Ltd., Gate City Ohsaki,
East Tower 16F 1-11-2 Ohsaki, Shinagawa-ku, Tokyo 141-0032, Japan
e-mail: takanobu.mizuta@sparxgroup.com

Kiyoshi Izumi
Digital Human Research Center, National Institute of Advanced Industrial Science and Technology, 2-3-26, Aomi, Koto-ku, Tokyo 135-0064, Japan
e-mail: kiyoshi@ni.mints.ne.jp

Q. Bai and N. Fukuta (Eds.): Advances in Practical Multi-Agent Systems, SCI 325, pp. 217–231.
springerlink.com

1 Introduction

As a consequence of the subprime mortgage crisis and the Lehman Brothers shock in the United States, stock markets around the world have crashed, revealing their instability. To stem the decline in stock price, short-selling regulations were implemented in a number of markets. Generally, these regulations are thought to inhibit the decline of stock prices. The effectiveness of such regulations has been evaluated in previous studies [2, 9]. A report on short-selling regulation in the United Kingdom, United States, and so on did not consider the market mechanisms for the case in which short selling is restricted [8].

An artificial market that is an agent-based model of financial markets is useful to observe the market mechanism. That is, it is effective for analyzing causal relationship between the behaviors of market participants and the transition of market price. The behavior of an artificial market can be observed when useful models of actual market participants are introduced to the market. Moreover, when institution models to stem market declines, such as short-selling regulation, are introduced to the market, we can observe the behavior of the market participants. In addition, we can observe market phenomena on which the above participant behaviors have a great effect. A number of studies have investigated phenomena in financial markets using artificial markets [1, 3, 4, 6, 7].

In this paper, we discuss the effectiveness of short-selling regulation using an artificial market. For simplicity, in Sect. 2, we constructed an artificial market with short-selling regulation and an artificial market that allows short-selling. We hereinafter refer to the market with short-selling regulation as the regulated market and to the market that allows short-selling as the unregulated market. The trading strategy models of these market participants (agents) are defined in Sect. 2.1 and Sect. 2.2. In Sect. 3, we analyze the market mechanisms based on the results of the artificial markets simulations. First, in Sect. 3.1, we observe the stock price variations of two markets. The unregulated market was found to be more stable than the regulated market, and a stock market bubble emerged in the regulated market. By comparing our proposed market to past actual market data (TOPIX), we demonstrate that our artificial market has some properties of actual markets. Next, we observe agent transactions in Sect. 3.2. We analyze how agent transactions influence our two markets and why a bubble emerges in the regulated market. Then, we discuss the performances of agents in our two markets. The proposed market considers agents with three trading strategies: fundamentalists, chartists and noise traders. We demonstrate that fundamentalist have the best performance among the three agent types. The theoretical price in above market model does not depend on the market price. In Sect. 3.3, we observe price variations under the market model in which the market price affects the theoretical price.

2 Construction of Artificial Markets

The proposed artificial markets consist of one hundred agents who each have one million yen and 10 shares as initial assets. All agents sell or buy only one company. Each of the agents is assigned a fundamentalist, chartist, or noise trader trading strategy. In the simulation, the ratio of fundamentalists, chartists, and noise traders is 45:45:10[1]. The maximum number of shares that each agent can hold is $S_{max}(= 1,000)$. In the unregulated market, the maximum number of short-sold shares is -$S_{min}(S_{min} =$ -$1,000)$, that is, each agent can hold the number of shares S that satisfies -$1,000 \leq S \leq 1,000$. However, agents are not allowed to straddle, but they can short-sell if they do not hold any shares. On the other hand, each agent in the regulated market can hold S that satisfies $0 \leq S \leq 1,000$.

2.1 Agent Models

In this section, the trading rules for the three trading strategies are presented.

2.1.1 Fundamentalist

In our markets, fundamentalists predict current market price based on theoretical price and control the number of shares that they hold in order to maximize their asset values. Theoretical price is a given value and is readjusted periodically. The initial theoretical price is set to 300 and readjusted according to a normal distribution $N(1, 0.1^2)$ every 1,000 periods.

Let the expected stock price of agent i at period t be $\tilde{P}_{i,t}$, and let the theoretical price be $\mathcal{P}_t$. Then, $\tilde{P}_{i,t}$ is decided according to $N(\mathcal{P}_t + \epsilon_{i,t}, (\alpha(\mathcal{P}_t + \epsilon_{i,t}))^2)$ and readjusted every 100 periods, where $\epsilon_{i,t}$ denotes the degree of bullishness[2] of i at t, and α is a coefficient that describes the degree of spread of expected prices, which is decided by each agent, and is set to 0.1. $Q_{i,t-1}$ denotes the amount of cash that i holds before a transaction at t. $q_{i,t-1}$ denotes the amount of shares that i holds before a transaction at t. P_{t-1} denotes the market price at $t-1$. The assets of agent i before a transaction at t are as follows:

$$W_{i,t-1} = Q_{i,t-1} + P_{t-1} \cdot q_{i,t-1}. \tag{1}$$

[1] The selected ratio is based on the intuitions of institutional investors that was further confirmed after getting properties of artificial markets similar to that of actual markets when the simulation was executed with the selected ratio (in Sect. 3.1). The values for other parameters used in this paper are selected on the basis of same criterion.

[2] Since a bullish agent expects a higher price than the theoretical price, this value is positive for a bullish agent. On the other hand, for a bearish agent, who expects a lower price than the theoretical price, this value is decided negative.

Let the amount of shares held by i at t that maximizes the subjective expected utility under condition (1) be $\tilde{q}_{i,t}$. Then, $\tilde{q}_{i,t}$ is as follows:

$$\tilde{q}_{i,t} = \frac{\mathcal{P}_t + \epsilon_{i,t} - P_{t-1}}{a(\alpha(\mathcal{P}_t + \epsilon_{i,t}))^2}.$$

Note that a constant $a(>0)$ is a coefficient of risk aversion and the larger the value of a is, the smaller the number of shares that fundamentalists hold to averse risk. Fundamentalists attempt to decide their trading rules based on $\tilde{q}_{i,t}$. If $\tilde{q}_{i,t} > q_{i,t-1}$, i places an order to buy $\tilde{q}_{i,t} - q_{i,t-1}$ shares at $\tilde{P}_{i,t}$, where the maximum number of shares that can be bought is $S_{max} - q_{i,t-1}$. If $\tilde{q}_{i,t} < q_{i,t-1}$, i places an order to sell $q_{i,t-1} - \tilde{q}_{i,t}$ shares at $\tilde{P}_{i,t}$, where the maximum number of shares that can be sold in the unregulated market is $q_{i,t-1} - S_{min}$, and the maximum number of shares that can be sold in the regulated market is $q_{i,t-1}$. If $\tilde{q}_{i,t} = q_{i,t-1}$, i does not trade at period t.

2.1.2 Chartists

In our markets, chartists trade based on the moving averages of stock prices. Chartists consist of market followers and contrarians. Let $n_{i,t}$ be a number that satisfies $1 \leq n_{i,t} \leq 25$. Let the $n_{i,t}$ period moving average at t used by agent i be

$$MA_{t,n_{i,t}} = \frac{1}{n_{i,t}} \sum_{j=1}^{n_{i,t}} P_{t-j},$$

and let $\Delta MA_{t,n_{i,t}} = MA_{t,n_{i,t}} - MA_{t-1,n_{i,t}}$. Market followers place orders based on the following rules.

- If $\Delta MA_{t,n_{i,t}} > 0$, then i places an order to buy $q_{i,t}^T$ shares at price $(1+\alpha_t)P_{t-1}$.
- If $\Delta MA_{t,n_{i,t}} < 0$, then i places an order to sell $q_{i,t}^{'T}$ shares at $(1+\alpha_t)P_{t-1}$.
- If $\Delta MA_{t,n_{i,t}} = 0$, then i does not trade.

Contrarians place orders based on the following rules.

- If $\Delta MA_{t,n_{i,t}} > 0$, then i places an order to sell $q_{i,t}^{'T}$ shares at $(1+\alpha_t)P_{t-1}$.
- If $\Delta MA_{t,n_{i,t}} < 0$, then i places an order to buy $q_{i,t}^T$ shares at $(1+\alpha_t)P_{t-1}$.
- If $\Delta MA_{t,n_{i,t}} = 0$, then i does not trade.

Let the initial $n_{i,t}$ be a random number that satisfies $1 \leq n_{i,t} \leq 25$, and let α_t be a random number according to $N(P_{t-1}, 0.1^2)$. Let $q_{i,t}^T$ be a random number according to $N(10, 0.1^2)$, where $0 < q_{i,t}^T \leq S_{max} - q_{i,t-1}$ when i places an order to buy. Let $q_{i,t}^{'T}$ be a random number according to $N(10, 0.1^2)$, where $0 < q_{i,t}^{'T} \leq q_{i,t-1} - S_{min}$ when i places an order to sell in the unregulated market and $0 < q_{i,t}^{'T} \leq q_{i,t-1}$ when i places an order to sell in the regulated market.

2.1.3 Noise Traders

Noise traders place orders to buy, sell, and wait with equal probabilities. At t, agent i places an order to buy $q_{i,t}^{N}$ shares at price $(1+\alpha_t)P_{t-1}$, where $q_{i,t}^{N}$ is a random number according to $N(10, 0.1^2)$ and satisfies $0 < q_{i,t}^{N} \leq S_{max} - q_{i,t-1}$. At t, i places an order to sell $q_{i,t}^{'N}$ shares at $(1+\alpha_t)P_{t-1}$, where $q_{i,t}^{'N}$ is a random number according to $N(10, 0.1^2)$ and satisfies $0 < q_{i,t}^{'N} \leq q_{i,t-1} - S_{min}$ in the unregulated market and $0 < q_{i,t}^{'N} \leq q_{i,t-1}$ in the regulated market.

2.2 Modeling of an Agent Influenced by the Trading Rules of High-Performance Agents

After a transaction at period t, all agents evaluate their performances. Each agent whose performance is lower than that of any other agent attempts to learn the trading rules of the high-performance agents. Moreover, high-performance agents who would like to improve their performance also learn trading rules.

However, agents cannot change their trading strategies to other strategies, because the ratio of the three strategies in our markets is fixed. For example, a low-performance fundamentalist attempts to learn a trading rule of a high-performance fundamentalist, and not a high-performance chartist. Noise traders do not evaluate their performance or learn the trading rules of other agents.

2.2.1 Fundamentalists

Let the ratio of change of the assets of fundamentalist agent i from period $t-1$ to t be $R_{i,t} = W_{i,t}/W_{i,t-1}$. When the mean ratio of change of the assets of i in the past $N(=5)$ periods

$$\overline{R}_{i,t} = \frac{1}{N}\sum_{j=1}^{N} R_{i,t-(j-1)}$$

ranks in the bottom $N_L(=20)\%$ of all fundamentalists, $\epsilon_{i,t}$ is changed with probability

$$p_i = \frac{\text{Rank of performance of agent } i}{\text{Total number of fundamentalists}}.$$

That is, i', whose $\overline{R}_{i',t}$ ranks in the top $N_H(=20)\%$ of all fundamentalists, is chosen at random, and $\epsilon_{i,t+1} = \overline{\epsilon}_{i',N}$ is set, where

$$\overline{\epsilon}_{i',N} = \frac{1}{N}\sum_{j=0}^{N} \epsilon_{i',t-j}.$$

To implement a change of trading rules of agents who would like to improve their performance, for agents whose $\overline{R}_{i,t}$ does not rank in the top $N_H\%$, their degree of bullishness is changed with probability 5% at random.

2.2.2 Chartists

When the mean ratio of change of the assets of chartist agent i in the past $N(=5)$ periods $\overline{R}_{i,t}$ ranks in the bottom $N_L(=20)\%$ of all chartists, the moving average period $n_{i,t}$ is changed with the following probability.

$$p_i = \frac{\text{Rank of performance of agent } i}{\text{Total number of chartists}}.$$

That is, i', whose $\overline{R}_{i',t}$ ranks in the top $N_H(=20)\%$ of all chartists, is chosen at random, and $n_{i,t+1} = n_{i',t}$ is set.

To implement a change of trading rules of agents who would like to improve their performance, for agents whose $\overline{R}_{i,t}$ does not rank in the top $N_H\%$, their trading rules (market followers and contrarians) and/or their moving average periods are changed with probability 5% at random, where moving average periods satisfy $[1, 25]$.

2.3 Equilibrium Price

Each agent determines the price and quantity of an order based on his trading strategy and places the order. The market price at period t is determined when the sell order prices and buy order prices of each of agent agree at t.

3 Discussion

3.1 Price Variation in the Markets

We observed market price variation in the unregulated and regulated markets (refer to Fig. 1 and Fig. 2). These figures reveal that prices in the unregulated market did not diverge significantly from the theoretical price. On the other hand, prices in the regulated market diverged greatly from the theoretical price and showed repeated sharp increases and decreases. In the regulated market, once the price rose sharply, it never returned to the theoretical price. These indicate that the short-selling regulation has the property that it not only stems the decline in the prices but also increases the prices excessively.

However, if our market models do not reflect the actual markets, the result is meaningless. Thus, we discuss whether our market models reflect the actual markets. The price variation statistics of our markets were first compared with those of past market data to analyze the problem. We used the Tokyo stock price index (TOPIX) from 16th May 1949 to 16th June 2009 (14,926

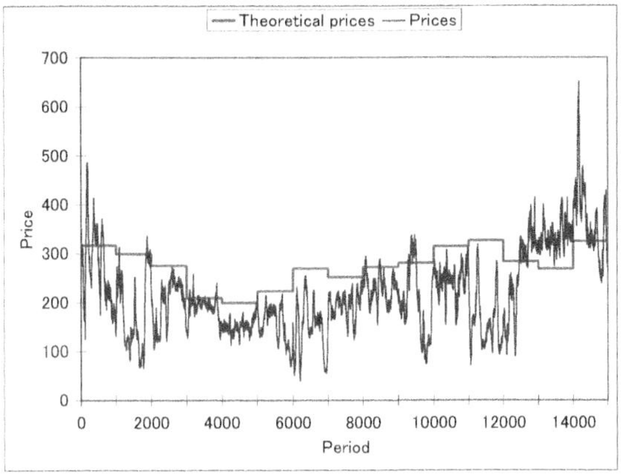

Fig. 1 Price variation in the unregulated market

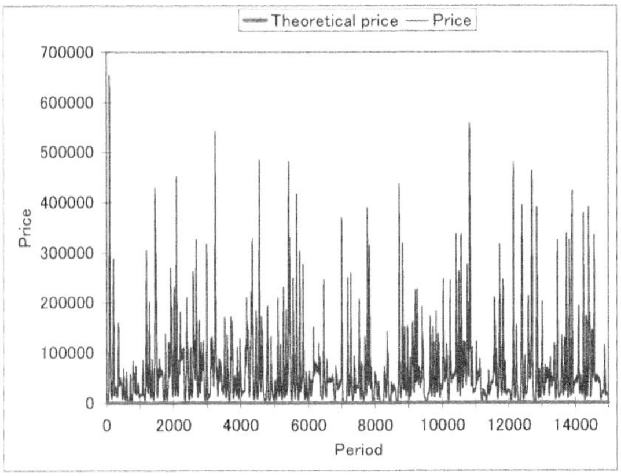

Fig. 2 Price variation in the regulated market

days) as past market data. Table 1 shows the parameters of the rate of return in these markets. Fig. 3, Fig. 4, and Fig. 5 show histograms of the rate of return in the unregulated market, the rate of return in the regulated market, and the rate of return of the TOPIX, respectively. The distribution of the rate of return of actual data is fat-tailed. (The kurtosis of the rate of return distribution of the TOPIX is 11.673.) Since the kurtosis of the rate of return distribution of the unregulated market is 128.65 and that of the regulated market is 0.50732, the distributions of the rates of return of our markets are also fat-tailed. The kurtosis of the unregulated market is larger than that of the regulated market, as the standard deviation (SD) of the rate of return in the unregulated market is smaller than that in the regulated market. Therefore, we can consider that the price variation in the unregulated market is more stable than that in the regulated market.

Since short-selling is seldom regulated in actual markets, a histogram of the rate of return in the unregulated market is more similar to that in the TOPIX than that in the regulated market. This indicates that our market models reflect the actual markets.

Table 1 Parameters of rate of return

	Unregulated market	Regulated market	TOPIX
Mean	0.00045	0.00317	0.00031
Standard deviation	0.03098	0.07745	0.01064
Kurtosis	128.65	0.50732	11.673
Skewness	3.8347	0.10937	-0.13232

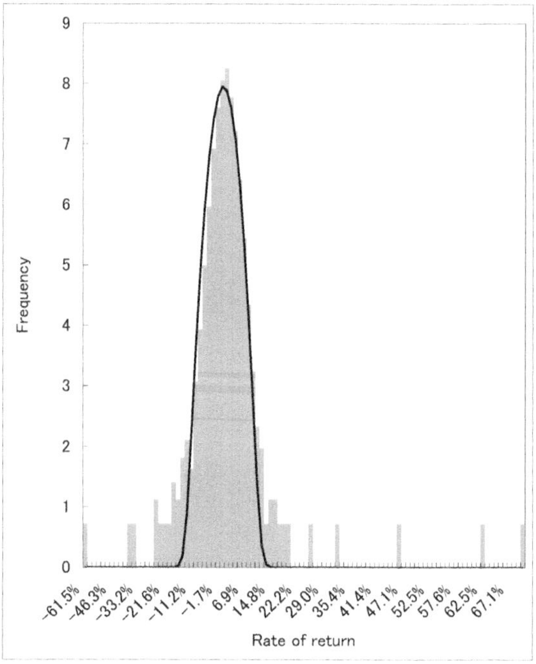

Fig. 3 Histogram of the rate of return in the unregulated market

We compare the autocorrelation of the rates of return of our markets with the TOPIX. The autocorrelation coefficient of the rates of return of the TOPIX converges to 0, and the autocorrelation coefficient of the squared rates of return the TOPIX is large and is not attenuated quickly (refer to Fig. 6 and Fig. 7). The autocorrelation coefficient of the rates of return in the unregulated market converges to 0, and the autocorrelation coefficient of the squared rates of return in the unregulated market converges to 0 faster than that of the TOPIX. However, the autocorrelation coefficient of the squared rates of return in the unregulated market converge to 0 slower than that of rates of return in it. The above properties are consistent with the properties of actual markets. The autocorrelation coefficient of the rates of return in the

Fig. 4 Histogram of the rate of return in the regulated market

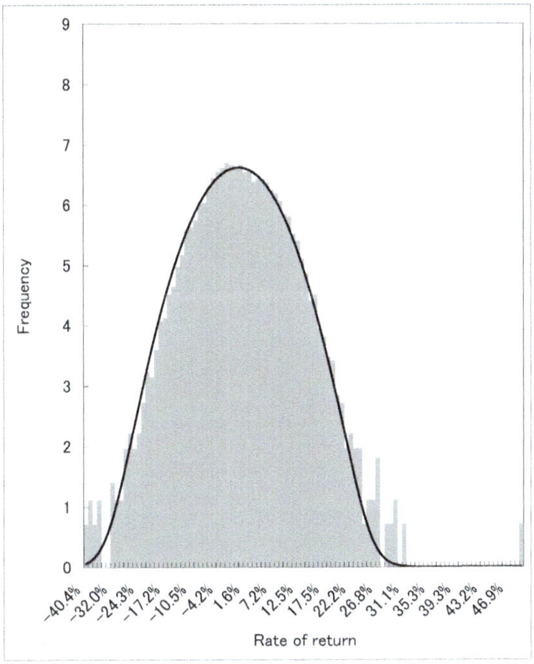

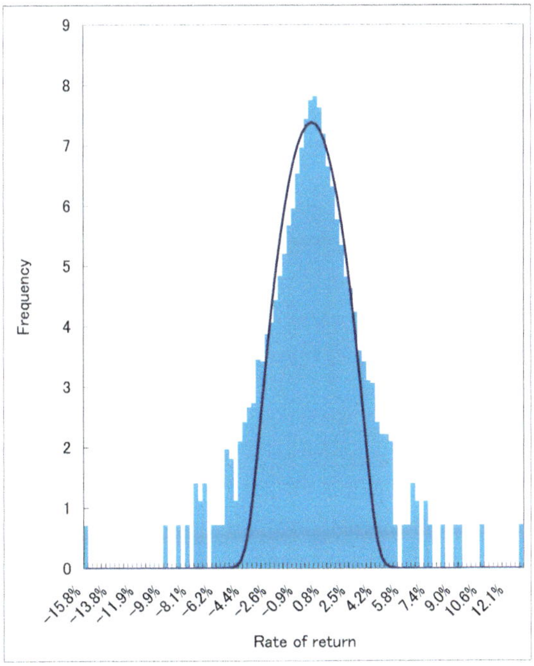

Fig. 5 Histogram of the rate of return of the TOPIX

Fig. 6 Autocorrelation of rates of return

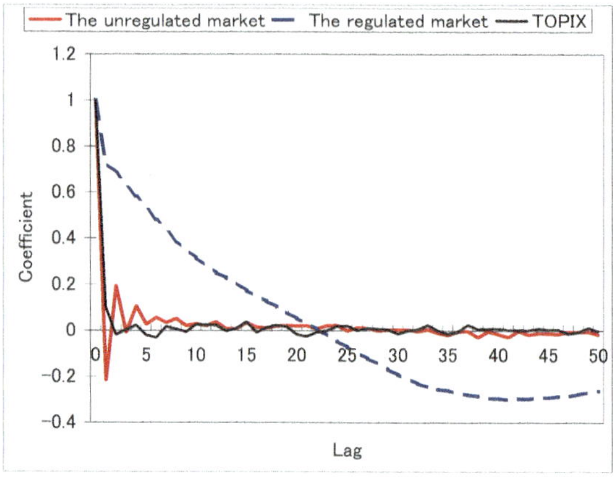

Fig. 7 Autocorrelation of squared rates of return

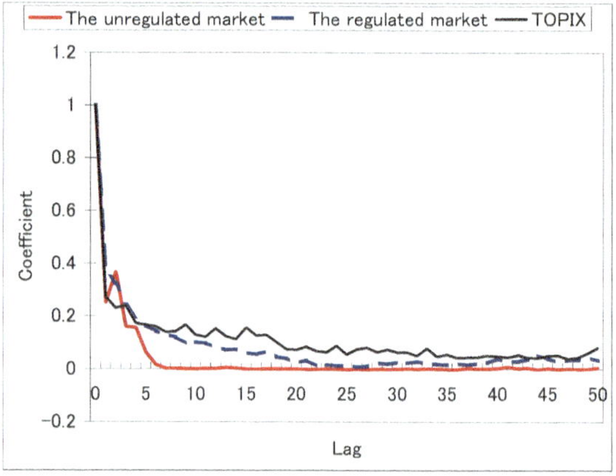

regulated market decreases slowly, which indicates that the rates of return depend on past rates of return. Therefore, the volatility of returns in the regulated market is considered to increase.

3.2 Agent Transactions

In this section, we discuss agent transactions in our markets. We analyze how agent transactions affect the markets and the difference in agent performance between the unregulated market and the regulated market.

3.2.1 Trends of Agent Transactions

Fig. 8 shows the transitions of the positions of agents in the unregulated market. The position rate of fundamentalists correlates inversely with the position rate of chartists (-0.902). Chartists can be considered as trend followers, because fundamentalists basically trade against the trend.

Fig. 8 Transitions of position rate of agents in the unregulated market

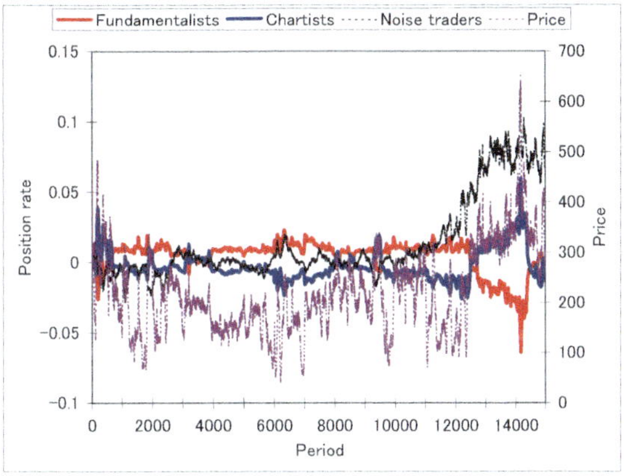

On the other hand, the positions of chartists correlate directly with market price (0.839) in the regulated market. In addition, in the regulated market, fundamentalists seldom trade is found. Fundamentalists sell all of their shares upon the emergence of a bubble. However, the price is seldom lower than the theoretical price, even if the bubble collapses. Therefore, they cannot buy shares after the first bubble. The reason for this is not analyzed in this paper.

3.2.2 Bubble in the Regulated Market

The mechanism of the emergence and collapse of a bubble in the regulated market can be considered as follows.

Generally, when price begin to rise, most chartists tend to place buy orders with the upward trend, because the trade rule of high-performance chartists is to place orders to buy with the upward trend. As stated above, in the unregulated market, the positions of fundamentalists is inversely correlated with the positions of chartists. Thus, price is prevented from increasing sharply, because many sell orders (including short-selling orders) placed by fundamentalists [3] match the buy orders of chartists, which may cause the price to increase sharply. However, in the regulated market, as the price begins to rise, fundamentalists sell all of their shares and do not trade, because they cannot place short-sell orders. Therefore, the price increases sharply, because the large number of buy orders from chartists causes the equilibrium price to increase.

As price increases, an increasing number of agents cannot place buy orders because of a shortage of cash. As a result, the number of buy orders decreases, buy orders with comparatively low prices match sell orders, and price begins to decrease (refer to Fig. 9).

As price begins to decrease, some chartists place orders to sell as the downward trend begin to increase their assets. Thus, most chartists begin

[3] Since fundamentalists basically trade against the trend, they tend to place sell orders if the price rises.

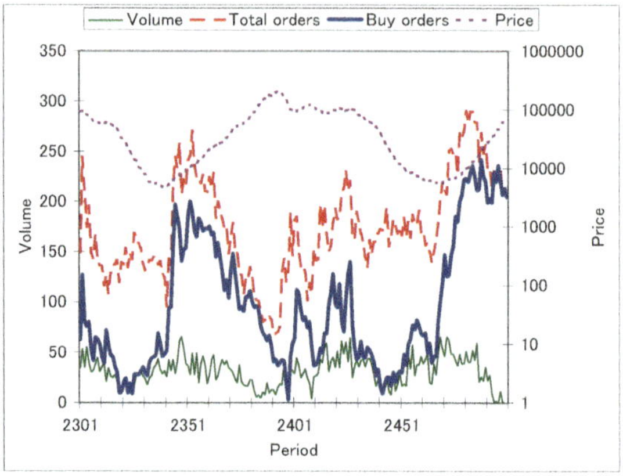

Fig. 9 Transitions of buy-sell ratio of agents in the regulated market from period 2,301 to period 2,500

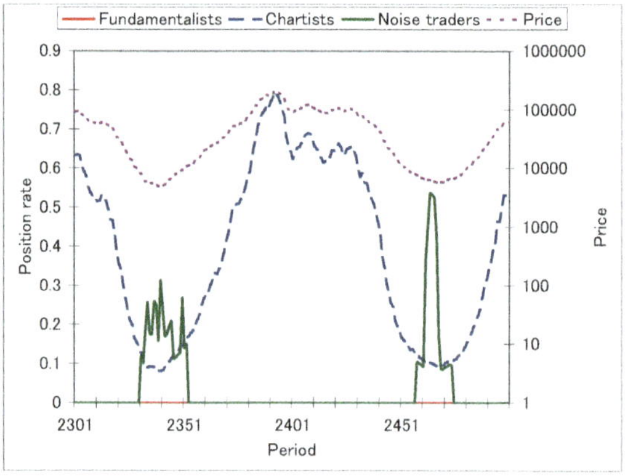

Fig. 10 Transitions of agent position in the regulated market from period 2,301 to period 2,500

to place sell orders as the trend and price decrease. When the price has fallen sufficiently, agents who were unable to place buy orders due to a cash shortage become able to buy some shares and price ceases to decrease (refer to Fig. 9). Fig. 10 indicates that noise traders place orders to buy and their positions begin to increase around the time prices cease to decrease.

Regarding the mechanism of the bubble emergence, in the theory of finance, there is reported to be a high possibility of bubble emergence if short-term investors or noise traders exist in markets [5, 10] and prices affect fundamental values [11].

3.2.3 Agent Performance

Table 2 shows performance of each agent type in the markets. These values are the arithmetic means of various values of SD obtained from 10 simulations in each market. In the unregulated market, fundamentalists and noise traders make profits, whereas chartists report losses. In the regulated market,

Table 2 Performance of each type of agent

	Unregulated market	Regulated market
Fundamentalists	14.670%	0.03906%
Chartists	-15.077%	187.54%
Noise traders	1.1808%	-99.752%

Table 3 Return ranking of chartists and noise traders in the regulated market

	Chartists		Noise traders	
	Top	Bottom	Top	Bottom
1	1604.8%	-99.238%	-99.602%	-99.921%
2	492.37%	-98.742%	-99.718%	-99.915%
3	389.51%	-98.524%	-99.789%	-99.889%
4	378.98%	-97.864%	-99.806%	-99.873%
5	264.84%	-96.644%	-99.830%	-99.832%

fundamentalists make slight profits, chartists make large profits, and noise traders lose most of their assets. Table 3 shows the five highest and five lowest returns of chartists and noise traders, respectively, in the regulated market. Not all chartists make large profits, and all noise traders lose most of their assets. The strategies of chartists in the regulated market can be considered to be high-risk-high-return strategies.

Therefore, in both markets, the strategies of fundamentalists have the most stable performance among three agent types.

3.3 The Market Model in Which the Market Price Affects the Theoretical Price

In the above market model, the market price does not affect the theoretical price. Fundamentalists in actual markets decide orders' price and the number of shares based on theoretical price that depend not only on fundamental elements but also on the relation of supply and demand of stocks, when the price diverges greatly from the theoretical price. In this section, we consider the market model in which the market price affects the theoretical price and observe price variances in both artificial markets. Let α_p be a coefficient to describe link between the theoretical price and the market price. The new theoretical price is decided using the following rule.

$$\mathcal{P}_{t+1} = \mathcal{P}_t + \alpha_p(P_t - \mathcal{P}_t)$$

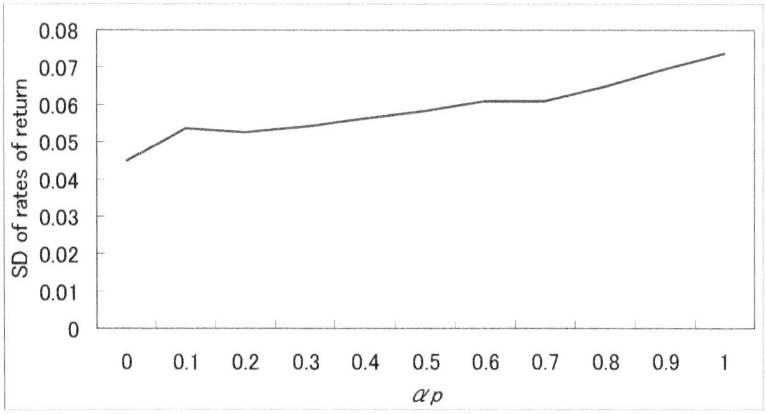

Fig. 11 The SD of rates of return in the unregulated market under the new theoretical price model

where P_t is more than three times of $\mathcal{P}_t$ or less than one third of $\mathcal{P}_t$. Fig. 11 shows the SD of rates of return in the unregulated market when α_p ranges from 0 to 1.0 at intervals of 0.1. These values are the arithmetic means of various values of SD obtained from the execution of 10 simulations in the unregulated market. The SD of the rates of return increases with increasing value of α_p. This means that the more market prices affect theoretical prices, the more instable markets are. Therefore, this property bears out Soros' prospect [11].

In addition, we do not have the scenario in which the market price is more than three times of the theoretical price in the unregulated market.

In the regulated market, we do not have any features under the new theoretical price model.

4 Conclusion

In this paper, we discussed the effectiveness of short-selling regulation using an artificial market. We first constructed regulated and unregulated markets. We found that the unregulated market was more stable than the regulated market, and that a bubble emerged in the regulated market. This indicates that the short-selling regulation has the property that it not only stems the decline in the prices but also increases the prices excessively. To confirm the adequacy of our market models, we compared the price variation statistics of our markets with those of past market data (TOPIX) and found that our market models reflected the actual markets. Next, we discussed agent performance in the two markets and showed that the performance of fundamentalists was the best among the three agent types. Moreover, we discussed the reason why a bubble emerges in our market models. Finally, we discussed the price variations under the market model in which the market price affects the theoretical price. In the future, we intend to clarify the reason why market price diverges from theoretical price in the unregulated market, prospect the price transition for the case in which the ratio of the three strategies (fundamentalists, chartists, and noise traders) in our markets change flexibly, identify whether properties of the artificial markets using other values for each parameter (e.g., the maximum number of shares) satisfy that of the actual markets, and discuss the effectiveness of other regulations (e.g., leverage regulation) using our market models.

References

1. Arthur, W., Holland, J., LeBaron, B., Palmer, R., Tayler, P.: Asset pricing under endogenous expectations in an artificial stock market. In: The Economy as an Evolving Complex System II, pp. 15–44. Addison-Wesley, Reading (1997)
2. Arturo, B., Goetzmann, W.N., Zhu, N.: Efficiency and the bear: short sales and markets around the world, Yale ICF Working Paper No.02-45 (2004)
3. Chen, S.-H., Yeh, C.-H.: On the emergent properties of artificial stock markets: the efficient market hypothesis and the rational expectations hypothesis. Journal of Economic Behavior & Organization 49(2), 217–239 (2002)
4. Darley, V., Outkin, A.V.: A NASDAQ Market Simulation: Insights on a Major Market from the Science of Complex Adaptive Systems. World Scientific Pub. Co. Inc., Singapore (2007)
5. De Long, J.B., Shleifer, A., Summers, L.H., Waldmann, R.J.: Noise Trader Risk in Financial Markets. Journal of Political Economy 68(4), 703–738 (1990)
6. Hara, A., Nagao, T.: Construction and analysis of stock markets model using ADG; automatically defined groups. International Journal of Computational Intelligence and Application 2(4), 433–446 (2002)
7. Izumi, K., Toriumi, F., Matsui, H.: Evaluation of automated-trading programs using an artificial market. NeuroComputing (forthcoming)
8. Marsh, I.W., Niemer, N.: The impact of short sales restrictions, The London Investment Banking Association (2008)
9. Saffi, P.A.C., Sigurdsson, K.: Price efficiency and short-selling, AFA 2008 New Orleans Meeting Paper (2007)
10. Shiller, R.J.: Bubbles, Human Judgement, and Expert Opinion. Financial Analysts Journal 58, 18–26 (2001)
11. Soros, G.: The Alchemy of Finance. John Wiley & Sons. Inc., Chichester (2003)

Learning a Pension Investment in Consideration of Liability through Business Game

Yasuo Yamashita, Hiroshi Takahashi, and Takao Terano

Abstract. While the importance of financial education is recognized in recent years, the technique for deepening an understanding to pension investment management is needed. In this research, we analyze learning method of the pension investment management in consideration of liability using the business game technique. As a result of analysis, interesting phenomena – the participant understood the learning method of the pension investment management in consideration of liability – were seen. This shows the effectiveness of the business game technique to learning the pension investment management.

1 Introduction

The company has introduced the pension system from the character of the complement of social security and the mitigation of the burden of individual security which retirement benefits[1] mean, or a personnel policy. In the pension system, although the organization which became independent of corporate management is usually performing investment and management, the sponsoring company takes the final risk of management. In Japan, it became clear that the shortfall to projected benefit obligation[2] comes to be added up

Yasuo Yamashita · Takao Terano
Department of Computational Intelligence and Systems Science,
Tokyo Institute of Technology
e-mail: yamasita.y.aa@m.titech.ac.jp

Hiroshi Takahashi
Graduate School of Business Administration, Keio University

[1] Retirement benefits say the thing of retirement lump sum grants or a business annuity in Japan.
[2] It is also called PBO. It is the current value of the Retirement benefits which a company will pay to employees in the future.

Q. Bai and N. Fukuta (Eds.): Advances in Practical Multi-Agent Systems, SCI 325, pp. 233–249.
springerlink.com

by the balance sheet of a sponsoring company as a debt, and the sponsoring company takes the risk of pension-assets investment by the Retirement-benefits accounting standards introduced from the 2000 fiscal year. Under the present accounting system, the measure called delay recognition is taken.

It eases the influence on an earning statement in which what is necessary is just to refund the unfunded liabilities by reduction in pension assets in a fixed period (about ten to 15 years) after a following term. However, it is in the tendency for the method that unfunded liabilities are appropriated for a generating fiscal year the total amount in the future to be adopted, and a possibility is increasing, that change of pension assets will affect the achievements of a sponsoring company.

In such a situation, the pension-assets investment persons concerned[3] think that they need the pension-assets investment technique in which the influence to the achievements of a sponsoring company is also taken into consideration. As opposed to the problem how to maintain the balance of pension assets and a pension liability regarding pensionary risk management in finance theory, it has tackled as the problem of pension-assets liability management (ALM), or a problem of pension surplus management [4]. By old research, it has been thought that the asset allocation method that a high return, a low risk, and the risk of pension liability is hedged, is important for a pension-assets investment. Therefore, the risk of pension liability was hedged by duration management[4] of pension assets, and the technique of considering realization of a high return and a low risk as an asset allocation problem in traditional pension assets has been chosen.

However, by this technique, the influence of the achievements on a sponsoring company is not taken into consideration. What is necessary is just to, lessen change of a surplus (difference of pension assets and pension liability) if possible, in order to lessen influence of the achievements on a sponsoring company as much as possible. For that purpose, it is required to choose the asset allocation method which minimizes the tracking error to the pension liability return of a pension-assets return. Since this asset allocation method differs from the conventional asset allocation method in business, an understanding is difficult for it for practitioners. The importance of different education from the learning method by classroom lecture regarding the pension asset management in consideration of liability is increasing.

Meanwhile, in the research through the business game which is one field of a gaming simulation [2], although research on management or marketing is done briskly [6] [7] [9], it is in the situation which is hard to be referred to

[3] They are not only the corporate pension fund which maneges pension system and the institutional investor to whom the pension-assets investment is entrusted, but the financial department of a sponsoring company and so on.

[4] There are simulation analyses as the technique of treating the risk of pension liability in detail [3] [8]. There is a fault which is going to express the structure of a complicated pension delicately of becoming complicated too much and becoming rather unclear.

as that research on a finance is done enough [1]. So far, some researches are made regarding learning through business game as the technique of the financial education in the finance field [10] [11]. However, it cannot say that the quantity of research is still enough, and research on learning of the pension asset management in consideration of liability is not done, either. Therefore, the approach of this research in consideration of liability on learning of pension asset management can be called meaningful thing. In this research, it is shown clearly by experiencing a business game that practitioners learn the difficult asset allocation method effectively.

In this research, in order to aim at showing the technique of learning the pension-assets investment in consideration of liability using the framework of business game, first, the model of a business game is built, next it experiments by using an actual human being as a player. After explaining the model used for analysis in the following chapter, results are shown in chapter 3. Chapter 4 is Conclusion.

2 Method

2.1 System of Business Game

The environment required for development of the system in this research, and execution of an experiment is constituted by business model descriptive language (BMDL) and business model development system (BMDS) [5]. BMDL is the programming descriptive language of a short form. HTML files, CGI files, and so on can be created in BMDS by describing BMDL for game managers (facilitator) and for game users (player). A business model written in BMDL is first translated into corresponding CGI scripts and C programs by BMDL translator. The programs are then run on a host computer with a WWW server and the model variable data in the form of spreadsheets. Finally, players execute the business game through browsers on the WWW environment. Fig.1 shows development and execution environment of an experiment. Players input decision-making in each round through WWW browsers, and a facilitator also advances a game through a WWW browser.

2.2 Business Game Model

In this research, business game regarding asset management in consideration of pension liability is performed. Fig.2 shows the outline of the business game in this research. A player was going to be a portfolio manager of a pension fund. He or she decides the asset allocation[5] of 5 assets of domestic long-term

[5] Asset allocation is determined that the ratio sum total of the assets will be 100%. The ratio of each asset is 0% or more of 100% or less. Short selling is an impossible.

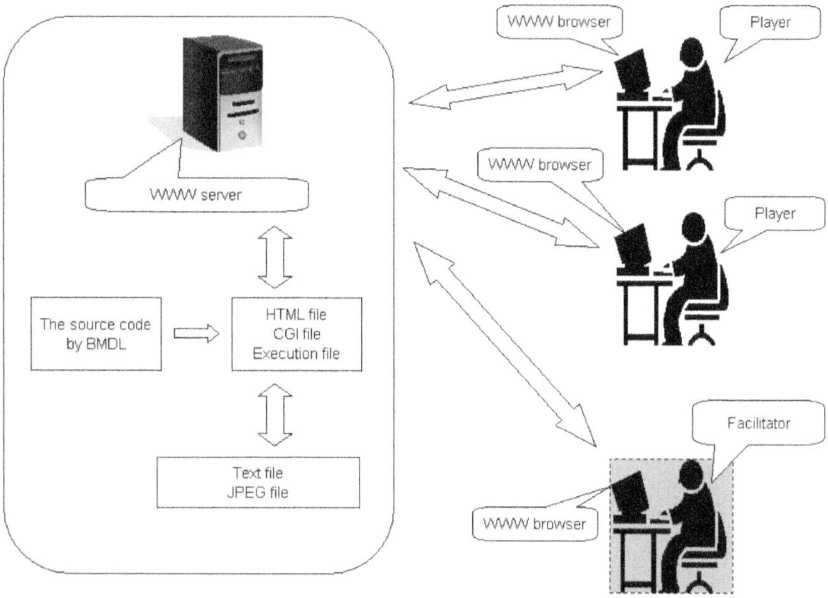

Fig. 1 Conceptual diagram of development and execution environment for the experiment

bond, domestic short-term bond, foreign bond, domestic stock, and foreign stock at the beginning of every term[6], and inputs into a model as "Input information".

A player refers to "Reference information" and decides asset allocation. "Reference information" is offered every term. The expected return of asset, the risk, the covariance, the duration, and so on are displayed on "Reference information". "Reference information" is provided also with the example (suppose that it is called a reference portfolio in this paper) of the portfolio used as reference which maximizes the expected return of a surplus and so forth.

As a reference portfolio, the case where the "expected return / risk" of a surplus[7] is maximized, the case where an expected return is maximized, and the case where a risk is minimized are raised according to the kind to optimize. The reference portfolio of a short period (the past one year), the middle (the past three years), and a long period of time (all the available data, in general five years or more) is raised regarding the period of the past return used for expected return estimation. The examples of the target duration regarding

[6] One term is set to one round, and the period of one term is considered as the period of three months.

[7] A surplus here is the difference of pension assets and a pension liability. In plus, a surplus means the savings surplus of pension assets, and in minus, it means the unfunded liabilities of pension assets.

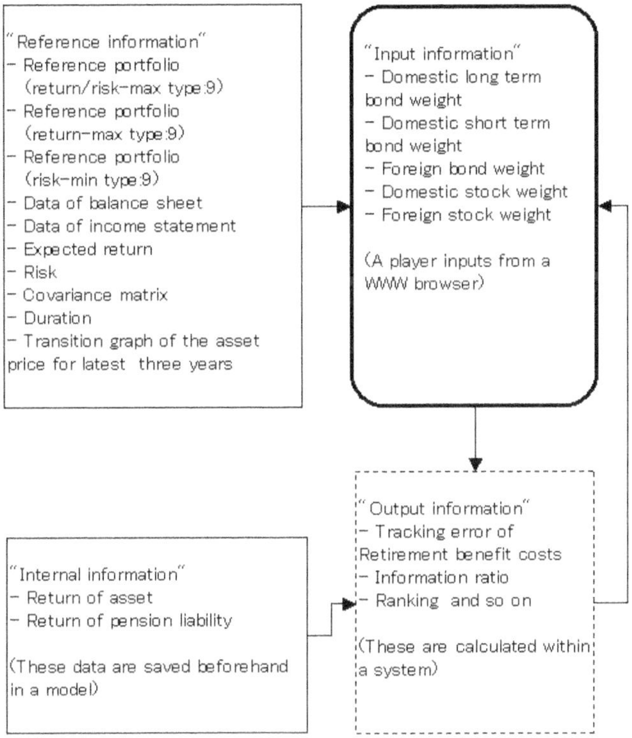

Fig. 2 Business game model

pension asset are given as for the cases where the percentage to the duration of pension liability is 80%, 100%, and 120%[8].

As a person in charge of a pension fund, the reservation of pension assets which provides pension liability is given to a player as a target. Although a savings ratio should just always change into not less than 100% of state (state of plus surplus), when a risk is taken too much in order to raise a return, sometimes, a reserve ratio may fall. When a reserve ratio falls, a sponsoring company compensates funds as retirement benefit costs, and a sponsoring company receives a loss by this costs donation. Therefore, a player must decide the asset allocation of pension assets, after also taking into consideration the influence which it has on the profit and loss of a sponsoring company. When a savings ratio becomes not less than 100%, it is a setup by which pension assets are balanced to pension liability by reducing the contribution of the pension which is inflow to pension assets, and a reserve ratio is adjusted to 100% at the time of every term and a start.

[8] Furthermore, in a Excel file which offers "Reference information", a program is able to calculate a portfolio by specifying the percentage of duration and expected returns (the short period, the middle, long period of time).

That is, when a reserve ratio is less than 100%, the insufficiency of pension assets is added up as a loss of a sponsoring company. When a reserve ratio is not less than 100%, it is adjusted so that an exceeded part of pension assets may become the same amount of pension liability, and there is not reduction to a sponsoring company[9]. It is explaining to a player being exposed to the situation of competing with the pension fund staff (other players in an experiment) of the other companies of a same sector at the time of an experiment start. It is explaining that the result of a pension-assets investment of a player may inflict a loss on a sponsoring company, and may make it the disadvantageous situation for competition with the other company.

The performance evaluation of a player is measured from the impact (tracking error of retirement benefit costs) given to the profit and loss of a sponsoring company. Ranking of the management capabilities of a player is carried out by this evaluation. In addition to the impact given to the profit and loss of a sponsoring company, players are provided also with information ratio (which is the average of excess return of pension assets to pension liability divided by the tracking error), the average of excess return of pension assets, and the average of excess return for every round to pension liability as "Output information" after the end of each round.

Messages are given to a player when every round finishes. The ranking based on the impact given to the profit and loss of the sponsoring company of a player is displayed on a message. Moreover, the links for acquiring the information for deciding the strategy of the next round are displayed. If equity ratio falls greatly, warning is emitted. As for a slight warning, a yellow card will be displayed, and, as for a serious warning, a red card will be displayed. Such a message is a device for giving the incentive which makes a player a competitive spirit. A sponsoring company may go bankrupt when defamation of pension assets continues and the loss done to a sponsoring company exceeds capital. Even when it goes bankrupt, a player can be made decisions succeedingly.

Experiments are two cases. One is a case where a player is a student, and the other is a case where a player is an institutional investor affiliation member. The experiment which uses a student as a player was conducted as a representative of the beginner who there is no knowledge not much regarding finance theory, and have not received professional training. Students are five Tokyo Institute of Technology graduate students (three first-year graduate students, two two-year graduate students). All five persons can be regarded

[9] Although adjustment by change of a premium and so on is performed only once in about five years, in this experiment, it shall be carried out every term. Moreover, it is assumed that a capital structure change by funding of a sponsoring company is not made every term, either. Since the pension premium increase for insufficiency stopgap of pension assets cannot be done by subscriber's strong opposition, either, it is assumed that there is only donation from a sponsoring company.

as beginners mostly regarding finance theory. On the other hand, the experiment which uses the three affiliation members in an institutional investor's investment section as a player, as a representative of those who it is knowledgeable regarding finance theory and have received professional training was conducted.

All three institutional investor affiliation members are financial analyst qualification holders. They hold the knowledge regarding finance theory, and since they have also experienced the training as a fund manager, it can be considered that they received professional training. The experiment was conducted 3 times[10] regarding each case where a player is a student, or an institutional investor affiliation member. One experiment took 10 rounds. It took about 1 hour. In order to avoid the end effect to a player, a player has not been told the end time beforehand.

The rational asset allocation method based on the finance theory in the experiment of this research is asset allocation which minimizes a prior risk[11]. Asset allocation which minimizes impact (tracking error of retirement benefit costs) given to the profit and loss of a sponsoring company ex post may occur besides the asset allocation which minimizes a prior risk. However, the rational asset allocation which can be made decisions in advance is asset allocation which minimizes a prior risk. Therefore, in the experiment of this research, it is judged that there is learning effect with the asset allocation of a player approaching prior risk minimization asset allocation.

3 Result

This chapter explains the result of having conducted the experiment regarding the pension-assets investment in consideration of liability , in order to find a difference between the case where professional training has been done and the case where professional training has not been done regarding finance theory. First, it explains that it was insufficient for bringing regarding learning effect as for the pension-assets investment which took liability into consideration in the case of the student, where the player has not received professional training. Subsequently, it explains that there was learning effect through this experiment regarding the pension-assets investment in consideration of liability in the case of the institutional investor affiliation member, where the player has received professional training.

---

[10] The 1st, the 2nd, and the 3rd experiment used respectively the actual market data from December 2001 to March 2004, from June 2004 to September 2006, from December 2006 to March 2009.

[11] "$TE = \sqrt{r^2 + \sigma^2}$" defines the prior risk. TE: Prior risk. r: The expected return presumed from a past return. σ: The standard deviation presumed from a past return.

3.1 Case Where Player Has Not Received Professional Training (Student)

This section explains that there was not much learning effect regarding the pension-assets investment in consideration of liability, when a player is a student. Fig.3 to Fig.8 shows transition of minimal risk asset allocation and the asset allocation of a player p3 in experiment A3 from experiment A1[12]. In the experiment A1, although the difference of the asset allocation of player p3 and minimal risk asset allocation is large, by the experiment A2 and A3 which are the 2nd and the 3rd experiment, the asset allocation of p3 is quite near minimal risk asset allocation. While there were those who come to do decision-making near minimal risk asset allocation when a student was used as a player, like a player p3, other players resulted in not coming to carry out decision-making not much near minimal risk asset allocation.

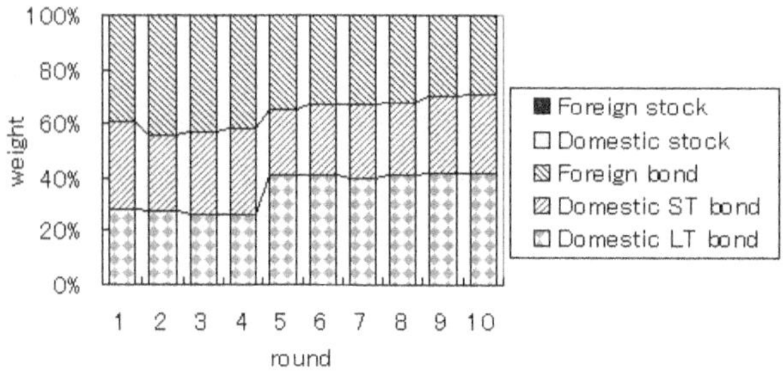

Fig. 3 Minimal risk asset allocation in experiment A1

Fig.9 to Fig.11 shows transition of the rate of deviation of the prior risk based on the asset allocation for which each player in experiment A3 from the experiment A1 in case a player is a student. The rate of deviation of a prior risk[13] is the difference of the prior risk of a player and the prior risk of minimal risk asset allocation divided by the prior risk of minimal risk asset allocation. It is shown that the rate of deviation of a prior risk is carrying out good decision-making, so that it is close to zero. The label of "p1" in a figure is a notation for distinguishing five players.

Even if it carries out comparison with Fig.9 and Fig.10, or comparison with Fig.9 and Fig.11, it is hard to find out the tendency for the rate of deviation of the prior risk of a player to decrease as it passes through an experiment. On the contrary, the round in which the rate of deviation of a prior risk is increasing can also be seen at a player p2 or a player p4.

[12] In this paper, the 1st experiment in case a player is a student, the 2nd experiment, and the 3rd experiment are called as experiment A1, experiment A2, and experiment A3, respectively. It is supposed that a player is called as a player p1 and so forth in order to distinguish five student players.

[13] A prior risk is measured as a presumed tracking error.

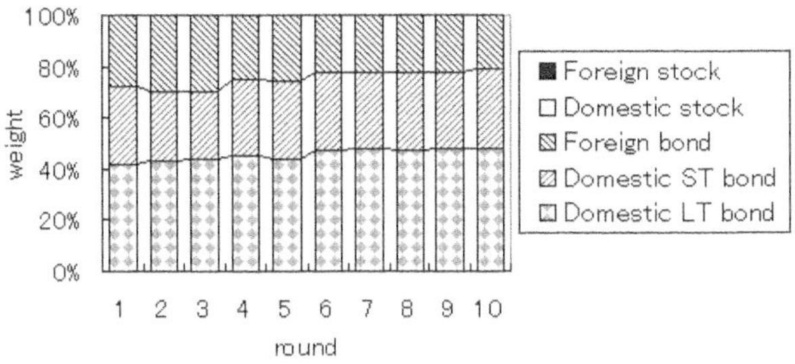

Fig. 4 Minimal risk asset allocation in experiment A2

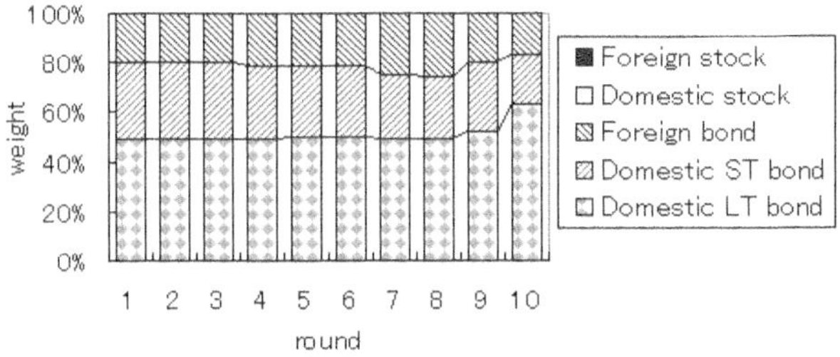

Fig. 5 Minimal risk asset allocation in experiment A3

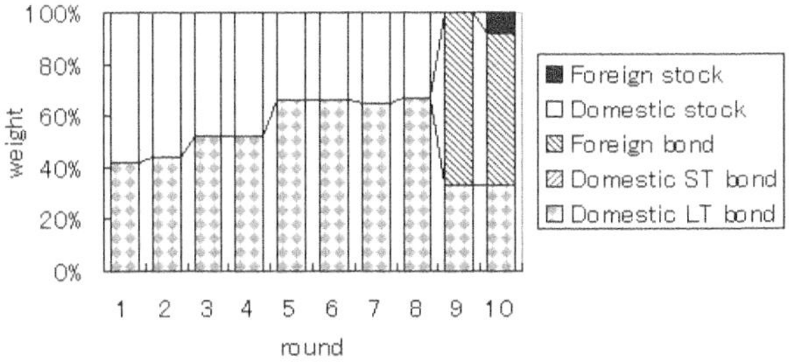

Fig. 6 Asset allocation of the player p3 in experiment A1

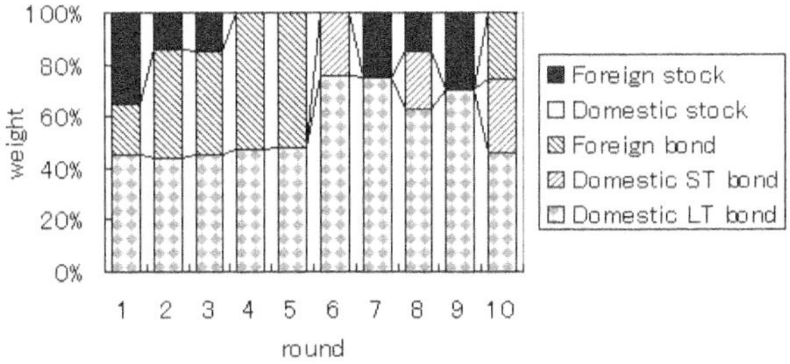

Fig. 7 Asset allocation of the player p3 in experiment A2

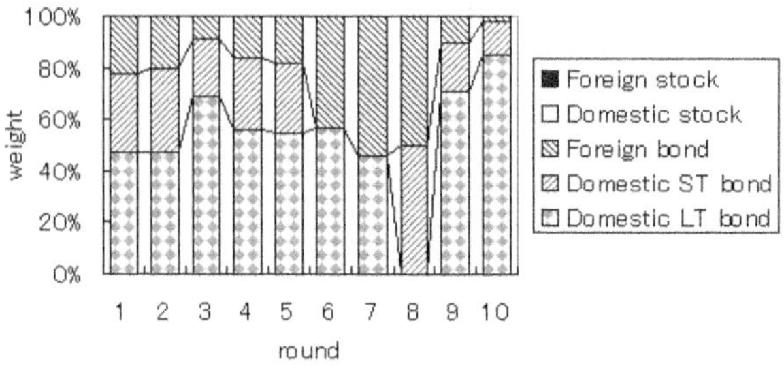

Fig. 8 Asset allocation of the player p3 in experiment A3

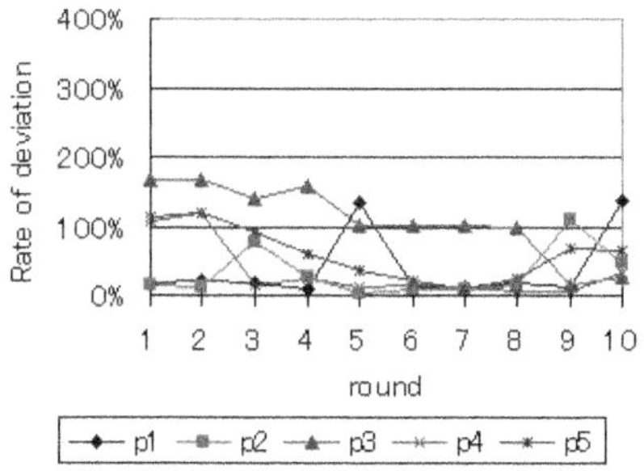

Fig. 9 The rate of deviation of the prior risk in experiment A1

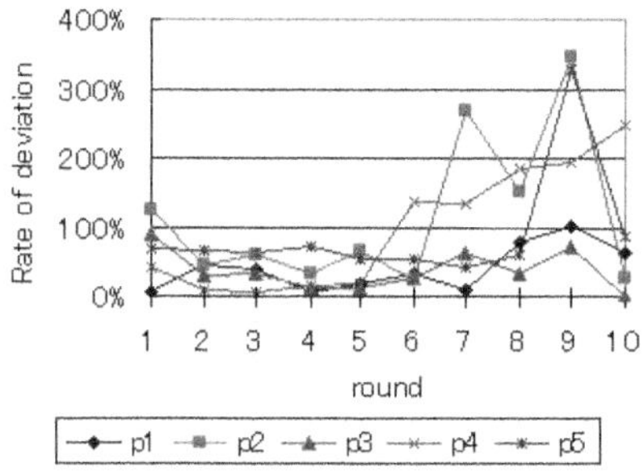

Fig. 10 The rate of deviation of the prior risk in experiment A2

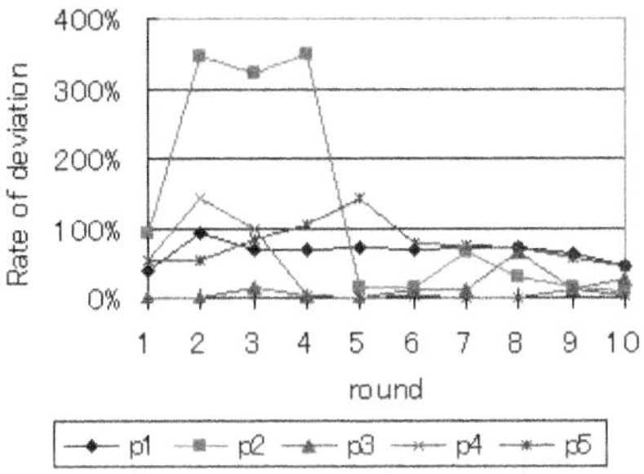

Fig. 11 The rate of deviation of the prior risk in experiment A3

Table 1 The average of the rate of deviation of a prior risk (student) (Unit:%)

Experiment	p1	p2	p3	p4	p5
A1	39	32	108	35	61
A2	40	115	37	99	90
A3	66	126	15	32	76

Table 2 The rate of the number of having thought as important on the decision-making (student) (Unit:%)

	A1	A2	A3
Reference portfolio(return/risk:max)	21	17	13
Reference portfolio(return:max)	2	1	1
Reference portfolio(risk:min)	2	1	7
Expected return and risk	17	20	19
Graph	15	14	10
Others	43	47	50
Total	100	100	100

Table 1 is the average of the rate of deviation of the prior risk in each experiment of each player. In the 2nd or the 3rd experiment, as for any players other than player p3, the rate-of-deviation average of a prior risk is not decreasing. As this reason, there is no notice that it must carry out decision-making that a player minimizes a risk in the asset allocation in consideration of pension liability.

Table 2 is a rate for every experiment which totaled the number of times of a reply of the item thought as important on the occasion of decision-making of a player. When a student is used as a player, the rates to which it is supposed that the reference portfolio which maximizes the return/risk of the 1st line was thought as important are 21% and a high value in experiment A1. Also in experiment A3, it has a high rate as compared with 1% of return maximum reference portfolios, and 7% of risk minimum reference portfolios. This shows not having recognized the importance of risk minimization, even if a player passes through an experiment. A reason which has not noticed that a player minimizes a risk is too much concern of player to change of an asset return. The reason has been presented in the high rate of thinking as important the transition graph of each asset current price for latest three years in the 5th line of Table 2. Because a player looks at a transition graph when deciding asset allocation, it is thought that decision-making of the player received influence by change of the latest asset price.

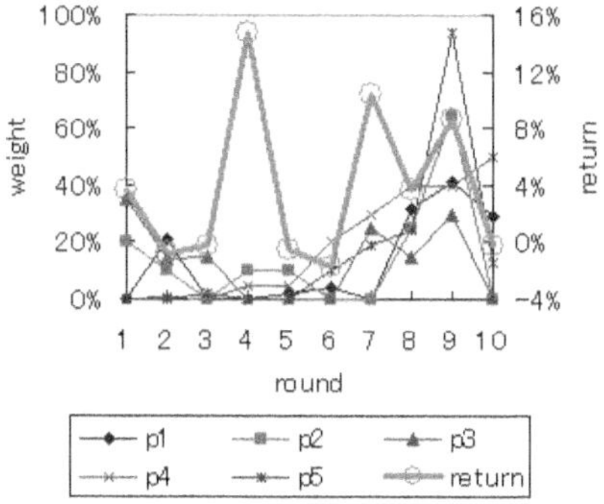

Fig. 12 Transition of the foreign stock weight (p1 - p5) and of the foreign stock return (right axis) in experiment A2

Fig.12 shows the transition of the foreign stock weight of each player, and the foreign stock return of a pre-round in experiment A2. Especially after round 6, when a return is high, it is shown that the player decide the raising the weight of the asset. Also from this, a player is considered to become difficult to notice better decision-making of minimizing a risk, by thinking an asset return as important.

As mentioned above, from this experiment, when a student is used as a player, it has checked that there is not much learning effect.

3.2 Case Where Player Has Received Professional Training (Institutional Investor Affiliation Member)

In this section, when a player is institutional investor affiliation member, it explains that there was learning effect through this experiment regarding the pension-assets investment in consideration of liability.

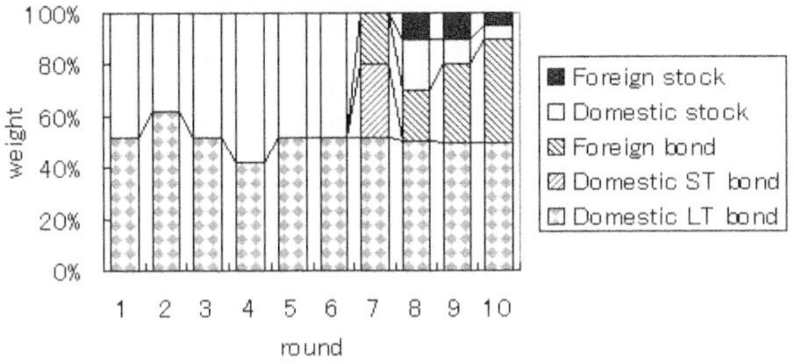

Fig. 13 Asset allocation of the player q1 in experiment B1

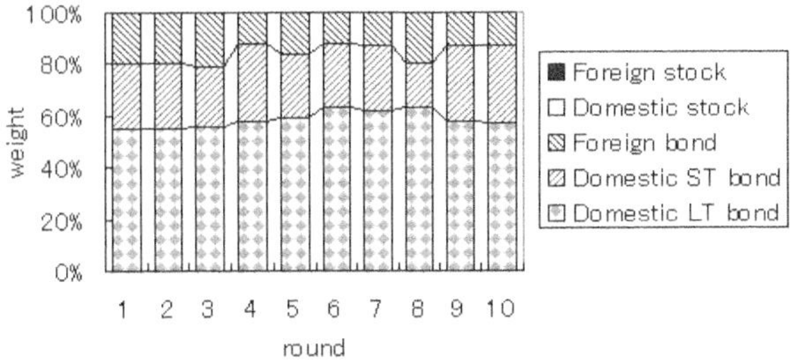

Fig. 14 Asset allocation of the player q1 in experiment B2

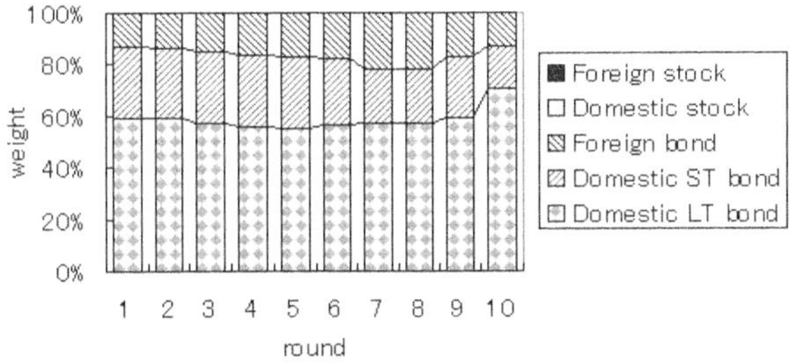

Fig. 15 Asset allocation of the player q1 in experiment B3

Fig.13 to Fig.15 shows transition of the asset allocation of the player q1 in the experiment B3 from experiment B1[14]. In the experiment B1, although the difference of the asset allocation of a player q1 and minimal risk asset allocation is large, by experiment B2 and B3 which are the 2nd time and the

[14] In this paper, the 1st experiment in case a player is an institutional investor affiliation member, the 2nd, and the 3rd experiment are called as experiment B1, experiment B2, and experiment B3, respectively. In order to distinguish three players, it is supposed that a player is called as player q1 and so on.

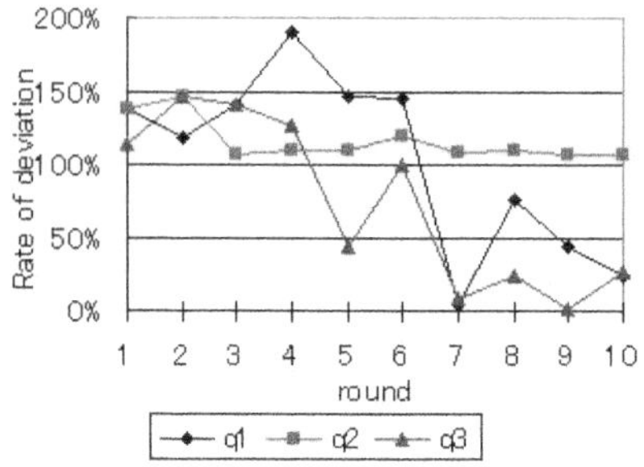

Fig. 16 The rate of deviation of the prior risk in experiment B1

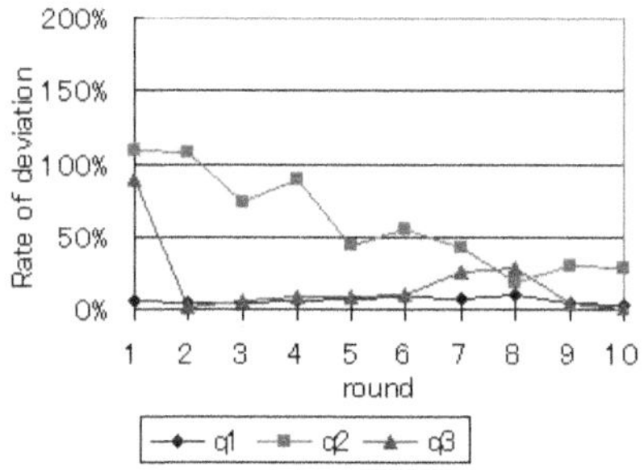

Fig. 17 The rate of deviation of the prior risk in experiment B2

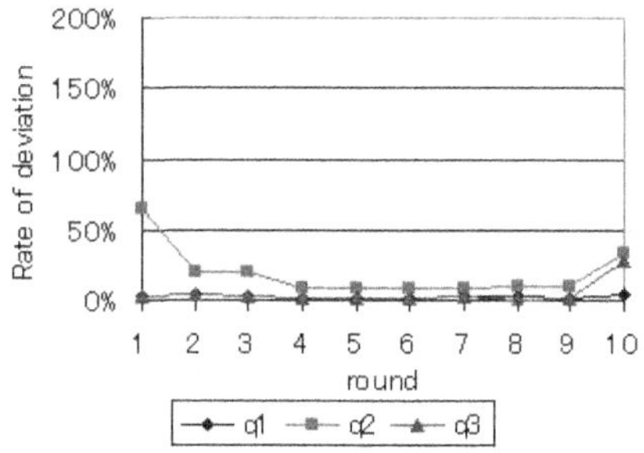

Fig. 18 The rate of deviation of the prior risk in experiment B3

3rd experiment, the asset allocation of player q1 is quite near minimal risk asset allocation.

Fig.16 to Fig.18 is transition of the rate of deviation of the prior risk based on the asset allocation as which each player in the experiment B3 from experiment B1 in case a player is an institutional investor affiliation member. With comparison of Fig.16 and Fig.17 or comparison of Fig.16 and Fig.18,

Table 3 The average of the rate of deviation of a prior risk (institutional investor affiliation member) (Unit:%)

Experiment	q1	q2	q3	p-value
B1	102	116	73	—
B2	6	60	18	0.036
B3	2	19	5	0.013

P-value of exp.B2 is calculated by paired t-test with exp.B1 and exp.B2.
P-value of exp.B3 is calculated by paired t-test with exp.B1 and exp.B3.

Table 4 The rate of the number of having thought as important on the decision-making (institutional investor affiliation member) (Unit:%)

	B1	B2	B3
Reference portfolio(return/risk:max)	31	6	0
Reference portfolio(return:max)	0	0	0
Reference portfolio(risk:min)	2	18	22
Expected return and risk	23	17	0
Graph	7	4	11
Others	37	55	67
Total	100	100	100

the tendency for the rate of deviation of the prior risk of a player to decrease will be seen as it passes through an experiment.

Table 3 is the average of the rate of deviation of the prior risk in each experiment of each player. In the rightmost column of experiment B2, p-value is tested for the difference of the average by paired t-test regarding the average of the prior risk of experiment B1 and experiment B2. In the rightmost column of experiment B3, p-value is tested for the difference of the average by paired t-test regarding the average of the prior risk of experiment B1 and experiment B3. All are significant at 5% of a level of significance. Through a player's experience of experiments, a player learns that a minimal risk strategy is better decision-making.

Table 4 is a rate for each experiment which totaled the number of times of a reply of the item thought as important on the occasion of decision-making of a player. When a player is an institutional investor affiliation member, it can check at the 3rd line that the rate to which it is supposed that the reference portfolio which minimizes the risk was thought as important is increasing as it passes through an experiment. By experiencing experiments, this shows that a player recognized the importance of risk minimization gradually. Also from this, when a player is an institutional investor affiliation member, it is shown that there is learning effect by experiencing an experiment.

3.3 Consideration

As a result of the experiment, between those who received the professional training regarding asset management, and those who have not received professional training, even if it experienced business game, the difference was found by the learning effect. In the case of the institutional investor affiliation member which is the representation of those who received training, better decision-making has been able to be done regarding the pension-assets investment in consideration of liability by experiencing business game. On the other hand, in the case of the student who is the representation of those who have not received training, it was the result of the ability to seldom say that there is learning effect even if it experiences business game.

As a reason of this difference, the existence of the professional training regarding asset management is raised. It is thought that the institutional investor affiliation member can apply now the rational view based on finance theory by having received professional training. From the experimental result of this research, the meaningful result that for those who received professional training it was effective for learning the decision-making method regarding the pension-assets investment in consideration of liability experiencing business game has been checked . On the other hand, since the student had not received professional training, it has checked that it was not enough just to have experienced business game to learn the decision-making method regarding the pension-assets investment in consideration of liability. In order to also make those who have not received professional training learn effectively the decision-making method regarding the pension-assets investment which took liability into consideration by experiencing business game, using a case method together can be considered. But I would like to consider it as a future subject regarding it.

4 Conclusion

In this research, through the business game technique, analysis aiming at learning of the pension-assets investment technique in which pension liability was taken into consideration was carried out. Regarding the participant who received professional training, interesting phenomena – an understanding progresses regarding the asset allocation technique at the time of taking pension liability into consideration – were seen as a result of analysis. This result shows the effectiveness of the business game technique to learning of the pension-assets investment technique. On the other hand, in the participant who has not received professional training, it has checked that learning effect was seldom acquired. Although the method which used the case method together can be considered to learning more effectively by experiencing business game to the beginner who have not received training, I would like to consider it as a future subject regarding it.

References

1. Faria, A.J.: Business simulation games: Current usage levels—an update. Simulation & Gaming 29, 295–308 (1998)
2. Greenblat, C.S.: Designing games and simulations. Sage Publications, Inc., Thousand Oaks (1988)
3. Kingsland, L.: Projecting the financial condition of a pension plan using simulation analysis. Journal of Finance 37(2), 577–584 (1982)
4. Sharpe, W.F., Tint, L.G.: Liabilities-a new approach. Journal of Portfolio Management 16(2), 5–10 (1990)
5. Terano, T., Suzuki, H., Kuno, Y., Fujimori, H., Shirai, H., Nishio, H., Ogura, N., Takahashi, M.: Understanding your business through home-maid simulatior development. Developments in Business Simulation and Experimential Learning (26), 65–71 (1999)
6. Tompson, G.H., Dass, P.: Improving students' self-efficacy in strategic management: the relative impact of cases and simulations. Simulation & Gaming 31(1), 22–41 (2000)
7. Walters, B.A., Coalter, T.M., Rasheed, A.M.A.: Simulation games in business policy courses: is there value for students? Journal of Education for Business 72(3), 170–174 (1997)
8. Winklevoss, H.E.: Plasm: Pension liability and asset simulation model. Journal of Finance 37(2), 585–594 (1982)
9. Wolfe, J.: The effectiveness of business games in strategic management course work. Simulation & Gaming 28(4), 360–376 (1997)
10. Yamashita, Y., Takahashi, H., Terano, T.: The development of the financial learning tool through business game. In: Lovrek, I., Howlett, R.J., Jain, L.C. (eds.) KES 2008, Part II. LNCS (LNAI), vol. 5178, pp. 986–993. Springer, Heidelberg (2008)
11. Yamashita, Y., Takahashi, H., Terano, T.: Learning a selection problem of investment projects and capital structure through business game. In: Proceedings of First KES International Symposium on Intelligent Decision Technologies, KES IDT 2009, Himeji, Japan, April 23-24 (2009)

An Agent-Based Implementation of the Todaro Model

Nadjia El Saadi, Alassane Bah, and Yacine Belarbi

Abstract. The problem of internal migration and its effect on urban unemployment and underemployment has been the subject of an abundant theoretical literature on economic development. However, most discussions have been largely qualitative and have not provided enough rigorous frameworks with which to analyze the mechanism of labor migration and urban unemployment. In this paper, we build up an economic behavioral model of rural-urban migration which is an agent-based version of the analytical Todaro model described by deterministic ordinary differential equations. The agent-based model allows to explore the rural-urban labor migration process and give quantitative results on the equilibrium proportion of the labor force that is not absorbed by the modern industrial economy.

1 Introduction

Economic development generates significant structural transformations such as changes in the demographic condition and in the production structure. The most important structural feature of developing economies is the distinction between rural and urban sectors and particularly, economic development is often defined in terms of the transfer of a large proportion of workers from agricultural to industrial activities [8]. The study of rural-urban labor migration has for long been an important area of research in development economics and a large body of literature has grown up in recent year around this topic in particular for less developed countries (see [1] for a survey on the theoretical literature).

Nadjia El Saadi · Yacine Belarbi
CREAD, Algiers, Algeria
e-mail: enadjia@gmail.com, yacinebelarbi@yahoo.fr

Alassane Bah
UMI 209, UMMISCO (IRD-Paris 6), ESP-UCAD, Dakar, Senegal
e-mail: alassane.bah@gmail.com

Q. Bai and N. Fukuta (Eds.): Advances in Practical Multi-Agent Systems, SCI 325, pp. 251–265.
springerlink.com

Economic historians agreed that a considerable part of the urban growth was due to rural-urban migration (see for example [11]). The towns offered new forms of employment opportunities and it was mainly the landless and the rural artisan who left and not the farmers [1]. When in the early 1950s economists turned their attention to the problems of population growth and economic development in the developing countries, it was natural to think that policies which emphasized industrialization would not only increase national incomes, but also relieve the overpopulation of the rural areas. However, during the 1960s, this view has led to a new orthodoxy in which rural-urban migration in less developed countries is viewed as "a symptom of and a contributing factor to underdevelopment". This orthodoxy is due to Todaro [10] and Harris-Todaro [6] whose models have provided a widely accepted theoretical framework for explaining the urban unemployment in less developed countries. The Todaro article [10] on urban unemployment in less-developed countries is an important advance in the study of this problem and although there has been some controversy on specific points, it has been widely applied to investigate various development issues. The key hypothesis of Todaro are that migrants react to economic incentives, earnings differentials and the probability of getting a job at the destination have influence on the migration decision. From these assumptions is deduced that the migratory dynamics in certain parametric ranges lead the economic system towards an equilibrium with urban concentration and high urban unemployment (the Todaro Paradox). The paradox is due to the assumptions that in choosing between labor markets, rural workers consider expected rather than current income differential . The expected income in the urban area is the fixed wage in the urban modern sector (formal sector) times the probability of obtaining a permanent job in this sector. This probability is defined as the number of opened jobs in the formal sector divided by the number of job seekers in the urban area. Since expected urban income is defined in terms of both wage and employment probability, in this model, it is possible to have continued migration in spite of the existence of sizeable rates of urban unemployment. Now assuming rural migrants respond to the employment probability, the Todaro model then demonstrates that an increase in urban employment may result in higher levels of urban unemployment. The repercussion of this simple set of assumptions is that contrary to received wisdom, the migration response which is factored in several policies aimed at reducing urban unemployment will raise urban unemployment rather than reduce it.

Todaro works have generated considerable discussions [2,4,5,7,9,12] from which it has been accepted in the literature that Todaro model (1969) predicts too high an unemployment rate in developing countries compared to observed rates and accordingly additional features have been incorporated to the basic model with a view to generate lower predictions.

In this paper, we propose to revisit the Todaro dynamic model from an agent-based approach. The migration agent-based model we conceive is the simplest version derived from the basic analytic model of Todaro [10]. It is formulated too close to the analytic model so that comparisons between the two approaches results can be made. The aim of such a formulation is to reproduce explicitly the migratory dynamics at individuals scale and analyze and visualize migration process. By simulating the agent based model, we explore the behavior of the model and hence check if from an agent-based approach the Todaro paradox holds or not.

The paper is organized as follows: in section 2, we recall the theoretical Todaro model. Section 3 is devoted to the agent-based model constructed from the Todaro model. We first present the simulator model and then describe the structure of the simulator. Section 4 presents the scenarios tests and the simulations results followed by a discussion. A conclusion is presented in section 5.

2 The Todaro Model

In the Todaro model [10], migration is viewed as a two-stage phenomenon: the first stage finds the unskilled rural worker migrating to an urban area and joining a large pool of unemployed and underemployed workers who arrived in town earlier and still are waiting for a modern sector job. This pool is the so-called "urban traditional sector" or "informal sector". It is modelled as an unproductive and stagnant sector serving as a refuge for the urban unemployed and as a receiving station for newly arriving rural migrants on their way to the formal sector jobs. The second stage is reached with attainment of a permanent job in the so-called "modern sector" or "formal sector". The decision to migrate from rural to urban areas will be functionally related to two principal variables: (1) the urban-rural real income differential and (2) the probability of obtaining a permanent urban job. Hence, the equation for growth of aggregate labor supply in the urban area is:

$$\frac{dS(t)}{S(t)} = \{\beta_u + \pi(t)F(\Delta(t))\}\, dt \tag{1}$$

with

$$\Delta(t) = \frac{Y_u(t) - Y_r(t)}{Y_r(t)}$$

where $S(t)$ is the size of the total urban labor force in period t, β_u the natural rate of increase in the urban labor force and $\pi(t)$ the probability of obtaining a job in the "formal sector" in period t. $Y_r(t)$ is the net rural real income in period t while $Y_u(t)$ is the net urban real income in period t. $F(\Delta(t))$ is a function such that $dF/d\Delta > 0$. $\Delta(t)$ is the percentage urban-rural real income differential and the product $\pi(t)F(\Delta(t))$ represents the rate of urban labor force increase as a result of migration.

As mentioned earlier, the probability $\pi(t)$ of obtaining a job in the "formal sector" is defined as being equal to the ratio of new employment openings in the "formal sector" relative to the number of accumulated job seekers in the urban informal sector at date t, that is:

$$\pi(t) = \frac{\gamma N(t)}{S(t) - N(t)} \tag{2}$$

with $N(t)$ the total employment in the urban sector in period t and γ the rate of job creation in this sector. The difference $S(t) - N(t)$ measures the size of the "informal sector".

The model considers that the number of new jobs created increases at a constant exponential rate over time, specifically

$$\frac{dN(t)}{N(t)} = \gamma \; dt. \tag{3}$$

The proportion of the urban labor force employed in the formal sector at time t (the employment rate) is denoted by $E(t)$, where

$$E(t) = \frac{N(t)}{S(t)}. \tag{4}$$

and the proportionate size of the urban informal sector (unemployment rate) is denoted by $T(t)$:

$$T(t) = 1 - E(t). \tag{5}$$

Todaro [10] solves for the equilibrium rate of employment E^* in the simple case that the income differential $D(t)$ remains constant over time ie $D(t) = D$ and $F(D) = D$. He solves for:

$$\frac{dE}{E}(t) = \frac{dN}{N}(t) - \frac{dS}{S}(t) = 0. \tag{6}$$

The solution for E^* is:

$$E^* = \frac{\gamma - \beta}{\gamma \Delta + \gamma - \beta} \tag{7}$$

and alternatively, the equilibrium rate of unemployment in the urban sector is simply:

$$T^* = 1 - \frac{\gamma - \beta}{\gamma \Delta + \gamma - \beta}. \tag{8}$$

Todaro claims that this equilibrium is stable. His intuition behind the equations above can be explained as follows:
for a developing country in the very early stages of industrialization such that almost the entire population resides in rural areas, when the urbanization process is just beginning, the pool of the urban unemployed is relatively

small so that the probability of obtaining a job is high. Therefore, for a significantly positive Δ and a positive rate of urban job creation exceeding the natural rate of urban population growth (ie $\gamma > \beta$), the resulting urban expected real income induces rural-urban migration such that the urban labor force grows at a faster rate than that of job creation, that is, $\beta + \pi(t)\Delta > \gamma$. This more rapid growth of labor supply results in an increase in the size of the urban traditional sector with the result that the probability of a rural migrant finding a job in the next period is lower ($\pi(t+1) < \pi(t)$). Assuming Δ and γ remain constant, this lower probability should slow down the rate of urban labor force growth although $dS(t)/S(t)$ may continue to exceed $dN(t)/N(t)$. Eventually, $\pi(t)$ stabilizes the urban unemployment rate at some level $1 - E^*$ depending upon the values of Δ, β and γ. If the unemployment rate falls below $1 - E^*$, equilibrating forces in the form of rising π will be set in motion to restore the equilibrium.

3 The Migration Agent-Based Model

3.1 The Simulator Model

In this section, we formulate an agent-based version of the Todaro model presented above. For this purpose, we consider an economic system formed by two sectors: a rural sector with a labor force of size N_r and an urban sector with a labor force of size N_u. In turn, the urban sector is divided into a formal sector of size N_f and an informal sector of size N_i so that $N_u = N_f + N_i$. As argued in Todaro [10], individuals in the rural sector take their decisions of migrating or not by considering the differential of their real income between their present sector and the sector they intend to go (the urban formal sector), and their probability of obtaining a job in the latter. So in our agent-based formulation, each individual i in the rural sector ($1 \leq i \leq N_r$) will evaluate at date t his percentage urban-rural real income differential $\Delta_i(t) = \frac{Y_u^i(t) - Y_r^i(t)}{Y_r^i(t)}$ and his probability $\pi_i(t)$ of obtaining an urban job in period t. If both $\Delta_i(t)$ and $\pi_i(t)$ are strictly positive, the rural worker i will migrate with probability $P_i(t) = \Delta_i(t)\pi_i(t)$, elsewhere, he will remain in his sector.

Similarly to Todaro, we consider the simplifying assumptions that the percentages urban-rural real income differentials for rural individuals are assumed to be constant over time and probabilities of urban employment are not individualized, namely, $\Delta_i(t) = \Delta$ and $\pi_i(t) = \pi(t)$ for $1 \leq i \leq N_r$. This is for being able to make a comparison between Todaro and our simulations conclusions. Nevertheless, a richer agent-based model which generalizes this version will be presented in a future work.

Based on the Todaro assumption that a rural migrant might spend a certain period of time in the urban informal sector before reaching the urban formal sector, in our model we consider no direct move from rural to the formal urban sector. The dynamics of individuals between sectors are:

migration from rural sector to informal sector, transition from informal sector to formal sector or inversely from formal sector to informal sector. From this, $\pi(t)$ can be regarded as the probability of transition from informal sector to the formal one. Using our notations, formula for $\pi(t)$ given in (2) becomes

$$\pi(t) = \frac{\gamma N_f(t)}{N_i(t)} \tag{9}$$

with γ the rate of job creation considered constant along a period of simulation. Employment and unemployment rates defined in (4) and (5) are given by:

$$E(t) = \frac{N_f(t)}{N_f(t)+N_i(t)} \quad \text{and} \quad T(t) = 1 - E(t).$$

To carry out the simulation of this economic system, workers are placed in the 3 different sectors according initial chosen values for N_r, N_f and N_i. Given the values of parameters Δ , β_u (rate of natural growth) and γ , the first step in the simulation is to add to urban informal sector the new seekers for jobs as a result of natural urban labor force growth at the rate β_u. Then, the probability of finding a job $\pi(t)$ is calculated using (9). If $\pi(t)$ is strictly positive, a fraction of individuals in the urban informal sector are selected randomly to join the formal urban sector, each with probability $\pi(t)$. If $\pi(t) < 0$ (due to a negative value of γ expressing jobs suppression), a fraction of individuals in the urban formal sector would be ejected to the informal sector with probability $-\pi(t)$. When $\pi(t) = 0$, no change takes place. Then, if both Δ and $\pi(t)$ are strictly positive, every individual i in the rural sector is a potential migrant with probability $P(t) = \Delta\pi(t)$. Otherwise, individual i in the rural sector does not migrate. To conclude the transition process from one sector to another one, a random number is generated from an uniform distribution on]0,1[. If this number is less than the transition probability assessed for an individual, than the transition of the corresponding individual holds, otherwise, no change takes place. Hence, a new sectorial configuration is obtained. Knowledge of the new populations allows the system to be reset. Therefore, the state variables of the sectors have to be calculated again and the whole procedure will be repeated as many times as we set in the simulation.

3.2 *Internal Structure*

The simulator proposed is a tool of experiment intended to represent virtually rural-urban migration by three sectors (rural sector, urban formal sector, urban informal sector) and their interactions. We use the platform Cormas for its implementation [3]. Generally, Cormas describes models by: spatial entities to represent the geographic space, social entities to describe the actors and passive entities for the rest of the models elements. For our migration model, the classes **Space** and **Sector** are spatial entities while classes **Individual** and **Group** are social entities. **Migration** class is the model class that is used

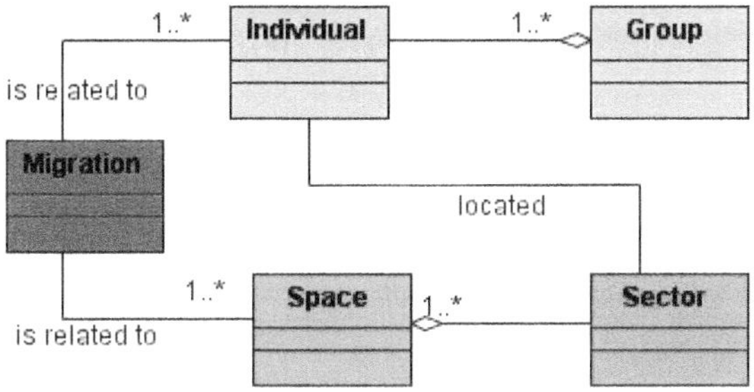

Fig. 1 Class diagram

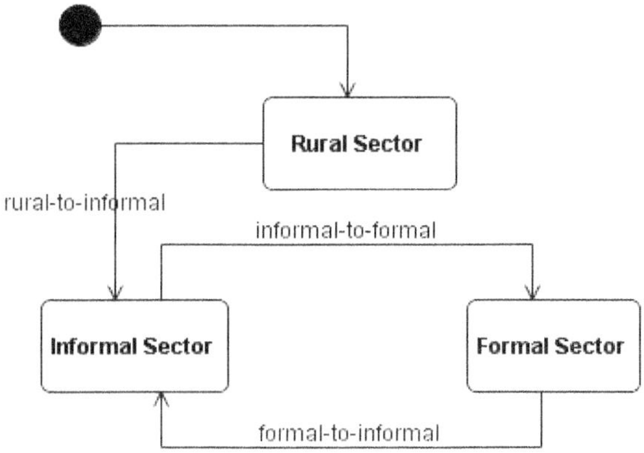

Fig. 2 The dynamic of Individuals

to manage the sequences of individual actions and the space (Fig. 1). The dynamic of individuals transitions between the three sectors is summarized in Fig. 2.

Our migration individual-based model is implemented in the Object-Oriented language Smalltalk using Visual Works Environment.Time evolves in a discrete way by steps $\triangle t$.

3.3 Interface

The simulator is provided with a three windows-interface:

(i) the first window allows to initialize, run and visualize evolution of the sectors and different indicators of migration (Fig. 3),
(ii) the second window allows the user the set up of the global variables and parameters (Fig. 4),
(iii) the third window is the simulation space, it shows the situation of the different sectors in time (Fig. 5).

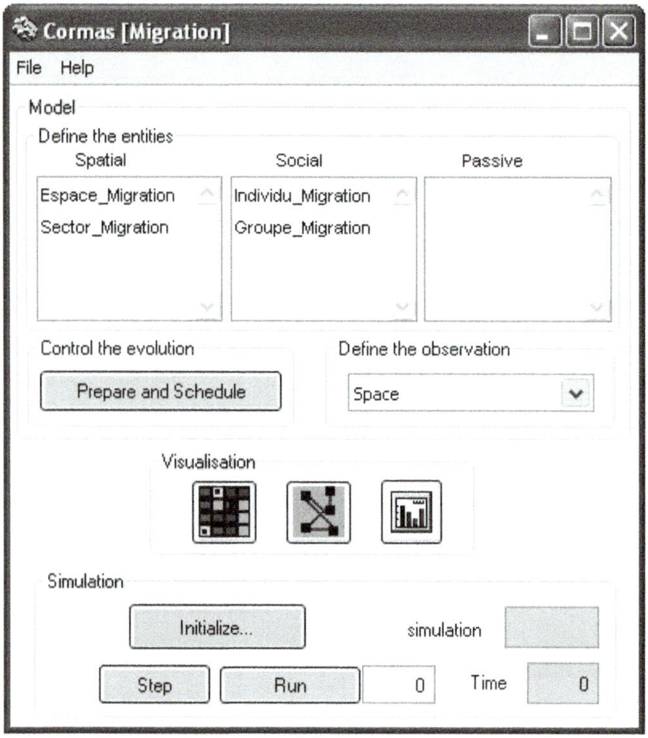

Fig. 3 The simulator interface: window 1

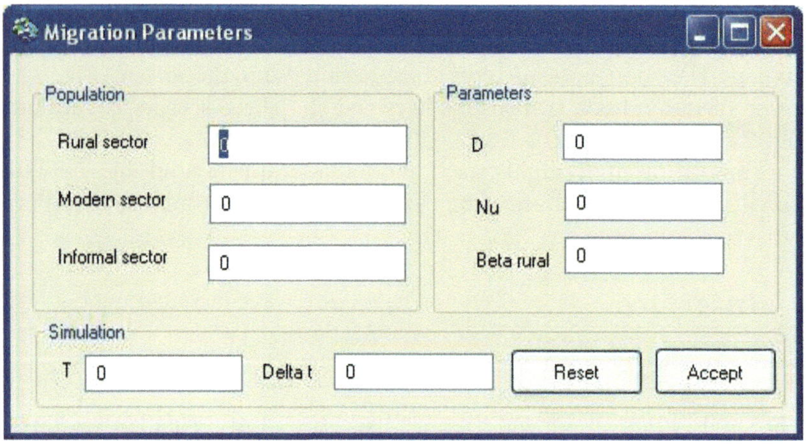

Fig. 4 The simulator interface: window 2

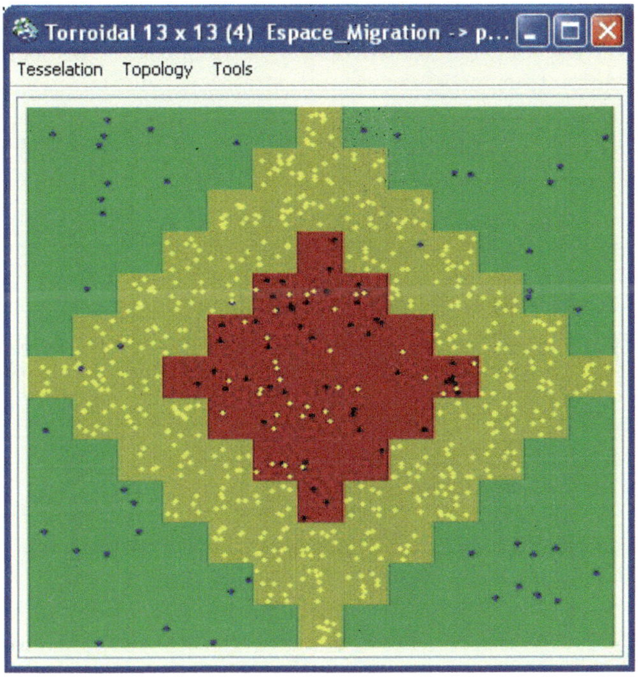

Fig. 5 The simulator interface: window 3. The rural sector is in green (its inhabitants are colored in blue), the urban informal sector is in yellow (its inhabitants are colored in yellow) and the urban formal sector is in red (its inhabitants are in black).

The simulator permits to undertake simulations for a large range of parameters. It has also been performed in order to visualize and quantify the evolution of some migration indicators (rural sector size, formal sector size, informal sector size, employment rate, unemployment rate).

4 The Simulations

4.1 The Scenarios Tested

The migration agent-based model has been conceived to perform simulations for a large parametric range. Here in this paper, we focus on the more interesting case from the economic point of view: the case of a significative positive percentage urban-rural real income differential ($\Delta > 0$) and a positive rate of urban job creation that exceeds the natural rate of urban population growth ($\gamma > \beta_u$). Our aim in simulating this case is double goal: first, we attempt to quantify the effects of the two factors Δ and γ on the migration process and unemployment and, second, we explore the behavior of the model to check if the Todaro Paradox is emergent property of our model. For these purposes, we perform simulations for different magnitudes of the parameters Δ and γ: $\Delta = 0.5$ that is $Y_u = 1.5Y_r$, $\Delta = 1$ ($Y_u = 2Y_r$) and $\Delta = 2$ ($Y_u = 3Y_r$), $\gamma = 0.1$ illustrating a high rate of urban job creation, $\gamma = 0.05$ illustrating an average urban job creation rate and $\gamma = 0.03$ an urban job creation rate proche to the natural urban labor growth β_u. These scenarios are tested for the following fixed initial condition $N_r = 5000$, $N_i = 500$, $N_f = 100$ and $\beta_u = 0.02$ a realistic value for labor natural growth rate in less developed countries. We point out that 10 runs of the simulation have been performed for each scenario to damp out randomness embodied in the simulator. Hence, each simulation result presented in the next subsection is an average of 10 repetitions.

4.2 Simulations Results

Simulations in Fig. 7 and Fig. 8 show the decrease of the initial unemployment rate and its convergence to 0. They also show the higher the rate of urban job creation, the faster the convergence. For instance, when $\gamma = 0.1$, unemployment rate reaches the equilibrium 0 after only 50 simulation steps while for $\gamma = 0.05$, the value 0 is reached after 100 simulation steps and later when $\gamma = 0.03$. Fig. 6 shows the final sectors situation when $\Delta = 1$ and $\gamma = 0.05$, it illustrates the fact that for the combination ($\Delta > 0$ and $\gamma > \beta$), the formal sector, in long run, absorbs all the labor force surplus in the economic system. Fig. 8 shows clearly the effects of the urban-rural real income

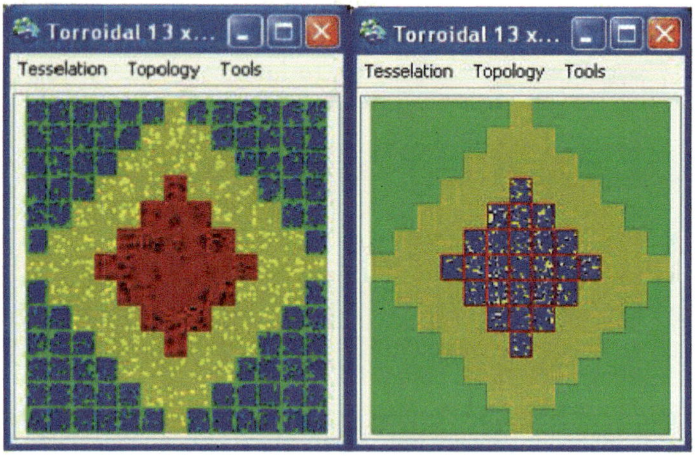

Fig. 6 Initial situation in the three sectors when $\Delta = 1$, $\gamma = 0.05$ and $N_i = 1000$ (on the left). The three sectors situation after $T = 200\Delta t$ (on the right).

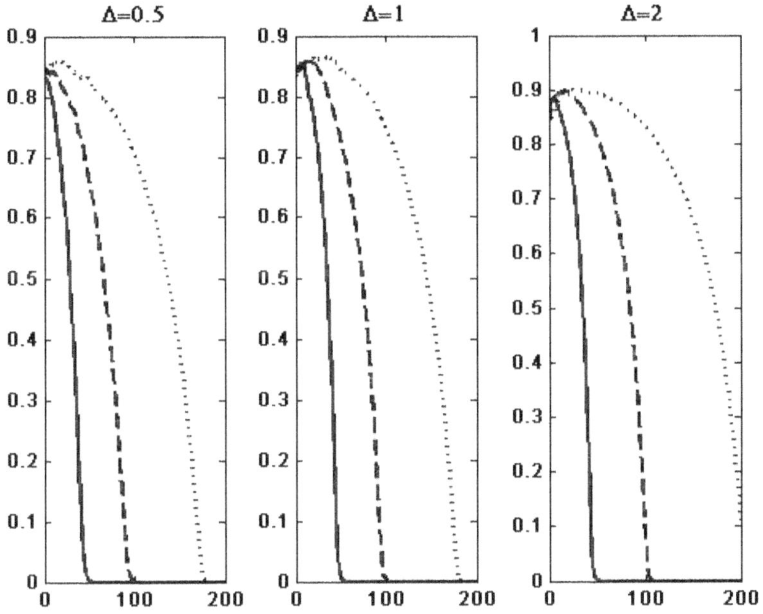

Fig. 7 Unemployment rate evolution in time for different values of γ: $\gamma = 0.1$ (solid line); $\gamma = 0.05$ (dashed line) and $\gamma = 0.03$ (dotted line).

differential Δ on the migration process, namely, an increase in Δ increases unemployment rate before converging to 0 and slows the convergence process towards equilibrium. These effects are better observed when Δ is significative and γ small. Fig. 9 shows the transitional dynamics and the long run equilibrium of the system through the three sectors evolution. Indeed, graphics in Fig. 9 show that for a positive urban-real income differential and a positive rate of urban job creation exceeding natural urban growth rate, there is a net migration towards urban sector. This migration is represented by a decline in the rural sector population and an increase in the informal sector population. It is clearly shown that the rate of decrease (respectively increase) of the rural sector size (respectively the informal sector size) is directly related to both Δ and γ. Graphics in Fig. 9 also show that in a first stage the formal sector grows but slower than the informal sector, this is due to migration which makes the urban labor force growing faster than job creation. This results in lower probabilities of migration and hence a decrease of the migration rate in the next stage. Indeed, we can observe after a certain time, the equalization of the formal and informal sectors sizes followed by a decrease in the size of the informal sector till extinction and a continuous growth of the formal sector size until it stabilizes. It is also observed that the formal sector size at equilibrium reaches a higher value and takes more time for such outcome for a higher urban-rural income and a smaller rate of urban job creation. Fig. 10

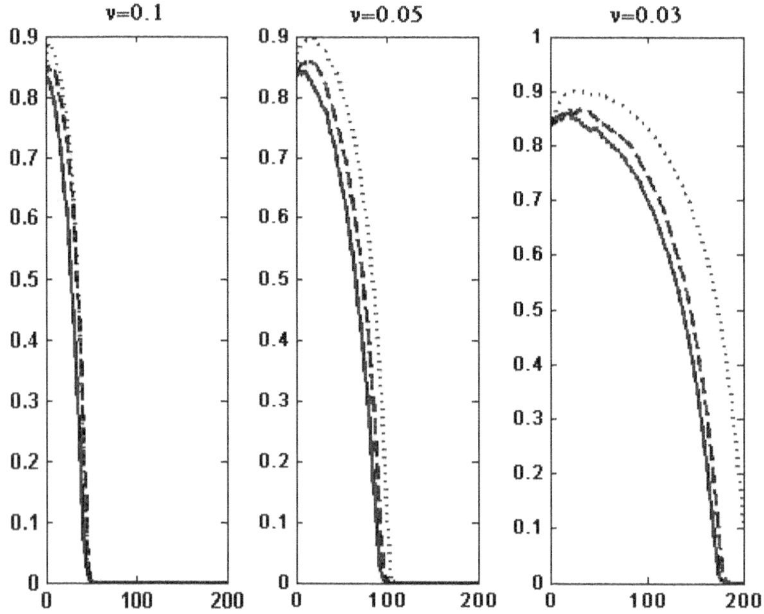

Fig. 8 Unemployment rate evolution in time for different values of Δ: $\Delta = 0.5$ (solid line), $\Delta = 1$ (dashed line) and $\Delta = 2$ (dotted line).

illustrates the case of a positive percentage urban-rural real income differential Δ with an urban job creation rate positive but less than the natural rate of growth in the urban labor force ($\gamma < \beta_u$). It is shown in this case that the unemployment rate increases progressively to converge to 1 (see (a) in Fig. 10). The graphic of the rural population size evolution shows at first a decrease in the population due to the rural-urban migration but after a certain time, migration stops and the rural population size stabilizes. Notice the large size of the informal sector in this case and the inability of the formal sector in absorbing the large labor force accumulated in the informal sector.

4.3 Discussion

The simulation of the migration agent-based model shows that a positive percentage urban-rural real income differential ($\Delta > 0$) and a rate of urban job creation exceeding the natural rate of urban population growth ($\gamma > \beta_u$) lead the economic system to an equilibrium characterized by a null unemployment rate (full employment). Indeed, simulations show that in the first stage of the urban development, the combination ($\Delta > 0$ and $\gamma > \beta_u$)

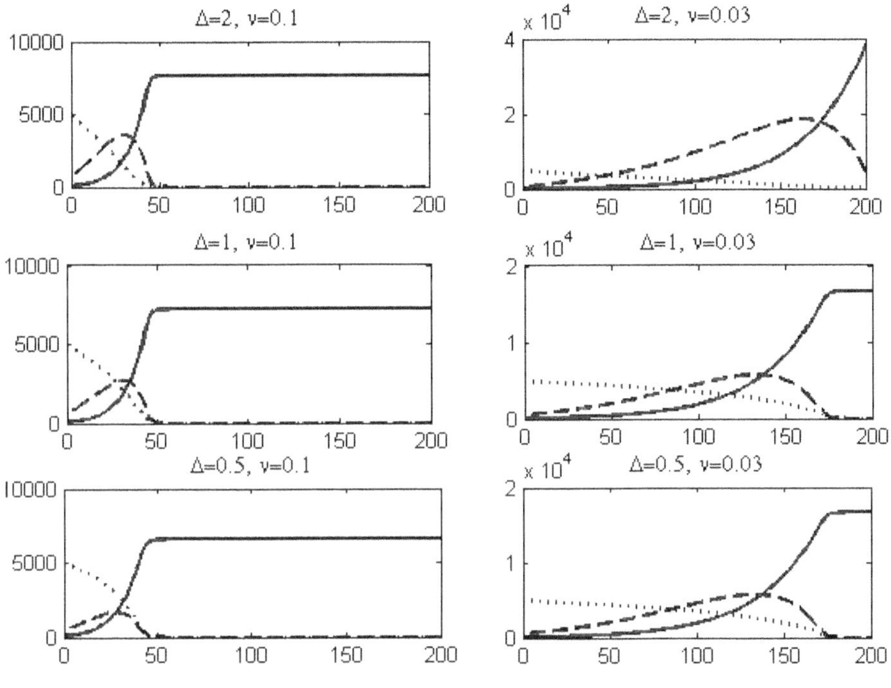

Fig. 9 Sectors evolution in time: urban formal sector (solid line); urban informal sector (dashed line) and rural urban sector (dotted line).

induces rural-urban migration such that the urban labor force grows at a faster rate than that of job creation, that is $\beta_u + \Delta\frac{\gamma N_f(t)}{N_i(t)} > \gamma$. This growth of labor supply results in an increase in the size of the informal sector with the consequence that the probability of finding a job in the next period is lower. Till now, our simulations results agree with the intuitive explanation of Todaro presented in section 2. But differently to Todaro intuition, simulations show clearly that the lower probability of obtaining an urban job slows down the migration rate and hence the rate of urban labor force such that from a certain threshold, the rate of urban job creation will irreversibly exceed the urban labor force ($\gamma > \beta_u + D\frac{\gamma N_f(t)}{N_i(t)}$) leading at long run to an equilibrium in which the formal sector absorbs all the urban labor force of the informal sector. But as soon as $\gamma < \beta_u$, the resulting equilibrium is unemployment 1.

On another hand, simulations have permitted to test implications of accelerated industrial growth and alternative rural-urban real income differentials. It has been shown that a high urban job creation rate has the effect of accelerating the convergence of the system to the full employment equilibrium while a high percentage urban-rural real income differential, through inciting to migration, has the opposite effect.

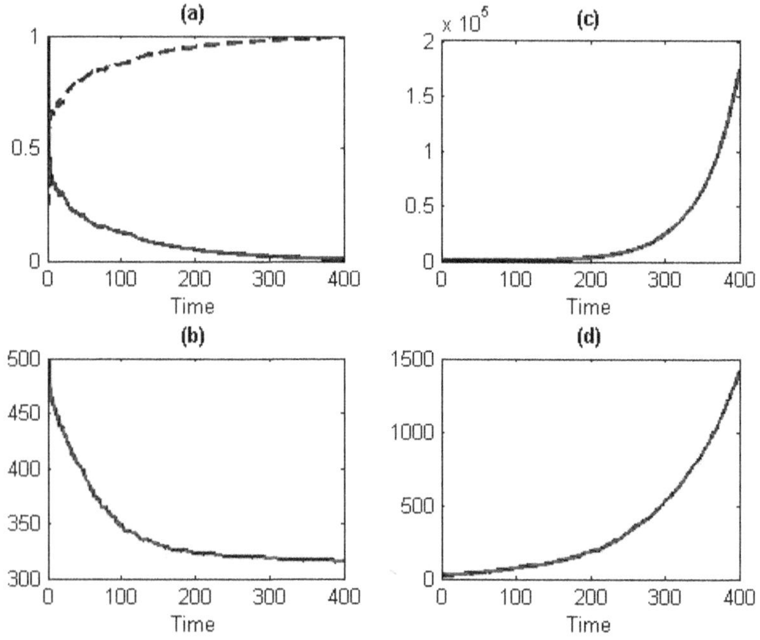

Fig. 10 (a) Employment (solid line) and unemployment (dashed line) rates evolution in time for $\Delta = 1$, $\gamma = 0.01$, $\beta_u = 0.02$, $N_r = 500$, $N_f = 20$, $N_i = 20$, (b) the rural sector evolution, (c) the urban informal sector evolution, (d) the urban formal sector evolution.

5 Conclusion

In this paper, we have conceived and implemented an agent-based model to study rural-urban labor migration in developing countries. This agent-based model has been derived from the analytic model of Todaro described by ordinary differential equations representing the links between rural-urban migration and urban unemployment. The agent-based model permits the analysis and visualization of the migration process for a large parametric range. It is also performed in order to give many quantitative results on equilibrium sectors sizes, equilibrium employment and unemployment rates.

Our main result in this paper is that the Todaro Paradox does not hold in the long run. Indeed, simulations results stemmed from our agent-based model show that as long as the urban job creation rate exceeds the natural urban labor force growth, the economic system converges to an equilibrium characterized by an unemployment rate 0 although the urban-rural real income differential is high and induces too much migration. Even if this finding is in contrast with the classical result of Todaro where inter-labor market (rural-urban) equilibrium mandates urban unemployment, it sustains many works (see for instance [7], [12]) reporting that Todaro model predicts too high an employment rate in less developed countries.

Acknowledgements. The authors would like to thank the anonymous referees for their pertinent and useful comments.

References

1. Bhattacharya, B.: Rural-urban migration in economic development. Journal of economic surveys 7(3), 243–281 (1993)
2. Blomqvist, A.G.: Urban job creation and unemployment in LDCs: Todaro vs Harris and Todaro. Journal of Development Economics 5(1), 3–18 (1978)
3. Bousquet, F., Bakam, I., Proton, H., Le Page, C.: Cormas: common-pool resources and multi-agent systems. In: Mira, J., Moonis, A., de Pobil, A.P. (eds.) IEA/AIE 1998. LNCS(LNAI), vol. 1416, pp. 826–837. Springer, Heidelberg (1998)
4. Collier, P.: Migration and unemployment, a general equilibrium analysis applied to Tanzania. Oxford Economic Papers 31, 205–236 (1979)
5. Fields, G.S.: Rural-urban migration, urban unemployment and underemployment, and job-search activity in LDCs. Journal of Development Economics 2(2), 165–187 (1975)
6. Harris, J.R., Todaro, M.P.: Migration, unemployment and development: a two-sector analysis. The American Economic Review 60, 126–142 (1970)
7. Mazumdar, D.: The theory of urban unemployment in less developed countries. World Bank Staff Working Paper 1980 (1975)
8. Ranis, G., Fei, J.C.H.: Development of the labor-surplus economy: theory and policy. Homewood (1964)
9. Stiglitz, J.E.: Rural-urban migration, surplus labour and the relationship between urban and rural wages. Easter Africa Economic Review 1(2), 1–27 (1969)
10. Todaro, M.P.: A model of labour migration and urban unemployment in less developed countries. The American Economic Review 59(1), 138–148 (1969)
11. Weber, A.: The growth of cities in the nineteenth century: A study in statistics, New York, pp. 184–186 (1899)
12. Zarembka, P.: Labour migration and urban unemployment: comment. American Economic Review 60(1), 184–186 (1970)

New Types of Metrics for Urban Road Networks Explored with S3: An Agent-Based Simulation Platform

Cyrille Genre-Grandpierre and Arnaud Banos

Abstract. The metric of the current road networks tends intrinsically to favour the efficacy of the routes which have the longer range, what leads to promote urban sprawl and automobile dependence. Indeed this metric ensures to individuals the possibility to travel even further, without necessarily increasing their transportation time in the same proportions. According to this assessment, we introduce a new kind of metric, the "slow metric", which amounts to invert the current ratios of efficacy between the different types of automobile travels, i.e. to favour the efficacy of the short range travels. This metric is reached thanks to traffic lights, under constraints regarding their location and duration. The calibration of the slow metric (number, duration and location of the traffic lights for different networks structures), its impact on traffic (fluidity versus congestion) and conversely the impact of traffic interactions on the slow metric, are simulated and evaluated with S3 which is an agent-based simulation platform.

1 Why Trying to Change the Current Road Networks "Metric"?

1.1 The Farther You Go the More Efficient Is the Road Network

From a planning perspective, road networks do not play fair game: the farther you go, the more efficient is your travel. Indeed, as road networks are highly

Cyrille Genre-Grandpierre
U.M.R. ESPACE, 6012/CNRS, Université d'Avignon, 74 rue Louis Pasteur, 84000 Avignon, France
e-mail: cyrille.genre-grandpierre@univ-avignon.fr

Arnaud Banos
U.M.R. Géographie-Cités, CNRS/Paris 1, Paris, 13 rue du Four, 75006 Paris, France
e-mail: arnaud.banos@parisgeo.cnrs.fr

Q. Bai and N. Fukuta (Eds.): Advances in Practical Multi-Agent Systems, SCI 325, pp. 267–285.
springerlink.com

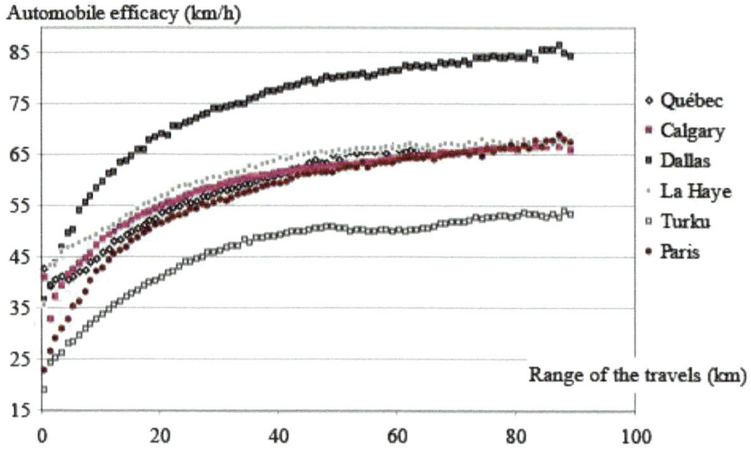

Fig. 1 Variations of the automobile efficacy with the range of the travels

hierarchised by speed, the farther you go, the more you stay on roads which allows you to drive faster when you are looking for a shortest path in time. This fact is rarely noticed because the performances of the road networks are almost always evaluated through global index (such mean speed) without making any differentiation according to the range of the travel.

Comparing the performances provided by several road networks for different range of travels in terms of efficacy, which is an average speed defined as the Euclidean distance between origin and destination of a trip divided by the duration of the trip [7], shows evidence of this phenomenon. If we plot the index of efficacy against the distance travelled (Fig. 1), we can notice that, on average, the level of performance increases non linearly with the distance travelled.

1.2 The Indirect Effects of the Current Metric

This "speed metric" ensures travellers the possibility to drive farther without necessarily increasing their transportation time in the same proportion. In other words, according to the ratio between the number of opportunities that can be reached and the duration of the travel, the speed metric encourages people to stay on the network with their car, as every additional second spent on the network provides a higher gain in terms of accessibility than the previous one. Moreover, this speed metric merely concerns cars, as public transportation modes (bus, tramway) are restrained by the frequent stops they have to make along their route. As a consequence, the structure of road networks intrinsically favours the use of car, especially for the longest distances.

So, this metric goes against the objectives of urban planning, as it allows and even encourages car use, separation between the various places of life (as home and work) and finally urban sprawl. As all the networks tested are more or less characterized by the speed metric, we can wonder if another type of metric is possible, in order to avoid the bad effects of the current one.

1.3 Traffic Lights for a "Slow Metric"

In previous work [4], it has been shown that it is possible to generate a different metric, which inverts the current ratio of efficacy between the different types of automobile travels, that is to say favouring the efficacy of short-range trips and therefore promoting higher densities and functional proximities in urban design, according to the hypothesis of the rational locator [9].

We call this metric the "slow metric". Our simulations with GIS show that imposing stops (traffic lights) on a network under certain constraints regarding their locations and durations which follow a stochastic distributions, produces the desirable effect (Fig. 2).

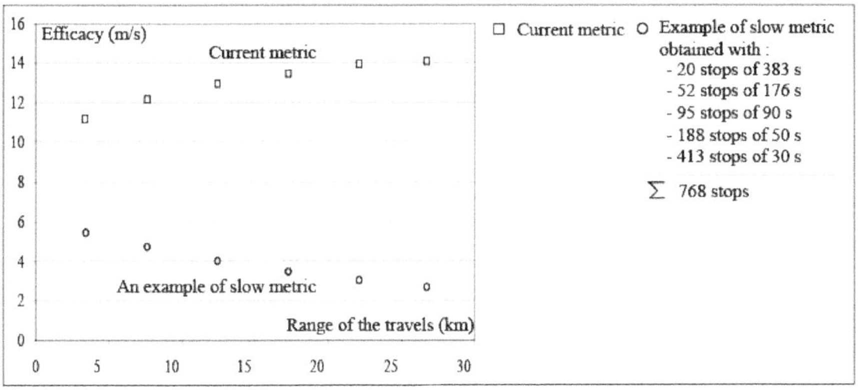

Fig. 2 An example of slow metric for the network 40km around the town of Carpentras

Traffic lights appears to be the only mean to change radically the logic of the current metric. Indeed, changing the topology of the network or the distribution of speeds (even if we homogenize circulation speeds) allows just to reduce the increasing of the performances of the trips according to their range, but not to invert the ratio of efficacy between short and long-range trips.

By modifying the number and duration of stops, we thus obtain various efficacy curves, favouring short-distance trips. These first encouraging results encouraged us exploring more dynamic and microscopic models, in order to address some keys issues identified so far such as: (a) the number, duration and location of stops, (b) the possible structural effect induced by networks topology and (c) the possible impact on traffic, including congestion and traffic jams and conversely the impact of the nature of traffic (density, types of interactions) on the slow metric. For this last point agent-based simulations seams to be the only way of exploration.

2 Smart Slow Speed (S3): An Agent-Based Simulation Platform

S3 has been designed as an interactive platform, allowing to explore with reactive agents the complex issues underlined previously. Therefore, the objective of S3 is not to develop a sophisticated traffic simulator as Aimsun (http://www.aimsun.com/site/) or Paramics (http://www.paramics-online.com/), but rather to have a relatively simple and robust tool available in order to understand under which conditions a slow metric can be obtained, to know how to calibrate it and how to analyse its effects in terms of traffic. To this aim, the model has to use a few input parameters for its output to remain under control. The model must also provide the adequate descriptors (i.e. efficacy of trips, percentage of vehicles stopped) in order to understand the emergence of the slow metric. In addition to problems of access to software, previous conditions are not met for true traffic simulators available on the market. Indeed, those aims at replicating real traffic conditions as closely as possible and thus use a lot of data and parameters. That is the reason why we have not developed S3 as an additional traffic simulator, but as a specific application which allows for the analysis of the slow metric and its induced effects.

S3 is composed of two interacting modules: the first one concerns both the creation of a road network differentiated by speed and the location of traffic lights, while the second one implements an agent-based traffic simulation model.

2.1 Traffic Lights: Choosing Location and Time Duration

The first module (network builder, Fig. 3) allows constructing regular (rectangular or octagonal), networks. Once the graph is generated, road links can be removed by hand or randomly, in order to test the impact of structural modifications on the global behaviour of the system. The graph generated is non-oriented but weighted by speed. Edges are indeed valued with a given speed v, such as $v \in \{30, 50, 70, 90, 110\}$, expressed in kilometers per hour. On that base, shortest paths are computed between any two distinct nodes, using Floyd-Warshall algorithm. As the graph is non-oriented, the total number of pairs is $(m * (m - 1)/2)$, with m the number of nodes.

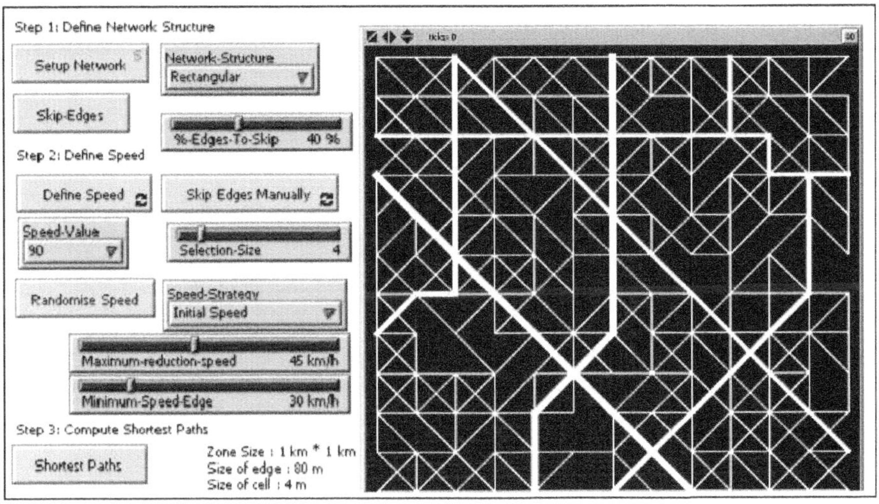

Fig. 3 The network builder of the S3 platform, developed in NetLogo

Then, a population of agents of size P is created, each agent being defined by an origin node, a destination node, and therefore the shortest path between these two end nodes. The P paths created are then used to create the traffic lights, in a four steps process (Fig. 4).

The main issue concerns the efficient localisation of a limited number of traffic lights. Let us assume two simplifications to begin. Firstly, let's assume there is no capacity constraint, which means that the flow on every edge of the network may overpass the edge's capacity: for all edges e, $F_e \geq U_e$ with F_e the flow on e and U_e its capacity. Secondly, let's assume our agents have no adaptation capacities, that is they strictly follow their allocated shortest path, whatever the context is.

Then, impacting the maximum number of agent with a limited number of traffic lights involves identifying target edges, that will be crossed by a large number of agents (Fig. 4, a): basically, the probability Π for a given edge e to host a traffic light L will be a function of the flow on e, $\Pi(L_e) = f(F_e)$. We can think about this process as a preferential attachment one [11]. What is more, given the assumptions previously formulated (no capacity constraints and no adaptation behaviour), it is evident that the flow F_e on a given edge e is the number of times that edge belongs to a shortest path between two given nodes. Therefore, the flow F_e of edge e can also be interpreted as a proxy of its "edge betweenness" [5], which is a generalisation of Freeman's betweenness centrality [2] to edges. However, introducing slow speed metrics in a network does imply targeting also the longest trips, as they benefit the most from the standard "speed metric" (Fig. 4, b). Therefore, we reduce the candidates to edges characterised by a flow superior or equal to a given threshold value γ:

$$\Pi(L_e) = f(F_e)\forall F_e \geq \gamma$$

Once traffic lights are localised (Fig. 4, c), we have to define their duration D. A power law is assumed, such that $\Pi(D_t) \propto D_t^{-\alpha}$. Increasing values of parameter α provide various distributions of traffic lights durations, once minimum and maximum durations are defined (Fig. 5).

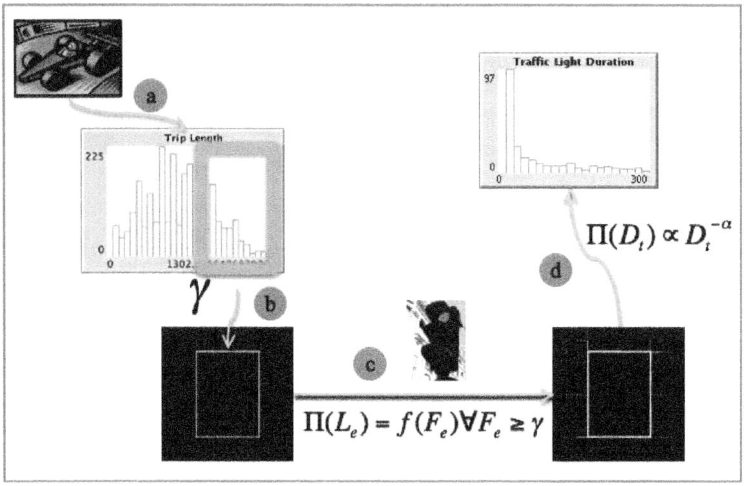

Fig. 4 Location and duration of traffic lights, a four steps process

Therefore, we assume that stopping probability at a certain stop is random, even we know that traffic lights are controlled in a systematic pattern by which a car can follow the pattern with rare stopping. Indeed, let us remember that our objective is to gain knowledge of the conditions necessary for the emergence of a metric different from the current one, rather than to copy the existing one.

Moreover, locating traffic lights with a random duration makes impossible the possible learning process of the agents which will allows them to avoid traffic lights with a long duration.

Lastly, this approach converges with Lämmer and Helbing's research which has shown that it was possible to contemplate a local self-regulation of traffic lights [8]. Indeed, that self-regulation provides outcomes as good as the ones resulting from a global regulation, which is increasingly more difficult to optimise due to a growing and more irregular traffic.

However, the possibility to define systematic pattern regarding the coordination of traffic lights is a possible evolution for S3.

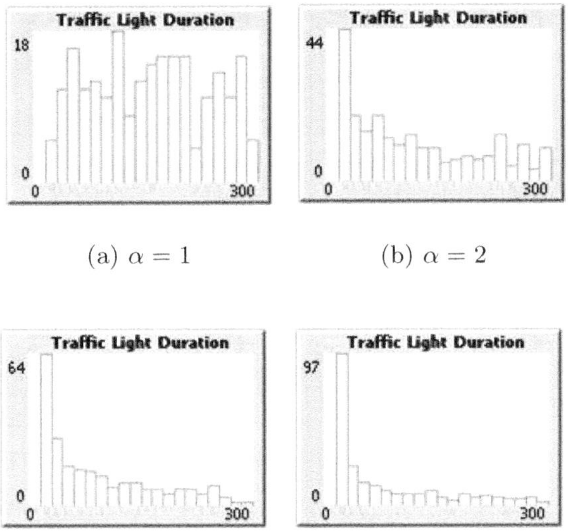

(a) $\alpha = 1$ (b) $\alpha = 2$

(c) $\alpha = 3$ (d) $\alpha = 4$

Fig. 5 Examples of distributions of traffic lights' durations obtained for various values of α

2.2 A Microscopic Traffic Model

The second module handles a microscopic traffic model, aimed at testing the efficiency of the network designed, as well as its impact on traffic fluidity. Before each simulation, n agents are created and localised at random on the nodes of the network, their destination being also chosen at random. During a simulation, each agent will have to reach its destination, following the shortest route computed previously, and taking into account speed links but also the presence of other agents in front, as well as the presence of red lights at intersection.

In order to do so, we use an underlying grid covering the 1 km * 1 km wide area, composed of a large number of small cells (length 4 m). Agents are then localised on cells underlying the network. They can occupy one and one cell at a time and only one agent can occupy each cell at the same time. On that base and following [1] [6], we then extended the NaSch model[1] [10], in order to introduce traffic lights. According to the prescription of the NaSch model, we allow the speed V of each vehicle to take one of the integer values $V = 0, 1, 2, ...Vmax$, $Vmax$ corresponding to the speed of the current link. At each discrete time step $t \to t + 1$, the arrangement of the n agents is then updated in parallel according to the following driving rules:

[1] A "probabilistic cellular automata able to reproduce many of the basic features of traffic flow" [13].

- *Step 1: Acceleration*
  If $V_n < Vmax, Vn \rightarrow min(Vn+1, Vmax)$, i.e. the speed of the *nth* vehicle is increased by one.
- *Step 2: Deceleration (due to other vehicles/traffic signal)*
  Suppose Dn is the gap in between the nth vehicle and the vehicle in front of it, and Dtn is the gap between the car under consideration and the red light in front of it on the road, then:
  if $dn \leq Vn$ or $dtn \leq Vn$, then $Vn \rightarrow min(Vn, Dn - 1, Dtn - 1)$
- *Step 3: Randomisation*
  If $Vn > 0$, the speed of the car under consideration is decreased randomly by unity (i.e., $Vn \rightarrow Vn - 1$) with probability $p(0 \leq p \leq 1)$. This random deceleration probability p is identical for all the vehicles, and does not change during updating. Three different behavioural patterns are then embedded in that single computational rule: fluctuations at maximum speed, retarded acceleration and over-reaction at braking.
- *Step 4: Movement*
  Each vehicle moves forward with the given speed i.e. $Xn \rightarrow Xn + Vn$, where Xn denotes the position of the *nth* vehicle at any time t.

Figure 6 illustrates the kind of traffic patterns generated by this simple thus powerful model.

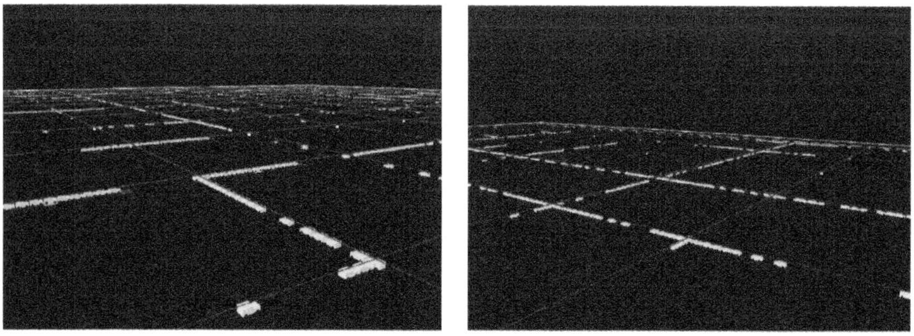

(a) A global view of the traffic

(b) Spontaneous formation of traffic jams

Fig. 6 Examples of traffic patterns obtained from the extended NaSch model

Once a specific hierarchised network is fixed, this traffic model allows exploring its efficiency as well as the impact of various strategies of speed reduction.

This quite simple but powerful model allows us to explore in what conditions we obtain a slow metric (number, duration, location of the traffic lights, effect of the network structure) and its potential effects on traffic regarding the number of agents and the parameters of the simulations. Exploring in depth these various issues in a systematic way is a challenge in itself because the combination of parameters remains huge. So in this paper we will present examples of results we obtained mainly related to the way a slow metric occurs and its effects on congestion, but without exhausting all the possibilities of analyzes, especially concerning the evaluation of the level of global accessibility provided by the different situations.

3 First Results

3.1 The Metric of the Road Network Depends on ...

3.1.1 ... the Number and the Duration of the Traffic Lights

Imagine a hierarchised star network, defined by high speed corridors converging towards a center. On that basis, we explore the influence of number and duration of traffic lights (determined in section 2.1) on network efficacy. As expected, the "speed metric" occurs in the absence of traffic lights. Introducing such equipments allows reducing the discrepancy between short and long trips, by penalising the long ones. However, the number of lights seems to play a secondary role compared to their duration, as expressed by the clustering of curves. This graph even suggests that a small number of traffic lights may do the job quite well, if we calibrate their duration, and location, correctly.

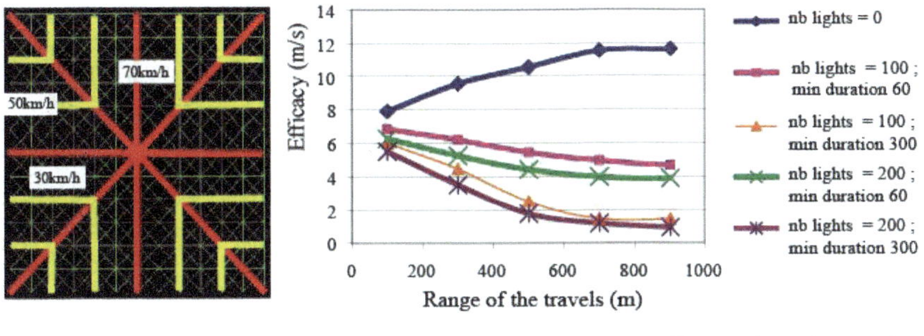

Fig. 7 Impact of number and duration of traffic lights on efficacy for a hierarchised star network

Beyond this particular case, it appears that the slow metric does not occur systematically when traffic lights are introduced. Indeed, traffic lights must have a sufficient density, which must be higher when the network is homogeneous in terms of speed. However, this density should not exceed a given threshold (of one traffic light / 2,5 edges in mean, this threshold varies according to the structure of the networks), since in this case we just decrease the level of efficacy of the network but without reaching a slow metric (the network becomes homogeneous again).

We also note that the slow metric is particularly obvious when the distribution of the traffic lights duration has high maximum and standard deviation, but with a high variability in the individual level of efficacy as a consequence.

3.1.2 ... the Structure of the Road Networks

Simulations on the effects of the structure of the road networks on their metric tend to separate two basic types: the connective and homogeneous networks in terms of speed on the one hand, and the less connective and hierarchised networks on the other hand.

Generally, under free flow conditions (no traffic interaction and no traffic light), the efficacy of the second type appears higher, in particular when the range of the trips increases. This is due to the fast lines which are used by

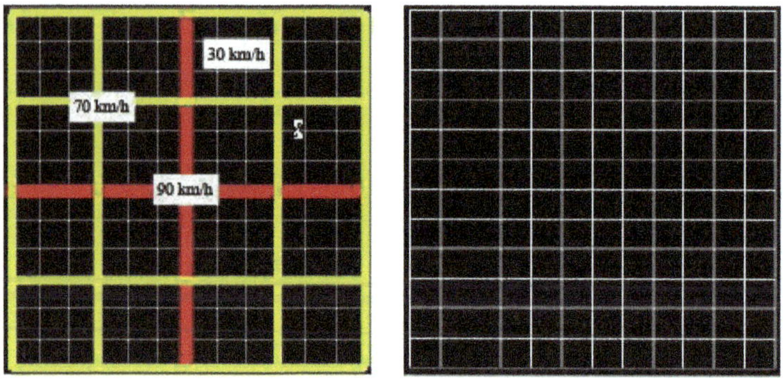

(a) A hierarchised network W1

(b) A homogeneous network with a speed of 50 km/h: homo50

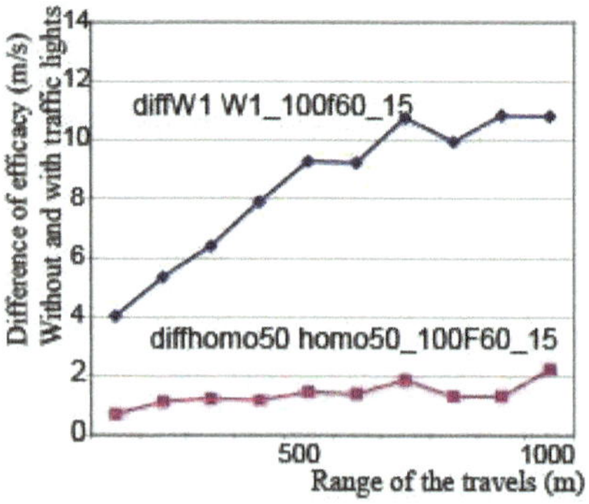

Fig. 8 Difference of efficacy for W1 (a) and homo50 (b) without and with 100 traffic lights.

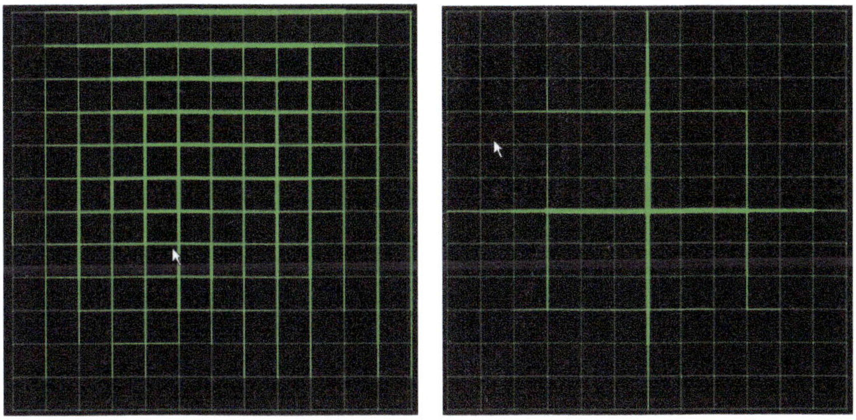

Fig. 9 Patterns of the flows (number of shortest paths per edge represents proportionnaly to the thickness of the edges) for 2000 random trips for a hierarchised network (W1) and a homogeneous network (homo50) without traffic lights.

many shortest paths when the range of the trips increases (cf. 1.1.). However, they appear on the other hand very sensitive to the introduction of traffic lights. For example we can see in the Fig. 8 that the difference of efficacy in the situations with or without traffic lights is higher for hierarchised networks (blue curve) than for homogeneous networks (red curve). Thus, for hierarchised networks, a limited number of traffic lights are enough to obtain a slow metric, especially if the networks are less connectives.

The sensitivity of the hierarchised networks can be explain by the fact that their "betweenness centrality" (2.1) is very concentrated on the edges with the higher speed, which belongs to a lot of shortest paths between origins and destinations. Thus, few edges participate actively to the routing of the traffic flows, which are very concentrated on particular roads (here the pattern of the flows depends strongly of the structure of the network as shown by Penn et al. [12] and Genre-Grandpierre [3]. Thus, if a traffic light is located on these edges, it will deeply impact the efficacy. Remember that agents do not adapt their shortest paths to the location of the traffic lights. Conversely, the homogeneous networks are intrinsically less effective, but they are less sensitive too to the traffic lights because the betweenness is more distributed on the whole network (Fig. 9).

3.1.3 ... the Connectivity of the Networks

If the pattern of the flows depends on the hierarchisation of the network, it depends on its connectivity too (connectivity expresses if the network is more or less complete, i.e. if there are a lot of edges for a given set of nodes). Indeed, the higher the connectivity, the less concentrated are the flow, because the

Fig. 10 Star network with 40% of the edges randomly deleted and square network with 30% of the edges randomly deleted.

shortest paths are more direct. In this case making a detour in order to drive on speed ways becomes irrelevant.

In order to test the effect of connectivity, we have successively and randomly deleted 10%, 20%, 30%, 40% and 50% of the edges of star and rectangular networks more or less hierarchised by speed (Fig. 10).

On this basis, we perform analyses of efficacy with and without traffic lights. The results shows that it is possible to delete numerous arcs with little impact on efficacity when connectivity is high. Indeed, in this case, there are a lot of alternative shortest paths which are very similar. For example, for a star network with homogeneous speed, it is possible to delete 40% of the edges with a very limited impact on efficacy.

However, for a less connective network or for a speed-based hierachised network, if strategic edges are deleted, good solutions of shortest paths which does not have any alternatives no longer exist. In this case the efficacy is much lower. For example for a connective and hierarchised network without traffic lights, we found that the efficacy decreases by 64% if we delete randomly 40% of the edges, but, if the deleted edges with low value of beetweeness centrality, the efficacy decreases by only 3%!

3.1.4 ... the Location of the Traffic Lights

Thus, if traffic lights are located regarding the betweeness centrality of the edges, it improves greatly their efficacy. We can even go further and target long trips thanks to the threshold γ which allows to locate the traffic lights on edges which are mainly involved in shortest paths concerning trips with long range. In this case, with the same number of traffic lights, the impact on efficacy is more important and it focuses mainly on long range trips (Fig. 11).

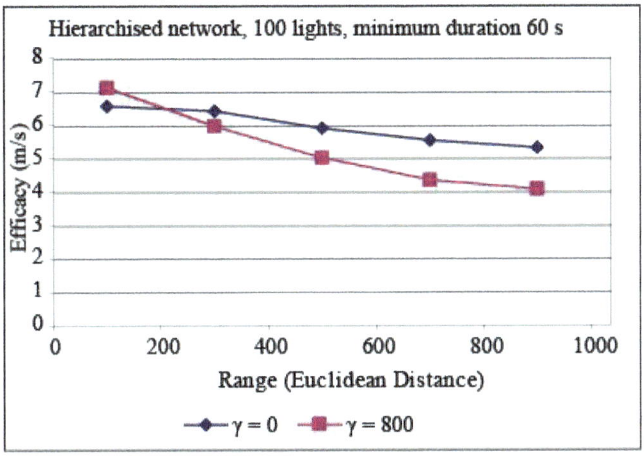

Fig. 11 Influence of parameter γ on network efficacy (network W1)

Here, for the same number of traffic lights, homogeneous networks can quickly become more efficient than hierarchised networks!

3.2 Exploration of the Traffic Model

S3 has been over all designed to explore the possible effect of a slow metric on the traffic and inversely the effect of the interactions between reactive agents on the slow metric, what is impossible with a GIS.

3.2.1 The Effects of Traffic Interactions on Efficacy

Introducing traffic interactions thanks to the NaSch model significantly lowers the efficacy. This decrease may be strong, in particular for hierarchised networks. For example, for the network W1 Fig. 8, we can see that the curve of efficacy taking into account the traffic interactions but without traffic lights is very close to the curve without interactions but with a hundred traffic lights of 60 seconds, in particular for short range trips (for long trips with a range superior to 500 m, we can see once again in Fig. 12 the impact of the γ parameter here equal to 500 meters).

Thus, taking traffic interactions into account translates the curves of efficacy at the bottom of the graph. This is particularly obvious when flows are concentrated as for hierarchised networks, i.e. when the probability to interact for agents is high. Here we can see that theoretical performances are very far from real performances when traffic interactions are taken into account...

The fluctuation of the value of the parameter p (2.2), which allows the simultaneous integration of fluctuations around maximum speed, retarded acceleration and over-reaction when braking, reinforces the previous findings. High values of p lower the efficacy, in particular for short range trips and if the

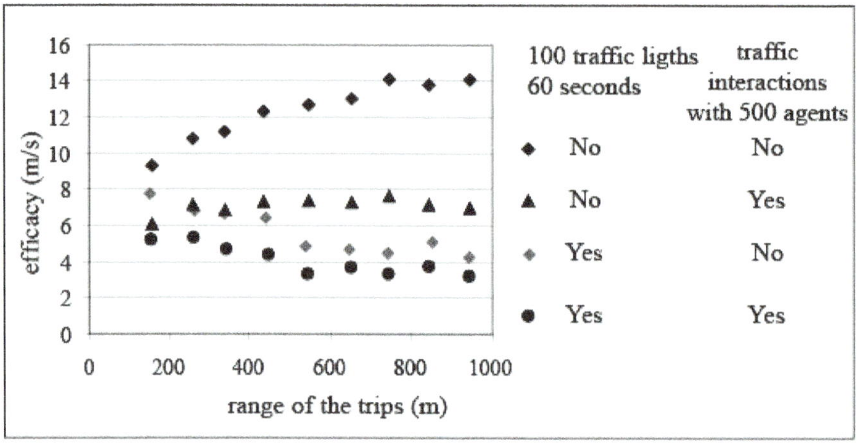

Fig. 12 Influence of traffic interactions on efficacy with and without traffic lights (network W1 Fig. 8)

interactions between agents are numerous (which is the case for hierarchised network with a lot of vehicles). In Fig. 13 we can see that it is even possible to reach a slow metric without traffic lights, for simulations with numerous agents (1,500) characterised by unhomogeneous behaviour (p = 0.9), but it remains an extreme simulation case.

Currently we are collecting data with GPS on efficacy of vehicles in traffic jam situations, in order to refine the calibration of the model and to see if the slow metric already exists in such situations.

This analyse focuses on the importance of driving habits on the global efficacy of the roads system. Thus, promoting homogeneous behaviours may be as important as investing in speed ways in order to improve accessibility. Concerning the impact of the number of agents in interactions on efficacy, it differs according to the type of networks. For homogeneous and connective

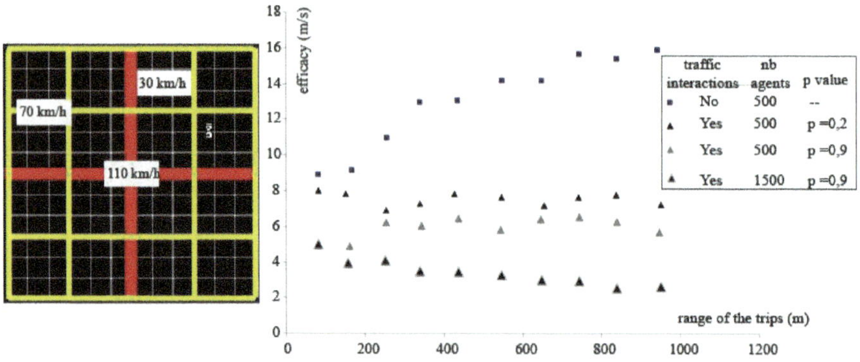

Fig. 13 Influence of parameter p on network efficacy (W2)

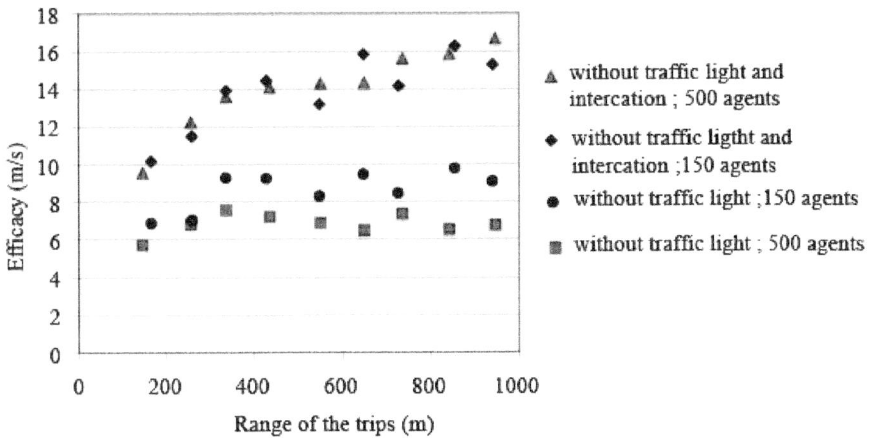

Fig. 14 Influence of the number of agents on network efficacy with and without traffic interactions for the network W2 Fig. 13

networks with 100 traffic lights, more than a thousand agents are required to really impact efficacy. For example for the network W1 in Fig. 8 efficacy decreases in mean of 0.8 m/s when the number of agents goes from 1,000 to 2,000 agents and the decrease of efficacy is of 1 m/s when the number of agents goes from 2,000 to 3,000. However, for hierarchised networks the impact of the number of agents is much more important, even if the number of agents is limited, since in this case the probability of interaction is very high (Fig. 14).

Here we can conclude that hierarchised networks are very efficient under free flow conditions, but that they are unable to assume an increase in traffic flows. At last, we can notice that the greater the number of agents, the higher the variance of the value of efficacy at the individual level. The convergence of individual situations seems to be reached in the case of free flow conditions or generalised congestion.

3.2.2 Efficacy versus Fluidity

However, equilibrating the efficacy for a given network between long and short trips, may not be an end in itself. Indeed, even if the goal of the slow metric is to lower the efficacy, especially for long range trips, the global system should remain overall efficient, i.e. it should provide quite a good accessibility and not lead to global congestion.

The decrease of efficacy and the congestion implied by any given solution have been respectively evaluated through the "fluidity index" and through the proportion of vehicles stopped during the simulations.

The fluidity index

The fluidity index compares a given state, characterised by some constraints (eg. traffic lights) with a previous one, free from such constraints. Let us define t_i as the duration of a simulated trip, under given traffic conditions. Let us define also τ_i the theoretical duration for that same trip, under free flow conditions (no traffic interaction and no traffic light). This theoretical indicator may be defined in a simple manner, for each trip, as:

$$\tau_i = \sum_{k=1}^{n} \frac{l_k}{v_k}$$

with l_k the length of edge k and v_k its speed. Given step 1 (acceleration) of the NaSch model, it is obvious that $t_i > \tau_i$. For each trip (and therefore each agent i), we can then define its loss in fluidity f_i as:

$$f_i = \frac{t_i - \tau_i}{t_i}, (0 \leq f_i \leq 1)$$

which varies between 0 and 1. From that point, we can then define an average indicator of fluidity loss F, which will also varies between 0 (no loss) and 1 (maximum loss):

$$F = \frac{1}{n} \sum_{i=1}^{n} \frac{t_i - \tau_i}{t_i}$$

Under free flow conditions, fluidity loss is null, as $t_i = \tau_i$, and it tends towards high values when traffic interactions (NaSch rules) and traffic lights are introduced. In extreme cases, F can go over 0.7. In these cases it may seem unacceptable to support such a decrease in the efficacy of the system. However, let us remember that F is a mean index whose value is lower for short range trips as the slow metric impacts long range trips above all. Moreover, for simulations taking into account the real frequency of the range of trips, which is very unsymmetrical (in urban areas trips lower than 3 kilometers represent more than 50% of the total), and if in the same time we improve the connectivity of the network for short range trips, we can reach a slow metric but with a global accessibility very close to the situation without traffic lights.

Here the question is, who should be encouraged through a high level of efficacy: people who made short range trips, or people making long range trips that urban policies try to restrain!

The proportion of vehicles stopped

As the fluidity index represents the loss of time due to news metrics compared to the theoretical situation of free flow conditions, it is highly correlated to the efficacy index and then it does not allows to really test the impact of the metrics on traffic jams. That is why we have added a graph to S3 which represents the number of vehicles stopped in real time during the simulation.

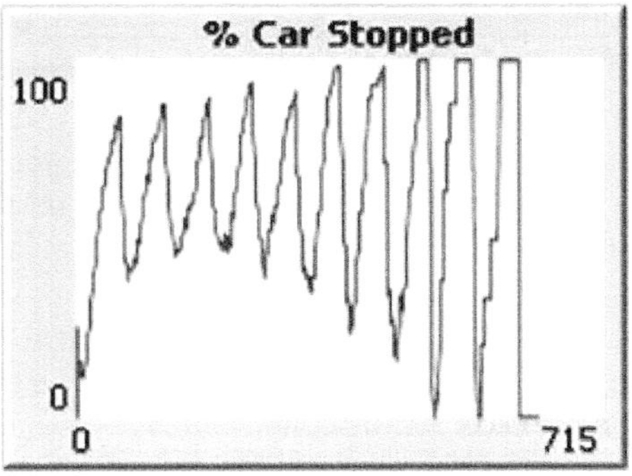

Fig. 15 Percentage of car stopped during a simulation

The more or less periodic shape of the graph is due to the traffic model and the traffic lights. The data shows that the percentage of vehicles stopped varies in mean from 5 to 40%. The values are once again higher for simulations involving hierarchised networks. For example, for the simulation in figure 13, for the hierarchised network W2 less than 30% of vehicles have come to a halt for only 13% of the duration of the simulation, whereas this value is of 45% in the case of a homogeneous network.

Theses values may seem very high, but one should not forget that the simulated traffic patterns are very important (a simulation with 500 agents represents one vehicle every 50 meters), and that vehicles stopped is not synonymous with congestion. Indeed, the vehicles stopped are not always the same. Thus, the simulations shows that congestion appears when the density of traffic reaches one vehicle every 8-10 meters. Until this extreme value the performance of the network decreases but without massive congestion.

Here we have a lack of reference data relative to real traffic conditions in order to know for example if a value of 30% of vehicles stopped represents a normal or unacceptable situation.

4 Conclusion

These first results are very positive as they suggest that slow speed metric could be reached with limited and well-targeted efforts, and this is particularly true for hierarchised networks as current networks are. However, several questions remain unanswered so far. The first one concerns the calibration of the slow metric which is very complex as it depends on the number, duration and location of the traffic lights, the structure of the network, the intensity of the traffic and the nature of the interactions. In order to optimize this calibration, we have to know more about what people will accept in terms of maximum duration of the traffic lights, fluidity and accessibility.

Concerning the simulation model, a second issue pertains to the assumption of non-adapting travelling behaviours: by imposing traffic lights with randomly chosen and ever-changing durations, we assume that no driver can predict a better solution to the shortest path (in distance) and therefore has no interest in modifying his route, thus reaching quite naturally an equilibrium. This very strong assumption needs further explorations, even before we imagine possible sources of heterogeneity introduced for example by real time traffic information.

Another issue concerns the regulation of the traffic jam. In the case of a traffic jam due to a long traffic light, we can imagine that the traffic lights automatically turn green when the charge on the edge is too important even if it has not reached its theoretical duration. Such a system of auto-regulation has been tested with very good results in terms of fluidity by Lämmer, Helbing [8].

The last issue concerns the social acceptance of such policy constraints. This issue raises a great variety of debates that we cannot cover here. For example, can the slow metric give more economic value to physical proximity in order to modify the current strategies of localization of households and firms. Such new strategies would be in line with sustainable development objectives. But what we can show from our simulations is that the "do nothing" policies that are usually adopted, i.e. letting traffic jams occur and "regulate" the system, lead to non-managed slow speed metrics. Indeed, we saw that increasing the number of agents in the systems without introducing traffic lights leads to the same kind of curves. Despite the complexity of the problem and its highly political dimension, our claim is that we can achieve better outcomes with a proactive and well-targeted strategy.

References

1. Banos, A., Godara, A., Lassarre, S.: Simulating Pedestrians and Cars Behaviours in a Virtual City: An Agent-based Approach. In: Proceedings of the European Conference on Complex Systems, p. 4 (2005)
2. Freeman, L.C.: A Set of Measures of Centrality Based on Betweenness. Sociometry 40, 35–41 (1977)
3. Genre-Grandpierre, C.: Forme et Fonctionnement des Réseaux de Transport: Approche Fractale et Réflexion sur L'aménagement des Villes: Thèse de Doctorat en Géographie. Université de Franche-Comté (2000)
4. Genre-Grandpierre, C.: Des Réseaux Lents Contre la Dépendance Automobile? Concept et Implications en Milieu Urbain. L'Espace Géographique 1, 27–39 (2007)
5. Girvan, M., Newman, M.E.J.: Community Structure in Social and Biological Networks. Proceedings of the National Academy of Sciences of the United States of America 99, 7821–7826 (2002)
6. Godara, A., Lassarre, S., Banos, A.: Simulating Granular Traffic Using Cellular Automata and Multi-Agent Models, pp. 411–418. Springer, Berlin (2007)
7. Gutiérrez, J., Monzón, A., Pinéo, J.M.: Accessibility, Network Efficiency and Transport Infrastructure Planning. Environment and Planning A 30(8), 1337–1350 (1998)

8. Lämmer, S., Helbing, D.: Self-Control of Traffic Lights and Vehicle Flows in Urban Road Networks. Journal of Statistical Mechanics: Theory and Experiment (2008), P04019
9. Levinson, D., Kumar, A.: The Rational Locator: Why Travel Times have Remained Stable. Journal of the American Planning Association 60(3), 319–332 (1994)
10. Nagel, K., Schreckenberg, M.: Cellular Automaton Models for Freeway Traffic. Journal de Physique I 2, 2221–2229 (1992)
11. Newman, M.E.J.: Power Laws, Pareto Distributions and Zipf's Law. Contemporary Physics 46, 323–351 (2005)
12. Penn, A., Hillier, B., Banister, D., Xu, J.: Configurational Modelling of Urban Movement Networks. Environment and Planning B: Planning and Design 25, 59–84 (1998)
13. Schadschneider, A.: Traffic Flow: A Statistical Physics Point of View. Physica A: Statistical Mechanics and its Applications 313, 153–187 (2002)

Impact of Tenacity upon the Behaviors of Social Actors

Joseph El-Gemayel, Christophe Sibertin-Blanc, and Paul Chapron

Abstract. The Sociology of the Organized Actions is a well-established theory that focuses upon the actual behaviors of the members of social organizations, and reveals the (to a large extent implicit) motives of social actors. The formalization of this theory leads to model the structure of an organization as a *social game*, including the Prisoners' Dilemma as a specific case. In order to perform simulations of social organizations modeled in this way, the *SocLab* environment contains an algorithm allowing the model's actors to play the social game and so to determine how they could cooperate with each other. This algorithm includes several parameters, and we study the influence of one of them, the *Tenacity*.

1 Introduction

In collaboration with sociologists, we have formalized the Sociology of the Organized Action (SOA) of M. Crozier and E. Friedberg [10]. This theory [3] [4] [5] is well experienced and it is the foundation of the methodologies applied by most consultants when they are called to analyze the very origin of dysfunctions within social organizations.

This formalization has led to the development of the environment *SocLab*[1] that allows the user to describe the structure of an organization, to explore its state space and properties, and to find, by simulating the rationality of social actors, how they could adapt their behaviors to each other [7].

This simulation algorithm, which is based on reinforcement self-learning rules [12], implements a model of rational actors as they are considered by the

Joseph El-Gemayel · Christophe Sibertin-Blanc · Paul Chapron
Université de Toulouse Capitole / IRIT, 2 Place du Doyen G. Marty, F-31042 Toulouse Cedex 09, 0033 (0)5 61 63 35 00
e-mail: {joseph.el-gemayel,sibertin,paul.chapron}@univ-tlse1.fr

[1] SocLab, available at : http://soclab.univ-tlse1.fr/

Q. Bai and N. Fukuta (Eds.): Advances in Practical Multi-Agent Systems, SCI 325, pp. 287–306.
springerlink.com

SOA: every one cooperates with others while trying to obtain the means to achieve his goals. This algorithm includes a set of parameters (ability to discriminate situations, memory, ...) that correspond to cognitive or behavioral properties of actors and have an impact on the expected simulation results: a fast convergence of simulations and a good cooperation between the actors.

Among these parameters, the *tenacity* of an actor determines his persistence, how hard he tries to get from others the capacity to reach his goals; tenacity is the inverse of resignation. Tenacity seems to be the most influential parameter, and this paper studies the influence of its value upon the obtained results, by means of a sensitivity analysis applied to different organizations. It appears that the more the actors are tenacious, the more the simulations last and the more the actors cooperate. Thus, the appropriate value of actors' tenacities in the simulation of an organization is a compromise between the cost of simulations and the quality of the cooperation. It also appears that this value depends on structural properties of organizations that are related to the difficulty, for every actor, to find how to cooperate. These results are useful for the study of organizations in the framework of the SOA, are they actual organizations or organizations envisioned in the context of organizational changes. They are also significant outside the SAO theory, since the social game may be applied in other domains, for example as a coordination model for societies of autonomous software agents [9].

The rest of this paper is organized into six parts. After a brief presentation of the SOA theory in Section 2, we present the main elements of the metamodel in Section 3 and the cases used in the experimentations Section 4. After a brief explanation of the actors' behaviors, Section 5 introduces the algorithm modeling this behavior and describes its various parameters. Then, we study its results on two social games by varying the value of the *tenacity* parameter in Section 6. Finally, we interpret these results in Section 7 before to conclude.

2 The Sociology of Organized Action

The French school of sociology of organizations has developed a research program whose fertility is well established and which attempts to discover the actual functioning of an organization beyond the formal rules that codifies it [3] [4] [5]. Organizations are "social constructs" binding the actors' behaviors, actors who are the heart of the organization. The actors of an organization have a strategic behavior framed by a bounded rationality [11]: they intend to achieve some goals (their own goals and also those of the organization), and each actor aims at having enough power to get from others the capacity of action he needs to achieve his goals. This power arises from controlling the access to one or more uncertainty zones needed as resources by other actors for their actions (each actor, at the same time, controls and depends on others through the resources they share). The controller of a resource sets the "terms of exchanges" for this resources: he makes his behavior more or less unpredictable for the actors that depend on this resource and so urge them to account for his own needs.

Therefore, the power relations structure the configurations of social organizations. These configurations feature a relative stability in a balance of social relationships, resulting from the fact that each actor both controls some resources and depends on others.

Describing an organization as a system of related Actors and Resources produces a context of interaction rather precisely delimited; this context structures the cooperation of the actors which, however, keep some level of autonomy, their room for manoeuvre.

3 The Meta-model of Organizations

The formalization of the SOA starts with the definition of a language for drawing formal models of organizations that is a meta-model of the structure of organizations as they are analyzed by sociologists.

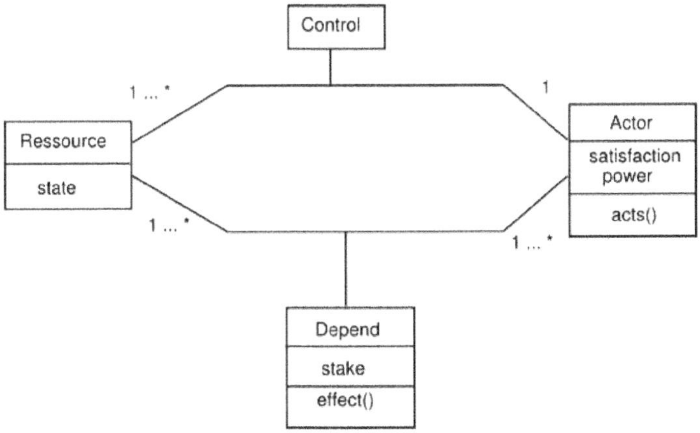

Fig. 1 The UML meta-model of an organisation

Fig. 1 is a UML representation of a simplified meta-model [10]. It contains the relevant *Actors*, and *Resources*, and various actors are related to each resource: one who controls it with some autonomy, and others who depend on it. *Actors* are the active entities, and *Resources* are both the means necessary for the actors to reach their goals and the supports of their interactions. As Friedberg wrote: "no power without relations, no relations without exchange" [5]. Therefore, the power supposes relations, which involves exchange, which itself requires objects of exchange: the Resources.

An organization can be modeled as a *social* game composed of:

- $A = \{a_1, \cdots, a_n\}$, the set of the actors.
- $R = \{r_1, \cdots, r_m\}$, the set of the resources of the organization.

Each resource has an attribute *state*, whose value is set by its controller inside a *space of choice*. This space of choice (chosen to be $[-1;1]$) characterizes the behaviors range of the controller: negative values correspond to uncooperative behaviors (-1 being the most uncooperative behaviors), positive values to cooperative behavior, and the close to zero values to neutral behaviors that may be assessed neither as cooperative or uncooperative. The state of the organization is thus the vector of every resource state.

- $controlledBy : R \rightarrow A$, a function that determines which actor controls each resource. Let's denote as $control : A \rightarrow \phi(R)$, the reciprocal function that determines the set of relations controlled by each actor.
- $stake : A \times R \rightarrow [0;10]$ a function that determines the stakes each actor puts on the resources he depends on. An actor makes this distribution according to the importance of the related resource towards the achievement of his goal; a null stake means that the actor does not depends on the resource. Each actor distributes a fixed quantity of 10 stakes units.
- $effect_r : A \times [-1;1] \rightarrow [-10;10]$, the effect function associated with each resource; it determines, for every actor that depends on this resource and according to the state of this resource, how well this actor can access the resource. Negative values correspond to difficulties to use the resources as whished by the actor, positive value to facilities, and the zero value as the normal, neither criticizable nor favorable cases.

The model of an organization may also include constraints between resources (the value of one resource state restricts the range of the space of choice of another resource), and solidarities between the actors. They introduce much more complexity into social games and are not considered here, to keep the paper concise and because they are not useful to analyze the impact of the tenacity parameter.

In order to assess the actors' situations at a state of the game, we consider, for each actor, the sum, over all the resources he depends on, of the product of his stake by the effect of the resource on him. This quantity, called the *satisfaction* of an actor, evaluates his possibility to access the resources he needs to achieve his goals, weighted by his need of these resources, and thus his global *capacity of action*.

$$Satis : A \times [-1;1]^m \rightarrow [-100;100]$$
$$(a, s_{r1}, \cdots, s_{rm}) \mapsto \sum_{r \in R} stake(a,r) \times effect_r(a, s_r)$$

Since the SOA requires that each actor controls resources, another significant quantity to be taken in consideration is the ability of an actor to influence the behavior of others depending on the resources he controls by his contribution to their satisfaction. This quantity, the *power*, is essential in SOA, and can be quantified as follows:

$$Power : A \times A \times [-1;1]^m \rightarrow [-100;100]$$
$$(a, b, s_{r1}, \cdots, s_{rm}) \mapsto \sum_{r \in R; a\ controls\ r} stake(b,r) \times effect_r(b, s_r)$$

In the social game, each actor seeks to obtain from others a high level of satisfaction and, to this end, he adjusts the state of the resources he controls. More precisely, at each step, every actor chooses moves of the values of the states of the resources he controls, and this change of the game state modifies the satisfaction of all actors. Let $(s_{r1}, ..., s_{rm})$ be a state of the organization and $(c_r)_{r \in control(a)}$ be the moves chosen by actor a such that $(c_r + s_r) \in [-1;1]$. Once each actor has chosen such an action, the game goes to a new state defined by

$$Transition : [-1;1]^m \times [-1;1]^m \rightarrow [-1;1]^m$$
$$((s_{r1}, \cdots, s_{rm}), (c_{r1}, \cdots, c_{rm})) \mapsto (s_{r1} + c_{r1}, \cdots, s_{rm} + c_{rm})$$

where c_r is chosen by the actor *controlledBy*(r). The game ends when a stationary state is reached: actors no longer change the state of the resource they control, because they accept the level of satisfaction that the current state of the game provides to them.

4 Illustrations

As an illustration, we present two types of organizations: a very simple example functioning as a Prisoners' Dilemma, and the Bolet case as the model of a complex social organization.

4.1 The Prisonner Dilemma

The Prisoner's dilemma [1] is a simple case consisting of two actors: *A1* which controls the resource *R1* and *A2* which controls *R2*. Each actor put 1 stake unit on the resource he controls and 9 units on the resource controlled by the other actor. The effect functions of the actors are linear and symmetric, with a slope of -1 on the resource controlled by himself and 1 on the other, as shown in Fig. 2.

Table 1 shows the respective satisfactions of the two actors for typical values of the resources. If we consider the *Global satisfaction*, that is the sum of the satisfactions of the actors, the best case is when both cooperate -they give up the capacity of action they could get from the resource they control to give the other full satisfaction-, while the worst case (which is the Nash

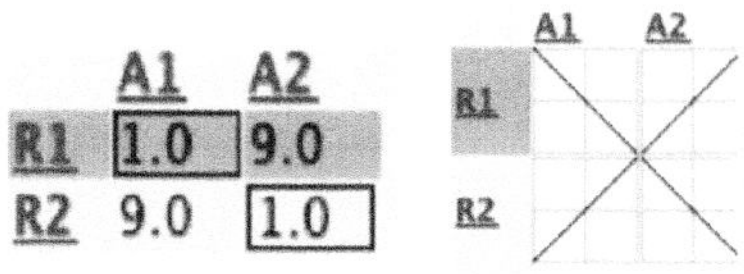

(a) (b)

Fig. 2 The stakes of the actors upon the resources (boxes with black borders show the controller) and the effect of the resources on the actors (x axis is the space of choice of the resource, y axis is the resulting capacity of action for the actor).

equilibrium) is when they do not. When one actor cooperates and the other does not, the former get -100 as satisfaction and the latter +100. (Due to the symmetry of the game, satis(A1) = - satis(A2) when R1.state = - R2.state).

Although this case is the model of no social organization, it fulfills an essential property of real social games: they are not zero-sum games, and high global scores first require the actors' cooperation and second are compatible with high individual scores [2].

Table 1 Satisfaction(A1) / satisfaction(A2) in characteristic states of the Prisoners' Dilemma

		Value of R1		
		-1	0	+1
Value of R2	-1	-80/-80	-90/10	-100/100
	0	10/-90	0/0	-10/90
	+1	100/-100	90/-10	80/80

4.2 *The Bolet Case*

The Bolet case is the model of a small familial firm. It contains four actors - the head shop (HS), Jean and the Bureau of Studies (Jean-BE), Andrew the chief of the production, and the father, founder of the enterprise - and six resources - the decision to purchase a new machine, the nature of the prescription given by Jean to the head shop, the supervising of the nature of the prescription, the application of the prescription, the supervising the application of the prescription, and the investment in the production. A careful analysis of the functioning of the firm leads to define the controls, stakes and effect function as shown in Fig. 3 and 4.

Since an organization state is the vector of every resource state that describes the behavior of every actor, each organization state corresponds to a way it can work. Since we are interested in understanding the actual behaviors of the actors of the organization, we can explore in an analytic way the

	HS	father	Andrew	Jean-BE
purchase-decision	4.0	1.0	1.0	4.0
application-presc	1.0	1.0	1.5	2.0
investment-production	2.0	5.0	3.0	0.0
control-application-presc	2.0	1.0	1.5	0.0
nature-presc	1.0	1.0	1.5	2.0
control-nature-presc	0.0	1.0	1.5	2.0

Fig. 3 The stakes of the actors upon the resources (boxes with black borders show the controller of the resource).

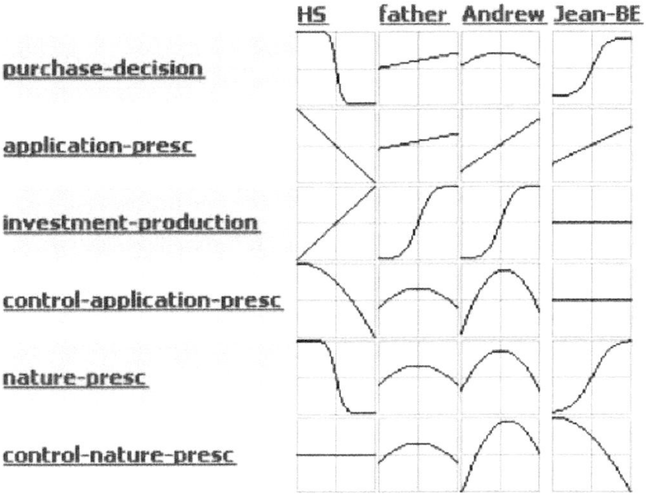

Fig. 4 The Effect functions of the Bolet case (x axis is the space of choice of the state of the resource, y axis is the resulting capacity of action for the actor)

whole set of all the possible states of the organization. Among them, some are of a high interest in that they maximize a given criterion, e.g. the Nash equilibriums, the Pareto optima or the extreme value of the satisfaction of a particular actor or the sum of their satisfactions. This analysis is global (the state space is sorted according to the criterion) and it is a good starting point when an organization has to be studied. The *SocLab* environment allows the user to see these particular states and to interactively get details upon all states.

As an example, Table 2 shows the states of the organization that maximize or minimize the satisfactions of every actor (in row), and the satisfactions that this state provides to every actor (in column). For example, the maximum satisfaction that the HS could get is 100, and in this case the global satisfaction is 87 while the Father's satisfaction is 42.

Table 2 Satisfaction values in particular states of the Bolet case that maximize or minimize the actors or global satisfactions

	HS	Father	Andrew	Jean	Global
HS max satisfaction	**100**	42.0	-13.9	-41.1	87.0
Father max satisfaction	-20.5	**65.8**	75.8	51.4	172.5
Andrew max satisfaction	17.3	63.5	**80.4**	14.9	176.1
Jean max satisfaction	-60.0	-49.5	-52.6	**81.1**	-81.0
Global max satisfaction	63.3	62.9	74.8	2.2	**203.2**
HS min satisfaction	**-100**	-49.5	-44.2	81.1	-112.6
Father min satisfaction	60.0	**-57.5**	-73.6	-41.1	112.2
Andrew min satisfaction	60.0	-57.5	**-73.6**	-41.1	-112.2
Jean min satisfaction	60.0	-57.5	-58.6	**-81.1**	-137.2
Global min satisfaction	-55.4	-55.1	-46.8	-32.1	**-189.4**

5 Simulation

Among the states of an organization, the question is to determine the ones that are socially feasible. If we consider a hypothetical organization where two persons regularly interact according to the pattern of the Prisoners' Dilemma, it is highly improbable that the one accepts to cooperate while the other does not! More generally, the analytic approach does not take into account the capacity of the actors to actually reach a specific state, for cognitive, psychological or social acceptability reasons. To get insights into the behaviors that the actors of an organization could adopt one to another, we can make them to learn how to behave, and thus make simulation of these join self-learning processes.

The design of the algorithm that gives rise to these self-learning processes can be grounded upon four principles of the SOA regarding the actors' behaviors. According to the SOA [3] [4] [5], the behavior of actors is *strategic*: each actor tries to protect and increase his action capacity and to this end, he handles the resources he controls. As a consequence, that behavior is quite stable, and a constitutive property of organizations is the *regularization* of the actors' behaviors. However, actors only have a *bounded rationality*, due to social relationship opaqueness, and to cognitive skills limitations [11], and they seek to obtain a satisfactory situation that is not necessary an optimum. Finally, this behavior is *cooperative*; indeed, cooperation is required for the organization to work in a proper way, and every actor recognizes his interest in this proper working.

5.1 A Bounded Rationality Algorithm for Actors' Cooperation

The assumption of rationality of social actors leads to base their behaviors on the basic cycle in which each player chooses an action according to his own objectives:

1. collect information about the state of the system,
2. select what action to undertake,
3. perform that action (i.e. modify the states of the resources he controls),

until the game is stationary, i.e. all actors are satisfied, everyone no longer wants to change the state of his controlled resources.

Our model of the actors' rationality [8] is a self-learning process by trial and error [12], where rules are expressed as (*situation, action, strength, age*):

- *situation*: the situation of the actor at the time of the creation of the rule, in the form of a list of the capacities of action he gets from each resource he depends on.
- *action*: a list of changes on the state of each resource controlled by the actor, chosen at random.
- *strength*: a numerical evaluation of the effectiveness of the rule, initialized at 0, reinforced according to the increase or decrease of satisfaction provided by the applications of the rule.
- *age*: the age of the rule initialized at 0 and incremented at each step.

Initializing the state of resources in an arbitrary manner, e.g. the value 0 corresponding to a neutral behavior, each player selects an action and updates his rule base at every step of the simulation as follows (see Table 3):

- He compares his current satisfaction with his satisfaction at the previous step. According to this assessment, the learning rate α and the respective reward *pr* of the last and penultimate rules, the *strength* of the previously and penultimate applied rules are increased or decreased proportionately to the improvement of the satisfaction. Then the *ambition*, the *action-range*, the *age* and the *learning rate* of the rules are updated.
- He selects rules, whose *situation* component is close to his current *situation* (inside a ball of a *scope* radius, according to a Euclidean distance), and chooses the rule with the greatest positive *strength*.
- If the set of selected rules is empty (at the beginning of simulation or because no rule is close to the current *situation*), a new rule is created, with the current *situation*, an action chosen at random (but smaller than the *action-range* parameter) and a *strength* and *age* initialized at 0.
- Finally, once every player has chosen an action, each player performs that action so that the resource state values are modified.

Any learning process requires an adjustment between the rates of exploration and exploitation of the state space [9]. If the exploration is excessive, the process will never reach a stable state because still another solution is tested, while if the exploitation is excessive, the process will retain the first acceptable solution and ignore possibly better alternatives.

In our model, each player has an *ambition* initialized to the maximum value of satisfaction he can reach, and this *ambition* will come closer to his current satisfaction in the course of the simulation. The bigger the gap between his current satisfaction and *ambition*, the more the actor explores, that is undertake vigorous actions (that is important changes of resource states), quickly forgets learned rules and rewards rules which increase widely his satisfaction. When each player gets a satisfaction greater than his ambition, the algorithm stops and then we consider that a regularized state is reached: every actor thinks that he has the best level of satisfaction to achieve his goals that he can obtain.

This algorithm requires very little knowledge about the state and the structure of the organization, as actors only perceive resources states and the satisfaction level they get from. That makes this learning process socially and cognitively likely, and it does not require any communication or deliberation skills.

5.2 The Main Parameters of the Algorithm

This section explains the role of the specific parameters of this algorithm. They correspond to psychological, cognitive, psychosocial or social capabilities of human beings that we do not discuss here.

5.2.1 Tenacity

This parameter determines the rate at which the *ambition* of the actor comes close to his satisfaction; it may vary from 0 to 999. The higher the tenacity, the slower the ambition decreases, and the longer the actor explores and has a chance to reach a higher satisfaction. As a consequence of a deeper exploration, the simulation is likely to last longer before a stationary state is reached. The value of this parameter is initialized according to the complexity of the organization to simulate, i.e. depending on the number of actors and resources and the distribution of the stakes of the actors: the more complex the organization is, the longer it will take to an actor to find how to cooperate with others, as there are more rules and more various situations to learn and to test.

5.2.2 Action Range

An action of an actor is a change in the state of every resource he controls, and the *actionRange* parameter defines the maximum amplitude of these changes in a new action. The value of actionRange can vary from 0 to 1.5. The more the current satisfaction is far from the actors' ambition, the higher is his action range in order to effectively operate the rate of exploration and exploitation. It can't be zero in order to maintain a minimum exploratory capacity.

Table 3 The actors' learning algorithm

/* *pr is the respective reward of the last and penultimate rules, α is the learning rate, SR_t is the Selected Rule at time t, M is the set of rules that are applicable at the current step, minSatis is the minimum satisfaction of the actor, e.g. as in Table 2.* */

```
While there is an actor such that satis < ambition
  For each actor a:
    updating assessment
    assessment = satis_t - satis_{t-1}

    updating strength of last and penultimate selected rules
    SR_{t-1}.strength=
      SR_{t-1}.α×SR_{t-1}.strength +(1-SR_{t-1}.α)×pr×assessment
    SR_{t-2}.strength=
      SR_{t-2}.α×SR_{t-2}.strength +(1-SR_{t-2}.α)×(1-pr)×assessment

    updating ambition
    ambition_t =
     ambition_{t-1} + (satis_t-ambition_{t-1})×(1000 - tenacity)/10^5

    updating the age and learning rate of each rule
    For each rule R in RuleBase
      R.age += 1
      R.α = R.age / nbSteps

    updating Action-Range
    if (satis_t<ambition_{t-1}) then
      actionRange=
        initialActionRange+0.5×(ambition_{t-1}-satis_t)/(ambition_{t-1}-minSatis)
    else
      actionRange = initialActionRange

    forgetting bad rules
    For each rule R in RuleBase
      if (R.strength < 0 and R ≠ SR_{t-1})
         then RuleBase.remove(R)

    selecting the set of matching rules M
    M.clear()
    For each rule R in RuleBase
      if(distance(R.situation , CurrentSituation())<scope)
      then M.add(R)

    selecting an action
    if (M.isEmpty()) then
      SR_t= (CurrentSituation(),(atRandom()*actionRange),0,0)
      RuleBase.add( SR_t)
    else
      SR_t= maximumStrengthRuleFrom(M)
      a.action = SR_t.action

    performing the actions
    For each actor a:
      performAction(a.action)
```

5.2.3 Scope

This parameter stands for the actor cognitive ability to discriminate the situations. A situation is a vector of the capacity of acting perceived from every resource the actor depends on. If the Euclidean distance (weighted by the stakes) between two situations exceeds the scope threshold, the two situations are considered as different. At each step, the applicable rules are the ones whose situation component is close to the current situation of the actor. A small value of *scope* (near 0), means that the actor makes a clear distinction between situations, so that rules have a small domain of application and the actor creates many specific rules. Whereas an actor with a very large value of *scope* (e. g. the amplitude of the range of the actor's satisfaction) is not able to distinguish the situations and all rules may be applied at every step.

5.2.4 Percentage of Distribution of Reward

This parameter determines the reward distribution on the last and penultimate rules. When a player applies a rule, he will be able to perceive its whole effect only two steps ahead, once the other actors have reacted to this state change, and consequently have modified the states of the resources they control. If only the last rule is rewarded, rules will be applied and rewarded that don't deserve to be rewarded because it is good for the actor himself, but bad for the others, implying a harmful reaction.

6 Sensitivity Analysis

As tenacity impacts the exploration / exploitation ratio of the social actors behaviors, we are looking for the values that provide actors with a good satisfaction level, without spending too many simulation steps to stabilize the social game.

To study the influence of the tenacity on the number of steps needed for convergence and the level of satisfaction gained by the actors, we analyze the results of the simulation algorithm on two types of organization models (the classical prisoners' dilemma, and the model of the Bolet case), and for each model, we carry out several simulations by varying the value of tenacity.

Each experiment was conducted with 50 simulation runs. The parameters of the algorithm were set to values (tested in previous simulations) needed for good convergence (scope = 16, InitialActionRange = 1.0, and percentage of distribution of reward = 30% for the last rule and 70% for the penultimate rule). For the tenacity values below than 900, the number of steps and the actors' satisfactions grow linearly, while significant variations appear beyond 900.

6.1 Prisoner's Dilemma

Fig. 5 and 6 show results from a sensitivity analysis of the PD model, including 100 experiences where the tenacity of both actors varies at random between 900 and 999.

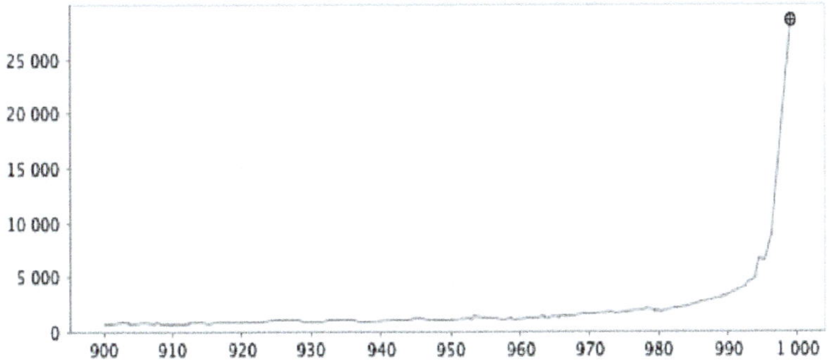

Fig. 5 The variation of the average number of steps needed for the convergence, depending on the variation of the tenacity of both actors.

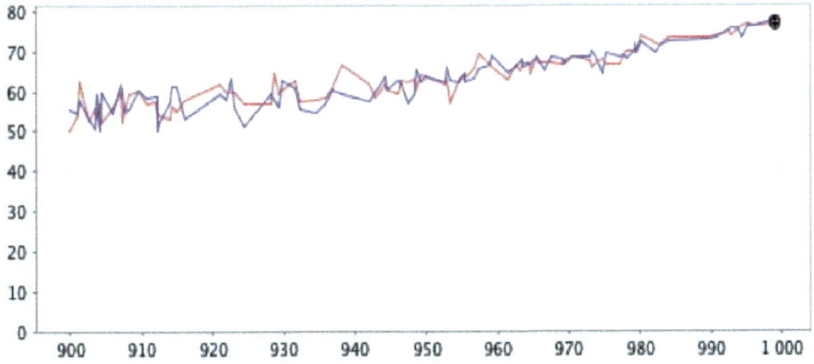

Fig. 6 The variation of the average satisfaction of each actor, depending on the variation of the tenacity of both actors.

The number of simulation steps varies from about 800 to 28000, with an intermediate value of 2000 for a tenacity of 980 (Fig. 5).The satisfactions of the two actors are quite similar, due to the symmetry of the game (Fig. 6) and they vary from 55 (that is 84%[2] of the maximum value) to 76 (97% of the maximum value), with an intermediate value of 70 (94% of the maximum value) for a 980 tenacity.

[2] This proportion of the maximum satisfaction is given by (value - min_value)/(max_value - min_value) = 55+ 80/(80 + 80).

It appears that for this game, making the tenacity parameter having a value about 975 is a good compromise between the quality of the cooperation and the length of the simulation, since a huge additional cost of the computation gives a very little improvement of the actors' satisfactions.

We can consider a variant of the standard PD consisting of four actors that depend the one on another in a circular way: A2 depends on A1 that depends on A4 that depends on A3 that depends on A2. In this case, it is much more difficult for the actors to discover how they could cooperate, since all of them must do simultaneously. The number of simulation steps varies from about 2000 to 30000, with a value of 10000 for a tenacity of 990, while the satisfactions of the four actors vary from 30 to 62, with a value of 50 for a tenacity of 990. These results are comparable to the ones of the two prisoners case, but the improvement of tenacity is not enough to produce a good cooperation.

6.2 The Bolet Case

For this organization, we ran two sets of experiments: in the first one, the sensitivity analysis is performed by varying every actor's tenacity, whereas in the second one, only the HS's tenacity varies.

6.2.1 Variation of All Actors' Tenacity

Figures 7 and 8 show results from a sensitivity analysis of the Bolet model, including 100 experiences where the tenacity of all actors varies at random between 700 and 999.

On Fig. 8, curve 1 represents Jean-BE's satisfaction, curve 2 represents Andrew's, curve 3 represents HS's, and the satisfaction of the Father is represented by curve 4.

Fig. 7 The variation of the average number of steps needed for the convergence, depending on the variation of tenacity for all actors.

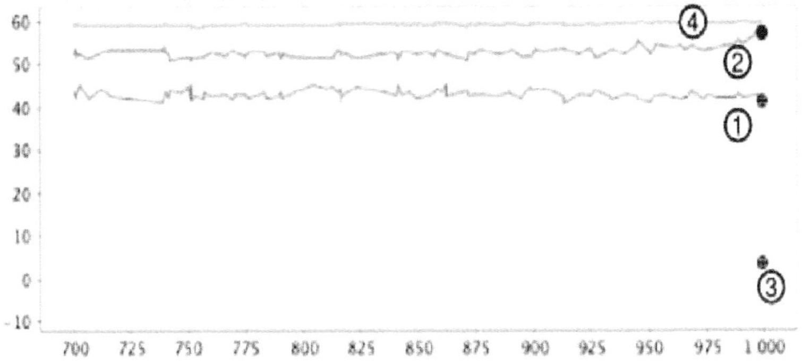

Fig. 8 The variation of actors' average satisfaction depending on their tenacity.

The number of steps of simulation varies from about 1500 to 265000. Up to a tenacity value of 960, the number of steps remains under 10000 (Fig. 7.). For a tenacity greater than 980, the number of steps explodes.

There is no stochastic variation in the (mean) global satisfaction that is stable at 38, that is 90% of its maximum value. The satisfactions of the Father (59), of Andrew (from 52 to 56) and Jean-BE (from 43 to 41) are also quite stable, while the HS's one slightly increases from -5 to 3. In this case, the improvement of tenacity is useless: it brings no satisfaction improvement. Compared with the states given in table 2, this situation is closer to the Father's maximum satisfaction than to the global maximum.

It is noticeable that the HS actor has a very low satisfaction with regard to others. Thus, it is of interest to proceed to a second experiment, where we plot the actors' satisfaction variations when only the HS's tenacity varies.

6.2.2 Variation of HS's Tenacity

The curves in Fig. 9 and 10 show results from a sensitivity analysis of the Bolet model, including 100 experiences. The HS's tenacity varies from 700 to 999 while the tenacity of other actors is fixed to 750.

An increase in the number of steps, beyond the tenacity value of 980, still occurs, but much less than in the previous case. The global (mean) satisfaction is stable until 900 tenacity, from 37 to 38, and then slightly increases to 43. The Father's satisfaction is very stable (57) as the Andrew's one (from 51 to 54). The HS's one grows from -5 to 45 while the Jean-BE's one decreases from 45 to 20.

The improvement of HS's tenacity reveals his conflict with Jean-BE that can be seen by examining the Table 2 (compare the states where they obtain their min and max satisfactions). In additions, they are the more sensitive actors to the behaviors of others (they have the larger amplitude of their ranges of satisfactions) and their best satisfactions are farer from the maximum global satisfaction.

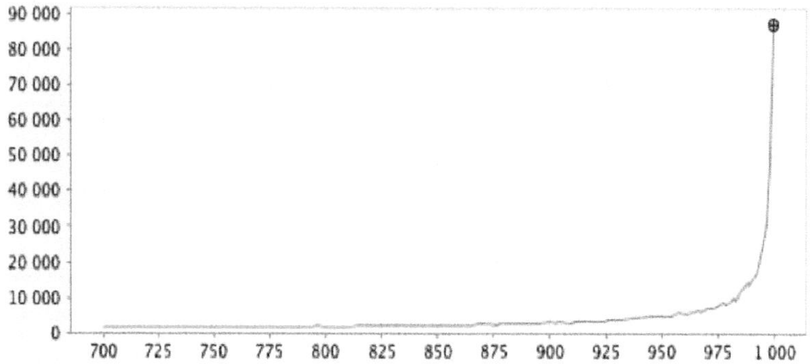

Fig. 9 Variation of the average number of steps needed for convergence depending on the variation of HS's tenacity.

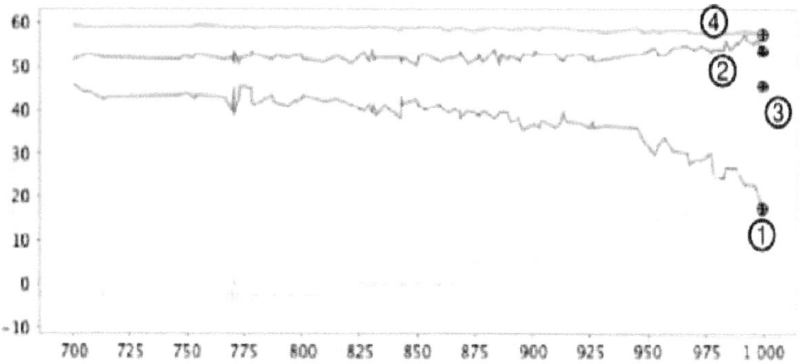

Fig. 10 Variation of the average actors satisfaction depending on the HS's tenacity.

To understand what happens, let us have a look at the detail of simulation results in both cases. First of all, the most relevant resources are the *decision to purchase* and the *investment in the production* that bear 10 stake units (see Fig. 3). At the 750 tenacity of all actors, the HS's investment is very high with a small standard deviation, he does not hesitate, while the Father's decision is to purchase, but with an important standard deviation: he hesitates. When HS is provided with a high tenacity, he still features a high investment, but he tests very often the inverse possibility. The decision of the Father to not purchase in this case (that produces the main part of the inversion of HS's and Jean-BE's satisfactions) is his reaction to the threat of the non-investment of HS.

7 Interpretation

For any organization, the level of the actors' satisfactions and the number of steps of simulations increase with the value of the tenacity parameter in a

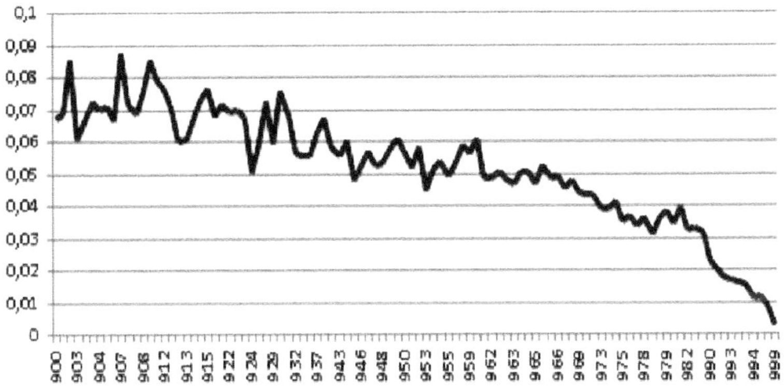

Fig. 11 The variation of the mean satisfaction divided by the number of steps, depending on the variation of the actors' tenacity in the Bolet case.

quite linear way until the 900 threshold, and they increase in an exponential way beyond this value. This is the reason to explore the influence of this parameter within the 900 -999 range.

The tenacity parameter has been introduced in the simulation algorithm to determine the level of exploration with regard to the level of exploitation: the more the current satisfaction of an actor is below his ambition, the more he explores, and his level of ambition decreases as the inverse of the tenacity. The increase in tenacity produces an increase in exploration which in turn produces an increase in the duration of simulations, and this increase is more and more important beyond a threshold which is about 960 - 980: this fact has been observed in the two cases presented in this paper and also in other organizations.

Regarding the influence of the tenacity upon the level of cooperation of actors, and their resulting individual levels of satisfaction, the two examples figure different situations: an effective linear improvement in both the two and four actors PD cases due to a better exploration and, in the Bolet case, no change when all actors have the same tenacity and an important inversion, surprising at first glance, of the HS's and Jean-BE satisfactions when HS improves his tenacity. To contrast the cases, the PD game is sensible to a characteristic of the actors, i.e. their tenacity, and the sparsity of cooperative states (in the four actors PD) requires a higher tenacity, while the Bolet game is not when all actors feature the same tenacity. The Bolet organization is strongly regulated and firmly constraints the behaviors of actors that are engaged in numerous relations. However, a singular actor, HS, can escape these constraints to obtain a much better situation to the detriment of a vulnerable actor.

In any case, a relevant value of the tenacity parameter has to be found, that is a compromise between the two effects of a high tenacity: a costlier simulation and a better cooperation. To this end we can consider the following function:

$$Tenacity\ Influence(tenacity) = \frac{Global\_Satisfaction(tenacity)}{number\_steps(tenacity)}$$

that compares the respective increases in the global satisfaction and the duration of the simulation (see Fig. 11). This function has a negative slope for any simulation: the number of steps increases more quickly than the global satisfaction while the tenacity is improved. A good value for the tenacity is when this function comes under some threshold that is determined by the relative importance of the cost of simulations and the quality of actors' cooperation.

8 Conclusion

This SAO-based model of coordination is relevant in cases where the cooperation of the actors is of first importance, notably in the contexts of Command and Control [6] or of crisis management [13]. On the one hand, the success of the operation requires the full cooperation of each member of the Headquarter; on the other hand, each member has to promote the own interests of his department, to protect the people acting on the ground, handle carefully the resources or in the perspective of further operations. The cooperation is not always so easy to achieve [14], [7], and a rigorous model of such interaction contexts can highlight structural drawback of the organization. Moreover, the tenacity directly refers to a psychological feature of the actors, and simulations could help to select the most appropriate representative of a department that is in a difficult position.

The distinctive feature of the algorithm presented in this paper is to allow the emergence of states featuring a high level of cooperation between the actors, a global property, while it is very parsimonious of the cognitive capacities of the actors and of their the knowledge of both the structure and the current state of the organization they belong to. It is a distributed optimization algorithm that provides good results while it demands very few information.

In this paper, we investigate the influence of the tenacity parameter on the quality of results of this simulation algorithm, that is the number of steps required for convergence and the actors' capacities to achieve their goals. It appears that the value of this parameter has a great influence on the simulation results. If some principle have been highlighted, a deeper study is necessary to relate the effects of this parameter with structural properties of games.

Since our model focuses on the behavior of autonomous actors and the value of tenacity influences the kind of regularization that is achieved, we have also to study how each actor could compute his own level of tenacity according to his satisfaction target, the best global satisfaction of the organization and of given maximum cost of the simulation.

The behavior of the algorithm is constrained by several other parameters, hence it will be necessary to study and analyze their influence on the algorithm's results (especially the Action Range's parameter that matters in exploration) and the effect of the join variation of several parameters.

Acknowledgements. This work has been supported by the ANR (French Research Agency) through the IsyCri project.

References

1. Axelrod, R.: The Complexity of Cooperation: Agent-Based Models of Competition and Collaboration. Princeton University Press, Princeton (1997)
2. Clegg, S.R.: Handbook of organization studies. Sage Publ., Thousand Oaks (2003), http://www.worldcat.org/oclc/249400076
3. Crozier, M.: The bureaucratic Phenomenon. Chicago University Press, Chicago (1964)
4. Crozier, M., Friedberg, E.: L'acteur et le système. Seuil, Paris (1978)
5. Crozier, M., Friedberg, E.: Organizations and Collective Action: our Contribution to Organizational Analysis. In: Research in the Sociology of Organizations, vol. 13, Jay-Press, Greenwich (1995)
6. Ioerger, T., Linli, H.: Modeling command and control in multiagents systems. In: 8th Int. Command and Control Research and Technology Symposium, Washington DC, June 17-19 (2003)
7. Mailliard, M., Amblard, F., Sibertin-Blanc, C., Roggero, P.: Cooperation is not always so simple to learn. In: Terano, T., Kita, H., Deguchi, H., Kijima, K. (eds.) Agent-based Approaches in Economic and Social Complex Systems IV, pp. 147–154. Springer, Heidelberg (2007)
8. Mailliard, M., Roggero, P., Sibertin-Blanc, C.: Un modèle de la rationalité limitée des acteurs sociaux. In: Chevrier, V., Huget, M.P. (eds.): Journées Francophones sur les Systèmes Multi-Agents, Annecy, 18/10/2006-20/10/2006, pp. 95–99 (2006), Hermès, http://www.editions-hermes.fr/
9. Sibertin-Blanc, C., Adreit, F., Chapron, P., Roggero, P.: A Model for the Social Regularization of Autonomous Agents (2009) (submitted to publication)
10. Sibertin-Blanc, C., Amblard, F., Mailliard, M.: A coordination framework based on the Sociology of Organized Action. In: Boissier, O., Padget, J., Dignum, V., Lindemann, G., Matson, E., Ossowski, S., Sichman, J.S., Vázquez-Salceda, J. (eds.) ANIREM 2005 and OOOP 2005. LNCS (LNAI), vol. 3913, pp. 3–17. Springer, Heidelberg (2006), http://www.springerlink.com
11. Simon, H.A.: The sciences of the artificial, 3rd edn. MIT Press, Cambridge (1996)
12. Sutton, R.S., Barto, A.G.: Introduction to Reinforcement Learning. MIT Press, Cambridge (1998)

13. Tahir, O., Andonoff, E., Hanachi, C., Sibertin-Blanc, C., Benaben, F.: A Collaborative Information System Architecture for Process-based Crisis Management. In: Lovrek, I., Howlett, R.J., Jain, L.C. (eds.) KES 2008, Part III. LNCS (LNAI), vol. 5179, pp. 630–641. Springer, Heidelberg (2008)
14. Van Doren, P.H.: Decision making and cognitive analysis: Track a historic failure in the social domain. In: 10th Int. Command and Control Research and Technology Symposium, Virginia Beach, VA (2005)

Part III
Agent Technology for Environmental Monitoring and Disaster Management

Recent advances in wireless communications and artificial intelligence have transformed the way people interact, trade, and entertain themselves. However, in spite of the current drive for sustainable development, the use of such technologies in critical domains such as disaster management and environmental monitoring has been limited, if not non-existent. To fill this gap, the scientific community has reacted by initiating a number of workshops dealing with the application of information technology to such important domains. In particular, these workshops have focused on the application of multi-agent systems technologies for the coordination of emergency responders and to manage sensor networks.

In particular, disaster management requires that a number of distinct actors and agencies, each with their own aims, objectives, and resources, be able to coordinate their efforts in a flexible way in order to prevent further problems or effectively manage the aftermath of a disaster. The techniques involved may necessitate both centralized and decentralized coordination mechanisms, which need to operate in large-scale environments, which are prone to uncertainty, ambiguity and incompleteness given the dynamic and evolving nature of disasters.

Against this background, and following from the success of the first edition, we organised the second edition of the International workshop on Agent Technology for Disaster Management (ATDM). Its aim is to help build the community of researchers working on applying multi-agent systems to disaster management, through either designing, modeling, implementing, or simulating agent-based disaster management systems. In this context, this collection consists of the papers accepted at the ATDM 2009 workshop and spans a number of topics including, but not limited to: (i) the design of coordination mechanisms for large disaster management organisations, (ii) the design and evaluation of simulation platforms involving software agents or virtual robots, and (iii) the design of intelligent emergency response strategies. Altogether, we believe these contributions represent an important step towards building better tools for emergency responders to prepare and respond to disasters in all shapes and form.

Turning to environmental monitoring, we note that advances in wireless sensor networks now enable real-time information retrieval from various environments with minimal human intervention. Since the need of the information varies from user to user, querying, analysing and publishing relevant information from heterogeneous sensor networks across various spatio-temporal scales is a challenging task. This is one of the hindrances for macro-scale sensing and monitoring.

Sensor Web is a concept that supports macro-scale sensing with web-enabled heterogeneous sensors. The agent paradigm may play a key role to scale the localised wireless sensor network to the sensor web and add much needed intelligence. It also helps to manage large-scale sensor networks, however, there are several challenges that need to be addressed from both the wireless sensor networks and agents perspective to make it a reality. The

First International Intelligent Agents in Sensor Networks and Sensor Web (IASNW) looks at the application of agent technologies in enhancing the capabilities of sensor web. The later chapters of part III describes: (i) the sensor network architecture based on web and agent technologies, where data gathering is managed by "pull-based" agents, (ii) agent-based autonomous data management system, and (iii) conceptual similarities between mulit-agent systems (MAS) and Open Geospatial Consortium - Sensor Web Enablement (OGC-SWE), this would give an opportunity to adopt MAS-based techniques to address some of the problems in OGC-SWE. The chapters will give an insight into the capabilities of sensor web concepts and also some of the limitations. We believe the chapters will stimulate more cross-disciplinary research to find effective and efficient way of monitoring our environment.

Sarvapali D. Ramchurn
Serge Stinckwich
ATDM Workshop Chairs

Siddeswara Guru
IASNW Workshop Chair

Dynamic Role Assignment for Large-Scale Multi-Agent Robotic Systems

Van Tuan Le, Serge Stinckwich, Bouraqadi Noury, and Arnaud Doniec

Abstract. In this paper, we introduce an approach for designing and deploying organizations on large scale Multi-Agent Robotic Systems. Lying on the well-known organizational meta-model AGR, we propose a role assignment protocol for automatically distributing the roles over the robots. We have run a series of simulations to validate the approach's feasibility. The simulations show that the protocol is well scalable as well.

1 Introduction

Traditional disaster relief operations involve a huge human resources, including different on-site rescue teams, and remote coordinators. On-site rescue teams are sent to explore the disaster area to find victims. Having located victims, they send victims situation to remote coordinators informing about the needs of medical cares or of any assistance else. Based on global information received, coordinators should give instructions to rescuers, who have usually a limited partial view. Hence, the most efficient cooperations between on-site teams is achieved. Moreover, as the rate of victims survivability drops drastically after 72 hours (thus called 72-golden-hours), and locating victims within three first hours is desirable the most, the rescuers have to find victims in very strict timely manner. Therefore, it is ideally required to have numerous teams in parallel and cooperatively exploring the disaster site.

In this context, the AROUND [3] project (Autonomous Robots for Observation of Urban Networks after a Disaster) aims at building a complete

Van Tuan Le · Serge Stinckwich
UMMISCO, UMI 209 (IRD, UPMC, IFI-MSI) and GREYC, UMR 6072 (CNRS, Université de Caen) and École des Mines de Douai, France
e-mail: {lvtuan@gmail.com, Serge.Stinckwich}@ird.fr

Noury Bouraqadi · Arnaud Doniec
Université Lille Nord de France and École des Mines de Douai, France
e-mail: {bouraqadi, doniec}@ensm-douai.fr

Q. Bai and N. Fukuta (Eds.): Advances in Practical Multi-Agent Systems, SCI 325, pp. 311–326.
springerlink.com

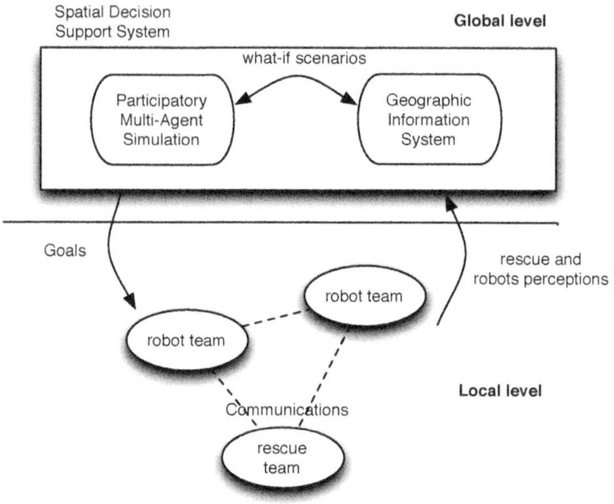

Fig. 1 AROUND project architecture

real-time deci- sion support system for natural disaster management in urban areas. The system is composed of 2 subsystems: (i) a local level with the deployment of autonomous mobile robots able to collect information in impacted urban sites and to dynamically maintain the communication links between rescuers, (ii) a global level based on a Spatial Decision Support System (SDSS).

The rescue operations impose a very quick exploration over disaster zone, yet the environment in this context is extremely complex and dynamic. Therefore, the number of the fleet of robots being deployed to assist the rescue operations is potentially very large, and *a priori* non-determined. We expect that the robotic system should be able to be re-organize for different rescue tasks. The proposed solution therefore should meet the following requirements:

- **Scalability:** we expect the solution to deal with a very large number of robots and this without performance degradation,
- **Robustness:** if there is one robot that can not function anymore, the others take over its task and the system can continue without any impact,
- **Openness:** system operators might easily add new robots to the system on-the-fly,
- **Flexibility and Reusability:** we expect to reuse as much as possible some parts of the solution for similar missions. Also, the set of robots might vary in technology, or individual capabilities.

In this paper, we consider a robotic system as a social organization [7, 11, 2]. The organizational approach of multi-agent systems has a long history since the 90s, when the term “agent society” was coined [7]. Organization is defined as a supra-agent pattern of emergent cooperation or predefined cooperation of the agents in a system, that could be defined by the designer or by the agents themselves, for a purpose [2]. To build such organizations, there are two possible points of view: an *agent-centered* approach and an *organization-centered* one.

While in the first case, the organization emerges from interactions between agents and in the second, rules of organization that must comply with agent to conduct a mission are determined *a priori* in the design phase of the system. The agent-centered point of view may cause many difficulties to ensure the overall coherence of a system composed of autonomous agents. The organization-centered point of view approach circumvents these problems by proposing a pre-defined setup which says nothing about the mental states of agents, but only on their ability to collaborate with other agents of the organization. The design of organizations still remains a difficult art. J. Hübner et al. [9] show that you can obtain a design where the organization is too flexible (the organization does not help achieve the overall objective) or too rigid (the organization prevents the expression of agents autonomy).

We propose in this paper an organizational approach applied to the multi-robot systems domain that guarantees the autonomy of agents. On one hand, system designers should specify the organization constraints that robots in the system have to follow (section 2). On the other hand, at the run-time the agents by themselves make decision (agents side, i.e. the robots) to whether take part in the organization while respecting the constraints of the organization or not (section 3). The decision process is totally distributed on the robots. We have run many simulations to validate the approach's feasibility (section 4). The simulations confirm also the good scalability of the approach. In section 5, we discuss the open issues and future work. Section 6 gives a overview over the work on role assignment in the literature.

2 Organizational Approach for Multi-Robot Systems Overview

Our proposal is based on the organizational model AGR [7] where we have made several changes and introduce the concept of *organization description*. This allows us to specify the target organization that can then be used by robots to build the effective deployed organization.

2.1 The AGR Model

The AGR model introduces three atomic concepts: Agent, Group, and Role.

- **An agent** *is an autonomous, communicating entity playing roles in groups. An agent might play many roles and thus becoming member of many groups.* At this level, no detail and constraints are imposed on the agent's mental capabilities.

- **A group** *is defined as a set of agents.* Groups are means to partition the system. They set up the boundaries between different sets of agents in system by imposing that only agents in the same group could communicate, and hence collaborate. As a result, they provide better security mechanism.
- **A role** *is an abstract representation of an agent's functionalities.* Defining a role is to indicate the services, identification within a group: the role encapsulates the way an agent should act within a group. Roles are local to a group. To play a role within a specific group, an agent must gain permission to do so. Once the required role has been assigned to it, an agent determines itself to continue playing role, or to resign the role and leave the group. The number of roles that an agent can play is determined only by its capability. Therefore, provided that it meets the constraints required by role's specification, an agent is free to take part in any groups, to play any roles.

2.2 Adapting AGR

In the context of multi-robot systems, there are some modification needed due to the physical constraints. In addition, the AGR model provides only the abstract concepts easing the design of a multi-agent system. The methodology stays purely at the systems modeling phase. Designers make use of the model to partition the system into roles, and groups in a comprehensive fashion. However, when and how the roles and agents should match together are all left open. In most of the cases[1], ones build the organization and the agents all from scratch. Agents then are told to play a role within the organization at specific run-point. The manners by which agents are assigned a role therefore vary from a system to the others. We argue that systems built in this style are quite closed, and do not suite very well with design of multi-robot systems, a truly open, distributed system.

This proposal requires that agents can reason about the organization when making their decisions about roles to endorse and groups to join. For this purpose, we introduce the concept of *organization description* that enable agents to know and manipulate the structure of the system. This extension of the AGR meta-model is depicted in figure 2.

The organization model serves as a tool for the system design phase. However, for implementation and deployment phases, the model need to be reified (i.e concepts should be defined explicitely) so that it can be deployed automatically over a real multi-robot system. Descriptions act as "blueprints" for systems, exposing their structures to agents who would like to dynamically join an organization. This meta-model is then used by the automatic role assignment process.

[1] Particularly in the MadKid platform, which is the reference implementation of the AGR model.

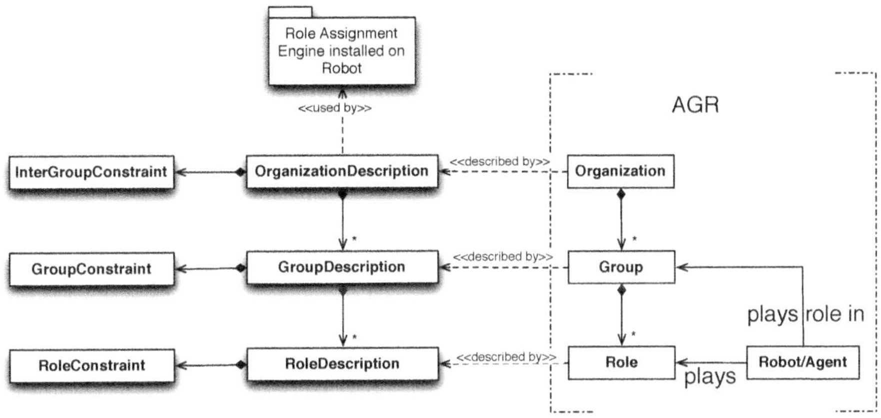

Fig. 2 AGR modifications for multi-robot system needs

We added to the AGR meta-model a description of organization, group and role:

- **Organization description:** *an organization description structurally specifies an organization; it consists of a set of group descriptions, inter-group constraints, and inter-group coordination protocols.* An example of inter-group constraint is that an agent from a group G_1 should not participate to a group G_2.
- **Group description:** *a group description is a structural specification of an effective group containing all descriptions of all roles in the group, group constraints (e.g. number of effective group of this kind) and an interaction protocol for the collaboration between groups entities.*
- **Role description:** *representing the constraints that a robot has to satisfy in order to play some role in a particular group. The constraints may be functional responsibilities and/or non-functional requirements imposed on the robot.*

3 Dynamic Role Assignment

Thanks to the organization description, we can design a dynamic role assignment to support deployment of organizations inside a multi-robot system.

3.1 Concepts

- **Neighbors** *: is a distance function whose goal is to compute which robots are neighbors of a given robot.* This function can be built in an ad hoc way, that varies from an application to the other. In its simplest form, two robots are neighbors if there is a communication link connecting them. The term neighbor in this paper refers to robots determined by this function.

- **Group Manager:** *The manager of the group controls the groups membership. This one does know all the robots within its group.* For each group, there is one and only one member holding this role. All robots of the group are neighbors of the Group Manager. In order to insure that there is a single manager in each group, the system designers have to specify this constraint in the role of group manager. In fact, expressing correctly the conditions for playing a role is a key point for the role assignment process. In the section 3.3, we give an example of role constraints.
- **Organization Manager:** *for the role assignment process, there is at least one starting point where we put the organization description. The robot holding the description organization then becomes the unique Manager for the organization. In an organization, the Organization Manager knows at least all the Group Managers.* To facilitate coordination between managers, they are gathered in a group led by the manager of organization.

3.2 Role Assignment Protocol

The protocol we present here is inspired by the Distributed and Mobility-Adaptive Clustering (DMAC) algorithm proposed by S. Basagni [1]. DMAC is an algorithm used to partition a Mobile Ad Hoc Network (MANET) into clusters. A cluster is made up from nodes of two kinds: a cluster head and some ordinary members attaching to the head. A robot will be a cluster head if it has the biggest number of "one-hop"[2] neighbors than any its neighbor does; otherwise, it will affiliate to a cluster head. In the protocol, we generalize the conditions to take a role in form of constraints. The constraints are then locally checked by a robot to determine its own role. With the same intention, the neighbors function has been introduced so that the formation of group can take place beyond the limit of one-hop neighbors. It is worth noting that, the original algorithm was designed with the presence of network nodes' mobility taken into account, the protocol is therefore adaptive to the robots' mobility as well.

Similarly to DMAC, we make the following assumption for the role assignment process:

- Messages sent by a robot will be correctly received within a determined period by all its neighbors.
- The lower layer provides infrastructural services so that robots are able to detect their neighbors, new coming neighbors, or the disappearance of "old" neighbors. Robots know their neighbors role.
- We assume that each robot has a unique ID.

[2] We use the term "one-hope" neighbor as its meaning in Mobile Ad hoc Networks, i.e. to specify that two robots are neighbors if there is a direct communicating link connecting each other.

Right after having started up, robots are in an initial state and try to join an organization. However, there should be at least one starting point : the robot which initialize the organization by assigning it the first role of the organization, along with the organization description.

- `Init`: every robot broadcasts the `OrgDescRequest` message to all its neighbors; the broadcast procedure is repeated for some interval of time, until they receive a response from an organization member.
- `Role selection`: received the `OrgDescResponse`, robot extracts the organization description to rebuild the organization's image. The next step is to verify role constraints to find an appropriate role. According to this, robot will broadcast `NewGroup` or `Join` message.
- `Disappearance of a neighbor`: whenever made aware of the disappearance of one of its neighbor, if the robot is a group manager and the newly disappeared robot is in its group, it removes the latter's ID from its group member list. If a robot is an ordinary member, and its group manager has disappeared, it needs to re-execute the `Role selection` procedure to adapt to the organization's changes.

In the implementation, the `Role selection` procedure can be re-executed on new events. For example, when robot detects that it is out of its group or when it finished its task attached to the role.

As long as robots have not joined the organization yet, they could communicate merely with their neighbors by sending these organizational setup messages. There are four messages; and all necessary information are supposed to be encoded in the message. Here are the messages and the procedure that robots will execute depending on the message received and on its own role:

- `OrgDescResquest`: sent out by the robots that are not yet members of the organization to their neighbors, and would like to join the organization. They send this message to ask for the description of the organization (so that it can select a role to play). A member, whatever its own role is, will answer with the `OrgDescResponse`, otherwise, nothing to do.
- `OrgDescResponse`: contains the organization description, with all group and role descriptions.This message is the response of the previous one. Receiving this message, robots do not thing if they have been organization members already. Otherwise, they execute the `role selection` procedure. This is the only message that a robot with non-determined role reacts upon.
- `NewGroup`: used to inform the creation of a new group. This message is created by a robot that decides to take the role of the manager of a new group. Because this message indicates changes in the organization, the robots that receive the messages need to check again the constraints of its current roles, if the constraints are violated; they need to re-execute the `role selection` procedure.

- `Join`: coming at the decision of taking a role in a group, a robot sends this message to the group manager making aware of its membership. In case the sender is already a group manager and decides to affiliate itself to another group, it broadcasts this message to all its members in the old group. Upon receiving this message, if robot is a group manager, it checks if the Group Managers ID in the message is its own, then it adds the sender into its member list. If the receiver robot is an ordinary member, it checks if the sender is its Group Manager, and only in this case, it will try to join to new group by sending out the message `OrgDescRequest`.

3.3 Example of Role Assignment

To illustrate the process of allocating roles, take the example of robots equipped with wireless network interface, which must conduct a rescue mission on unknown terrain.

Neighbors: two robots are said to be neighbors if both of them are in one-hop communication coverage zone of each other.

Constraints: we use another utility function other than the neighbor function. The weight function is to determine the number of robots neighbors. A robot will be the cluster head if none of its neighbors is a cluster head, and it has the biggest number of neighbors than any neighbor. Otherwise, it will join to a cluster head with the biggest weight. For the group of the manager, the constraint for the member role in this group is that a robot must be a cluster head in some cluster. There are only local constraints.

Now let us consider the group formation for system of robots deployed on the terrain as depicted in figure 3(a). Without loss of generality and for the simplicity, we illustrate the algorithm for static robots. The organization starts with the first member - the starting point (figure 3(a)).

At time step 1, all robots execute the init procedure to broadcast the `OrgDescRequest` message. After this step, only two neighbors of the starting-organization robot, robots A and B receive the `OrgDescResponse` with the encoded organization description. The robot A extracts the description from the received message, finding out that it has bigger neighbors count than the starting point, the only member of the organization so far, then robot A will take the role of the group manager and broadcast `NewGroup` message. On receiving this message, the first organization member react by re-executing the `Role selection` procedure, it finally take the ordinary member role, and affiliates to robot A's group. In the same time step, robot B follows the same procedure, and it might first of all join the group of the first robot, but when the robot A decides to create a new group and announces the group's creation, robot B re-executes the `Role selection` procedure, and then joins to the robot A's group. After three steps, the organization is formed as shown in figure 3(b).

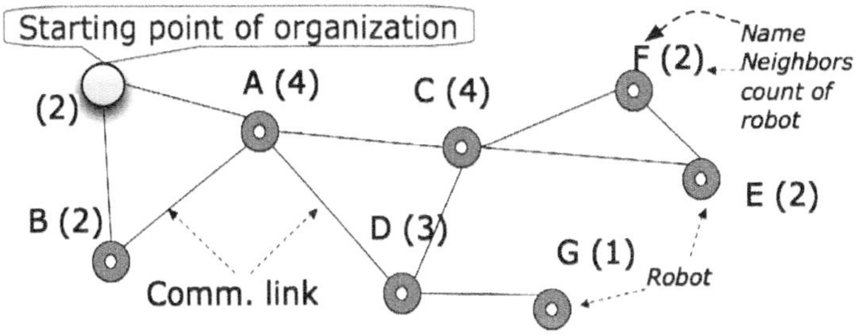

(a) The organization's initial status

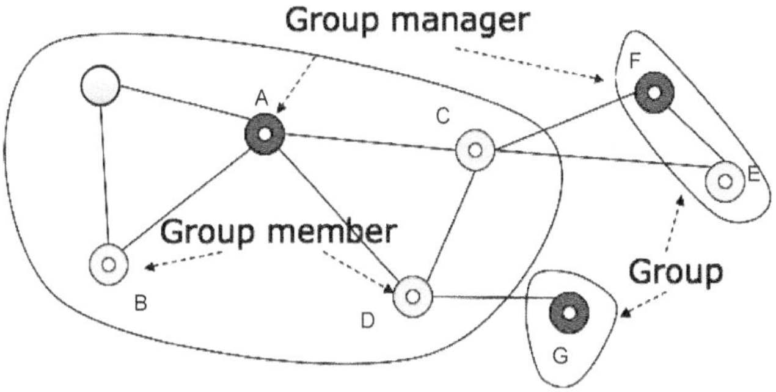

(b) Complete formation of the organization

Fig. 3 Example of organization formation

4 Validation

In this section, we sum up and analyze the results of different simulations we have made to validate our proposal.

4.1 Validation Scenarios Design and Setup

In the simulations, robots are randomly dropped on the terrain before the simulations run. The infrastructure services help robots to detect their neighbors. At the beginning of each simulation, we randomly take a robot and initialize it as the first member of the organization. This robot then becomes the manager of the organization, and hold the organization description as well.

In the simulations, all messages sent by robots will be correctly received within a finite period of time. Therefore, an organization formation step refers to the amount of time required to have all robots at one-hop to at least one

Table 1 Number of step to finish a simulation

Number of steps	10	11	12	13	14	15	16	17	18
Frequencies over 1000 simulations	33	170	111	180	163	153	122	65	3

organization member, join to the organization. This amount of time includes the duration required to perform a change in the organization due to the presence of these new comers. This assumption was made without loss of generality thanks to the parallel computation in each robot. We have run a series of 1000 simulation with 1000 robots, randomly dropped on terrain of 1300x900m. The radio coverage range was set to 100m.

Number of steps: as shown in the table 1, the minimum number of steps to finish such a simulation is 10 with the frequency of 33 times, while there are only 3 times the simulations finished after 18 time steps. The number of steps required to finish an organization deployment ranges –most frequently– from 11 to 16 steps.

Role changes: so far, we have not yet implemented any exploration strategy so the robots are all motionless. However, at each step, there are new members added to the organization causing changes in roles. Thus the re-organization implicitly takes place at each step. This effect is similar to that caused by the robots' mobility, or robots' failures, etc. The chart in the figure 4 shows the relationship between the number of new members, and the corresponding number of role changes due to these modifications of the

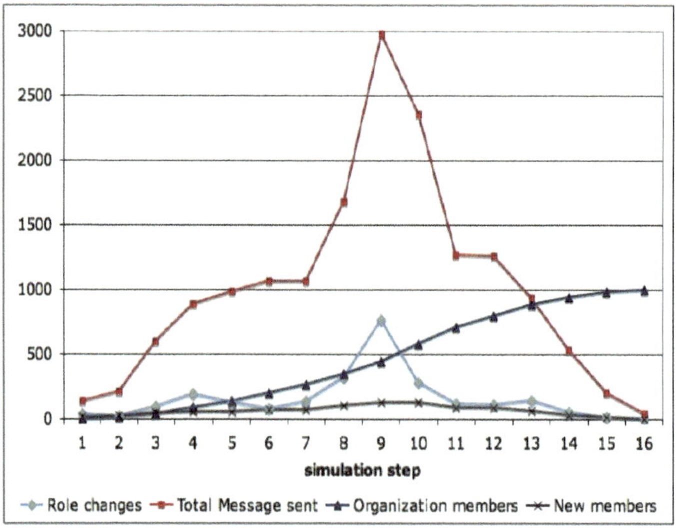

Fig. 4 Number of members, new members, role changes and messages sent in each step.

organization. In average of 1,000 simulations, each robot has changed its role from 2 to 3 times.

Number of messages sent: the average rate of number of messages sent in the simulations is around 18 times higher than the number of the robots. There are totally from 15,000 to 20,000 messages sent with 1,000 robots in the simulation.

The graph in figure 4 shows the relationship between the number of the robots who have just join the organization, the number of role changes, and the number of the messages of all kinds issued at each step. We might notice that the variations of the number of role changes and the number of messages sent are relatively close. This observation can be explained by the fact that, a new "comer" to the organization will send one `OrgDescRequest` message, it then decides to take part in the organization, and this might cause the changes for the organization member close to it. For this, they have to exchange the messages, mainly of kind `NewGroup` or `Join`.

4.2 Push vs. Pull and Optimizations

The propagation of the organization description we have just described follow the *pull* principle, where the non-member agents take the initiative to join the organization by *broadcasting* the `OrgDescRequest` message (the `Init` procedure). We have realized through the simulations that this naive tactic flood much the network with useless message when robots are far from the organization. As an alternative, we can use the *push* approach: if an organization member finds a neighbor that is not the organization's member, it sends the description to this one. In the simulation, we choose to minimize the number of messages to send by making the robots aware of an organization member among their neighbors, and only when this is the case, it send a request to the member. The non-member then sends the `OrgDescRequest` message to this member. The simulation then shows that the total number of sent messages are reduced, on average, by 6 times. We end up with a nice factor of only 3, i.e. the number of sent messages is only 3 times the number of robots. This significant reduction do not come from sending `OrgDescRequest` messages, but from "saving" the number of response to this messages.

4.3 Impacts of Robots' Density

The density of the robots has a big impact on the formation process of the organization. We have varied the robots' density by changing the distance of the radio coverage. Each time, the same configuration (set of robots and their positions on the terrain; and the starting point of the organization) was used in 5 simulations with different radio range of 70m, 170m, 270m, 400m, and 500m on the area of 1300x900m, with 1,000 robots. By changing the radio range, the robots' density has been varied as well. The simulations are

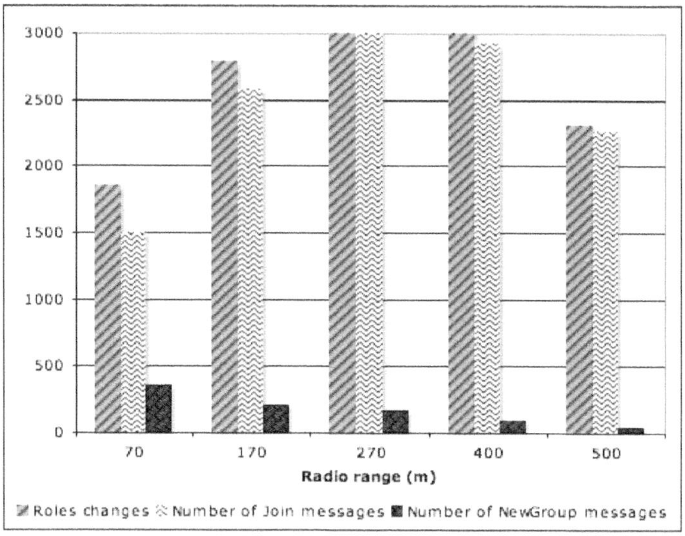

Fig. 5 Impacts of robots' radio range on number of messages sent and number of role changes

optimized as discussed in section 4.2. We have run 5,000 simulations in total, and the results we got lead to some interesting remarks. Figure 5 represents the correlation between these variations of radio range.

- Messages sent: the number of messages `OrgDescRequest` and `OrgDescResponse` is similar for all cases, as each robot sends out exactly one of these messages. However, the number of messages `NewGroup` and `Join` varied case by case. Note that we run *optimized* solution and on the graph shows the *total* messages of each kind in the simulation.
- Changes of role: the number of messages sent and the role changes are quite close. At radio range of 70m, we find this qualification is close to that of the simulation with the radio range of 500m. The number of role changes peaked when the radio range was set to 270m.
- Steps: the wider the radio zone, the fewer the simulation steps: around 20 steps with 70m, whereas the simulations finished after around 2 steps with robots' radio range of 500m.

From the graph, we can draw the conclusion that a good radio range setting can make significant optimization.

5 Discussion and Open Issues

5.1 Toward a Generic Solution

We call the presented approach *optimistic strategy*, as each robot is expected to be responsible for selecting an appropriate role while respecting the imposed constraints. Provided that the constraints can be locally correctly expressed, we have not to worry about the global integrity. However, this solution is inappropriate for applications where security is a critical issue.

Considering now a soccer team in the context of an organizational system, it is easy to identify several global constraints, including: no more than four groups: defend, attack, mid-field, and coach. Depending on the strategy, e.g. a team formation of 3-5-2 or 4-4-2 that the team employs, each group has a specific number of members. In such a case, the optimistic approach is not very appropriate as the global constraints are difficult to be ensured. In fact, there is no unique solution for all cases. Therefore, we propose two other strategies, namely pessimistic and hybrid ones providing to system designer as alternatives.

- *Pessimistic strategy*: each robot sends out a request to be assigned a role. The process of role assignment is totally controlled by the system.
- *Hybrid strategy*: staying between two extremes above, the third strategy might be a harmonized solution where all the local constraints will be validated by the robot itself, whereas the global constraints are controlled by some specialized robots in the system.

The optimistic strategy offers the advantage of being distributed and open, whilst the security and integrity of global constraint are difficult to solve. On the contrary, security and global constraints will be more easily to ensure with the pessimistic strategy, while this one suffers many hassles from centralized approach in physically distributed systems.

The key point in the role assignment process is how to correctly express the role constraints so that each robot can check these constraints based on its view over the whole system. Depending on the nature of system to build, one needs to carefully find out and express the constraints and use the suitable approach. In term of perspectives, we will focus on the hybrid strategy so that this approach becomes a really generic solution.

5.2 System Re-organization

Considering a rescue robot system again, there are many heterogeneous robots with differents capabilities deployed on the disaster area. Among other, some robots are more appropriate for assisting victims, whereas others are more suitable for locating victim positions. However, before a victim found, we should organize all robots to search around in the damaged area. Once a victim has been located, the most appropriate robot switches itself to take the role of supporter. In this case, a dynamic role assignment schema helps the organization to adapt to surrounding changes in optimized way.

All that the system designers should deal with is to take into account such stimulus, then the `role selection` procedure will be triggered in responding to these events. Furthermore, the dynamic role assignment is not only triggered by the external changes, but also by the internal changes of robots themselves.

6 Related Work

In the literature, the role assignment in the domain of multi-robot systems is a well-known problem already investigated in several works. Role assignment is also known as *task allocation* [10] where the roles in the system is identified as a set of (related) tasks. The most popular approach for task allocation lies on a market-like protocol, namely the ContractNet Protocol [12]. Whenever there is the task to be solved, a corresponding task proposal is generated and submitted to all available robots. Robots, based on the information provided along with the proposal, evaluate their appropriateness for the task to refuse or accept it. In the latter case, robots submit their bid, and the one whose bid value is the highest will be assigned for the task. This market-based task allocation is implemented in [12], and particularly in the domain of RoboCup [10, 6]. C. Candea et al. [4] have implemented the role assignment mechanism to coordinate the RoboCup soccer team. The process consists of two steps: (1) *formation selection*; and (2) *role assignment*. In the first step, all robots vote for the best team formation. Then passing to the second step, all roles are sorted in their order of importance. Using a greedy market-based task allocation algorithm, the role is then assigned for each robot in the fixed, total order. This is to insure that in case of robots' failure, the remaining member(s) will be always assigned the most "important" roles.

This market-based approach is useful to deal with task allocation where the ordering of tasks, the availability of robots, and the suitability of a robot for a particular task are not know. However, it does not scale well as the space to search for robots includes all the robots in the system. To address this shortcoming, C. McMillen and M. Veloso [10] implemented in their soccer robots team a mechanism to limit the search space by introducing the region-based task allocation: each robot is assigned to a region of the field. A robot is primarily responsible for going after the ball whenever the ball is in that robot's region. In our approach, we view a role from the higher level of abstraction than that of a task [11]. A role is not only a set of (related) tasks, but also represents the relationship between different system entities. We expect that robot enacting a role in an organization knows how to fulfill the task associated to that role (robot's autonomy) in cooperating with other robots in the system (thanks to the organization constraints, and role relationship).

Chaimowicz et al. [5], proposed an architecture for tightly coupled multi-robot cooperation - a system in which always exist a leader, and the others follow this one. The role of leader and that of the followers are not fixed, but might be changed on-the-fly. This work can be seen as a particular case in our approach, where the organization is composed of a single group.

Similarly to the approach presented in this paper, the role assignment and the coordination between robots are guaranteed by the communication protocol and control algorithms.

Automatic allocation of role is also largely investigated in pervasive computing systems such as sensor networks. C. Frank et al [8], proposed a generic algorithm to assign role to a node in a sensor network. They have developed a framework allowing the system designers to specify the *role specification* - a list of role-rule pairs in form of predicates. For each possible role, the associated rule specifies the condition for assigning this role. These specifications are then used by node in the network to determine its proper role. However, their solution is optimized for sensor networks, thus the mobility is not taken into account. Furthermore, whilst the evolution of the organization might take place during its execution if we change the organization description with our approach, it is not applicable for their algorithm.

7 Conclusion

We presented in this paper an organization-based approach for structuring large-scale robotic system. This can be used to design systems that need to be self-organized through the dynamic process role assignment.

Our proposal is built as an extension of the AGR meta-model. It unifies the design process and deployment of multi-agent robotics systems. We call our proposed approach "optimistic" because each agent immediately endorses roles he choose. Potential conflicts are handled when they are encountered (e.g., choosing the best robot to play the role of manager of a group). Another possible strategy, known as "pessimistic", is to anticipate potential conflicts. Assigning a role to an agent is subject to validation by an another agent. Another alternative strategy is to combine the "optimistic" and "pessimistic" approaches. A future perspective for our work is to study these different strategies and classes of problems they are best appropriate for.

References

1. Basagni, S.: Distributed clustering for ad hoc networks. In: Proceeding of the 1999 International Symposium on Parallel Architectures, Algorithms and Networks (I-SPAN 1999), pp. 310–315 (1999)
2. Boissier, O., Hübner, J.F., Sichman, J.S.: Organization Oriented Programming: From Closed to Open Organizations. In: O'Hare, G.M.P., Ricci, A., O'Grady, M.J., Dikenelli, O. (eds.) ESAW 2006. LNCS (LNAI), vol. 4457, pp. 86–105. Springer, Heidelberg (2007)
3. Boucher, A., Canal, R., Chu, T.Q., Drogoul, A., Gaudou, B., Le, V.T., Moraru, V., Nguyen, N.V., Vu, Q.A.N., Taillandier, P., Sempé, F., Stinckwich, S.: The AROUND project: Adapting robotic disaster response to developing countries. In: Proceedings of 2009 IEEE International Workshop on Safety, Security, and Rescue Robotics, IEEE Computer Society, Los Alamitos (2009)

4. Candea, C., Hu, H., Iocchi, L., Nardi, D., Piaggio, M.: Coordination in multi-agent robotcup teams. Robotics and Autonomous Systems 36, 67–86 (2001)
5. Chaimowicz, L., Sugar, T., Kumar, V., Campos, M.F.M.: An architecture for tightly coupled multi-robot cooperation. In: Proceedings of the IEEE International Conference on Robotics and Automation (ICRA), vol. 3, pp. 2992–2997 (2001)
6. Emery, R., Sikorski, K., Balch, T.: Protocols for collaboration, coordination and dynamic role assignment in a robot team. In: Proceeding of the IEEE International Conference on Robotics and Automation (ICRA), pp. 3008–3015 (2002)
7. Ferber, J., Gutknecht, O., Michel, F.: From agents to organizations: an organizational view of multi-agent systems. In: Giorgini, J.O.P., Müller, J. (eds.) Agent-Oriented Software Engineering (AOSE) IV, pp. 214–230 (2004)
8. Frank, C., Römer, K.: Algorithms for generic role assignment in wireless sensor networks. In: Proceedings of the 3rd international conference on Embedded networked sensor systems. ACM Press, New York (2005)
9. Hübner, J., Sichman, J., Boissier, O.: S-moise+: A middleware for developing organized multi-agent systems. In: Boissier, O., Dignum, V., Matson, E., Sichman, J. (eds.) International Workshop on Organizations in Multi-Agent Systems: From Organizations to Organization Oriented Programming (OOOP 2005), pp. 107–120 (2005)
10. McMillen, C., Veloso, M.: Distributed, play-based role assignment for robot teams in dynamic environments. In: Alami, R., Chatila, R., Asama, H. (eds.) Proceedings of Distributed Autonomous Robotic Systems 6, pp. 145–154. Springer, Heidelberg (2007)
11. Odell, J.J., Parunak, H.V.D., Fleischer, M.: The Role of Roles in Designing Effective Agent Organizations. In: Garcia, A.F., de Lucena, C.J.P., Zambonelli, F., Omicini, A., Castro, J. (eds.) Software Engineering for Large-Scale Multi-Agent Systems. LNCS, vol. 2603, pp. 27–38. Springer, Heidelberg (2003)
12. Ulam, P., Endo, Y., Wagner, A., Arkin, R.: Integrated mission specification and task allocation for robot teams - design and implementation. In: Proceeding of the 2007 IEEE International Conference on Robotics and Automation, pp. 4428–4435 (2007)

Multi-agent System for Blackout Prevention by Means of Computer Simulations*

Miroslav Prýmek, Aleš Horák, and Adam Rambousek

Abstract. The paper presents a dynamic simulation model of power flows in a power distribution network using communication in multi-agent systems. The model is based on local interchange of knowledge between autonomous agents representing the network elements. The agents use KQML as an inter-agent knowledge interchange language. The main purpose of the work is to develop a scalable distributed simulation model, flexible enough for incorporation of intelligent control and condition-based management of a power distribution network. This model is designed to optimize the electric distribution networks and prevent failures in the power grid (blackout). The network's reliability is affected by the parameters of connected power sources (both traditional and renewable sources), but also by the unexpected failures. The simulation is using the parameters calculated from the real database of electric distribution network failures.

1 Introduction

Multi-agent systems bring new possibilities into many research areas. The main idea of decomposing the given problem into separated independent task is not new and is already used in many other software design areas. But multi-agent paradigm moves such an approach even further. In the classic software design paradigms (e.g. object-oriented programming, OOP), the decomposition into separated units mainly helps the designer to decrease the (logical) complexity of the problem but usually does not automatically make

Miroslav Prýmek · Aleš Horák · Adam Rambousek
Faculty of Informatics, Masaryk University
Botanická 68a, 602 00 Brno, Czech Republic
e-mail: {xprymek,hales,xrambous}@fi.muni.cz

* This work has been partly supported by the Czech Science Foundation under the project 102/09/1842.

Q. Bai and N. Fukuta (Eds.): Advances in Practical Multi-Agent Systems, SCI 325, pp. 327–337.
springerlink.com

the (physical, real) solution decomposed, modular and scalable. Even if the designer uses the OOP approach well and extensively, the final software system remains monolithic unless the explicit element borders and interfaces are designed and implemented.

On the other hand, the multi-agent systems force the designer to think about the software system units as really independent, self-contained and autonomous entities. The overall system is then viewed as a community of autonomous "organisms" rather than one complex with many subsystems.

In OOP, the main principle constituting the software system is (usually synchronous) *message sending* or *method calling*. We can consider it to be an instruction: *you* [the object] *should do this* [the message] *and give me the result of the action*. The object is not instructed on how the action should be taken but is supposed to perform the action. If the action is not performed, the whole system will probably get broken.

The multi-agent approach is a little bit different. The "organisms" (agents) constituting the "community" (system) are regarded as self-controlled autonomous entities with a *knowledge* about their surrounding world and a *decision-making algorithm* which controls the actions of the agent. The problem of the system design is then a problem of the appropriate agent design and configuration. The binding between the agents should remain as loose as possible – it is based mainly on the interchange of the particular agents' knowledge about the surrounding world. A typical communication between agents will be not an instruction but rather a question-answer pair followed by the decision-making process:

Agent A: What is your current condition?
Agent B: Pre-failure. Probability of the failure in the near future is high.
Agent A decision-making: prepare for taking over the B's functions

The capability to predict electric power network failure is an important research topic since the first power networks delivered electricity to wide public [5]. Before 1980s, control and prediction systems utilized global computational techniques [4, 10], which are not scalable enough to support large networks. Systems with the architecture of intelligent agents [9] and their applications for power networks simulation [8, 1] are able to react flexibly on changing network conditions.

In this paper we will try to illustrate how to use autonomous agents for modelling electric power network processes and for constituting a dynamic simulator of the power distribution network.

2 The Rice Power Distribution Network Simulator

The multi-agent simulation system called *Rice* is a result of research of the NLP Centre at Masaryk University in Brno, Czech republic, in cooperation with the Technical University of Ostrava, Czech republic. The main aim of the research is to develop a simulation tool for the explorations in the field of a condition-based maintenance of the power distribution network facilities. The multi-agent approach is suitable for highly flexible, modular and distributed simulations. The flexibility of the system makes it possible to implement simulations of the highly dynamic processes involving intelligent decision-making.

The Rice simulator is based on autonomous agents implemented as independent programs or processes [6]. The inter-agent communication is based on the CORBA standard (Common Object Request Broker Architecture [3]) which enables the agents to communicate through local or even wide-area network. Thus it is possible to deploy the system as highly distributed.

The meaning-layer of the communication is based on the KQML (Knowledge Query and Manipulation Language [2]) in cooperation with two possible content languages: python as a highly-expressive but expensive language and a simple one-purpose declarative language as a standard communication instrument. For details about the communication languages see [7].

2.1 Event-Driven Agents

The Rice basic agents are strictly reactive ones[1] which means that the behavior of the agents is composed of the reactions to the events of internal (e.g. knowledge change or timer alarm) or external (e.g. message arrival) origin.

The main behavior components are centered on the knowledge manipulation and communication:

- *Subscription* – mechanism used for subscribing to some piece of knowledge of another agent. When the piece of knowledge changes, the subscriber is informed. E.g. agent B subscribes for the knowledge about the status of power output of agent A. This means that agent B will be automatically informed about every change of characteristics of the output voltage of the agent A. When outage comes to the agent A, agent B can decide to propagate the outage or to switch to another input. *Unsubscription* of the knowledge means that agent is no more interested in its value and its propagation is stopped. Such an action can be done e.g. if the network topology is to be changed and agent A is no more the input of the agent B.
- *Message handlers* – these functions are triggered when selected messages arrive at the agent.
- *In-reply-to handlers* – used for filtering the arriving messages, preferably for reacting to answers to the previously sent questions and subscriptions
- *Missing value generators* – function used for generating values which are not present in the agent knowledge base

[1] For the detailed description of the distinct agents architectures, see e.g. [11].

- *Knowledge generators* – function generating a particular knowledge value. It is triggered each time the knowledge is accessed.
- *Knowledge change handlers* – function triggered whenever the value of the associated knowledge changes.
- *Timer event* – single or recurring time-based events
- *Event database* – the handlers can use built-in event database methods to save knowledge values for off-line data post-processing (statistical analysis, chart generations, etc.) The gathered data can then be exported to standard formats.

These behavioral components are designed to be implemented as independent pieces of code and thus the overall agent behavior is composed of the set of features assigned to the particular agent. The usual event-reaction code of such a feature is a small primitive consisting of a few lines in the Python language code.

The simulation is then implemented by combinations of these knowledge and interface primitives that are assigned to agents implementing particular power network nodes and elements. An illustration of such simulation implementation will be given further in the paper.

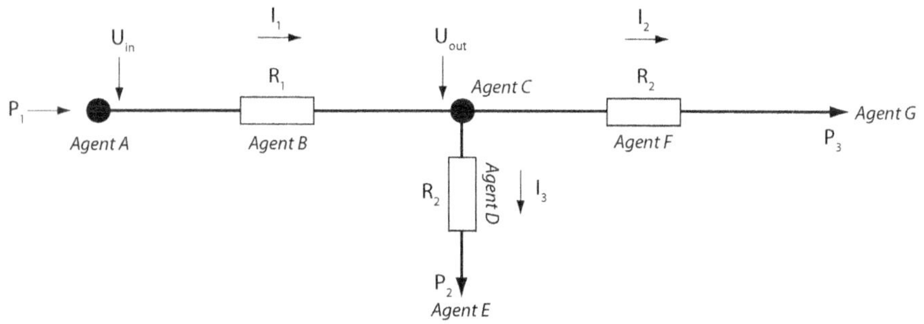

Fig. 1 Overview of the watched indicators

3 Model Situation Implementation

Our illustrative simulation will be based on the direct-current circuit model. The indicators involved in the simulation are displayed in Figure 1. There are four types of agents: the source node (agent A in Figure), the power line (agents B, D and F), the distribution node (agent C) and the consumer nodes (agents E and G). Using these network elements, there are two processes/network features demonstrated: the energy flow and its characteristics and the network topology.

The main knowledge interchange messages which build up the simulation are concluded in the sequence diagram Table 1.

Table 1 Agent communication

Source (S)	Line (L)	Consumer (C)
Initial state		
output voltage U=10kV output current I=0A	line resistance R=1Ω input node=S output node=C	power demand=0W
Network set-up		
S → L: inform me about power demands	L → S: add me as output L → C: add me as input	C → L: inform me about power available
Energy flow		
S → L: I=50A	L → S: current demand = 50A I = 50A U output = U-R*I = 9950V L → C: I=50A L → C: U=9950V	power demand = 500kW current demand = 50A C → L: demand I=50A P = U*I = 497500W

3.1 The Network Topology

As mentioned above, the multi-agent approach motivates to implement really independent software components. The central-controlled functionalities are kept to the necessary minimum. Even the topology of the network is implemented as an *agreement* between agents. The main advantage of this approach is that the network topology can be changed dynamically according to the observations and decisions of the agents.

The agents which make the first step to establish the topology, are the ones representing the power lines. For each agent there are two pieces of knowledge pieces inserted into their knowledge base: the name of the agent representing the line's input and the name of the output agent.

Immediately after the launch of the simulator, a line will send these agents a message telling "I am your input" and "I am your output." All the actions related to establishing a power flow channel will be taken as a reaction to the arrival of these messages as will be illustrated further.

3.2 The Source Node

The source node is a source of electricity with voltage U_{in}. The immediate overall consumption of the network connected to this source is:

$$P = U \cdot I \tag{1}$$

There is also the maximal amount of electric power the agent can deliver to the network as a pre-set value. When this amount is exceeded, the special event is triggered.

The agents knowledge base will contain these items:

1. U – voltage (volts)
2. I – current (amperes)
3. P – load (watts) – computed according to the relation (1)
4. P_max – maximal allowed load
5. *overloaded* – true/false, meaning that the source is actually overloaded

To make all the knowledge pieces correct and make the agent cooperate with other agents in the network, we must define these behavioral reactive primitives:

1. an agent tells me that it is connecting to my output $\longrightarrow$ subscribe for its I_demand knowledge
2. subscribed I_demand changes $\longrightarrow$ change the current I accordingly
3. the current I changes $\longrightarrow$ recalculate the value of the load P
4. the voltage U changes $\longrightarrow$ recalculate the value of the load P
5. if the load $P > P\_max$ $\longrightarrow$ put `true` into *overloaded*
6. if the load $P \leq P\_max$ $\longrightarrow$ put `false` into *overloaded*

Note that the rule 3 if separated from the rule 5. If the classic functional paradigm were used, this behavior would be implemented in a function `changeI(i)` which takes I, computes P and triggers an alarm if P exceeds P_max. But with the behavior-primitives approach, each of these steps is a distinct behavioral component. This approach allows to implement nodes which do not monitor their overloading with just removing this specific rule 5 from the set of their attached behavioral primitives.

Let us now present the details of implementing the model behavior using the functionalities of the Rice simulator (section 2.1).

The electric line which is going to be connected to the source node sends a message

```
{tell
    :sender   <line-agent id>
    :receiver <source-agent id>
    :content output=<line-agent id>
}
```

which changes the value of the knowledge `output` in the agent representing the source. The rule 1 is then implemented as a *change handler* of this knowledge. The Rice's XML definition of this behavior looks like this:

```
<handler valueName="output" name="subscribe-I_demand">
  <action type="sendKqmlMessage">
     <kqml performative="subscribe" language="Kqml">
       <kqml performative="ask" language="Trivial"
```

```
        reply-with="I_demand of output">
        I_demand=?
        </kqml>
      </kqml>
   </action>
 </handler>
```

Whenever an agent connects to the source output, the source will send him the subscription for the *I_demand* knowledge. Whenever the *I_demand* value changes in the output, it will inform the source with the message:

```
{tell
    :sender  <line-agent id>
    :receiver <source-agent id>
    :in-reply-to "I_demand of output"
    :content I_demand =<value>
}
```

When this message arrives to the source, the appropriate *message handler* is triggered and will set the source's I (rule 2).

The meaning of this messages passing chain can be interpreted as: "Whenever an agent becomes my output, I will be interested in its power current demands. I will saturate this need." Note that the source node will not send the message about changed I automatically. If the output agents need this piece of knowledge, they must subscribe for it.

All the other rules are implemented as the *change handlers* defined by the Python code. They ensure that the values of all the knowledge about the power flow in the node remain consistent to the given relation (1). For instance, the code of the rule 3 will look like this:

```
<handler valueName="I" name="I-handler">
<action type="python">
def action(data):
 agent = data['agent']
 agent['P'] = agent['U'] * data['newValue']
 return True
</action>
</handler>
```

3.3 The Power Line

The power line agents initialize the network topology establishment. Each such agent automatically sends the only two unsolicited messages in the system thus interconnecting and starting the whole simulation.

A power line agent has some features common with the source node, so we will describe only the different ones. The power line has one specific knowledge

R which is the *resistance* of the line, and two voltage indicators: the *input voltage* U and the *output voltage* $Uout$.

The line resistance causes the power loss on the line. The loss is:

$$\Delta U = R \cdot I \tag{2}$$

The output voltage is reduced by the loss:

$$Uout = U - \Delta U \tag{3}$$

The behavior primitives of the line are as follows:

1. simulation start ⟶ send notifications to your input and output
2. simulation start ⟶ subscribe for the voltage U and the current I of the input
3. simulation start ⟶ subscribe for I_demand of the output
4. the voltage U has changed ⟶ recalculate the output voltage $Uout$ according to the relation (3)
5. the current I has changed ⟶ recalculate the output voltage $Uout$ according to the relation (3)

All these primitives can be implemented by the same means as it was described in the case of the source agent – message sending, change handlers and subscriptions.

The *distribution node* (agent C in Figure 1) is a special case of the line with no resistance, thus copying the input voltage to its input. Its main purpose is to split the incoming current according to the demands of its outputs. All the outputs subscribe for the value of the knowledge of the current. The distribution node uses the *knowledge generator* for replying with the value corresponding to the particular output.

3.4 The Consumer Node

The consumer agent represents one of the endpoints in the power distribution network. It initializes the flow of the electric power according to its current power consumption demands. This is implemented by setting the I_demand knowledge (which is automatically subscribed when line connects as an input to this agent) according to the P_demand knowledge and the relation (1).

The pieces of knowledge of the agent are:

1. I – the input current
2. U – the input voltage
3. P_demand – the actual power consumption requested
4. P – the actual power delivered by input
5. $P\_threshold$ – the desired accuracy of the computation
6. I_demand – the current corresponding to the P_demand

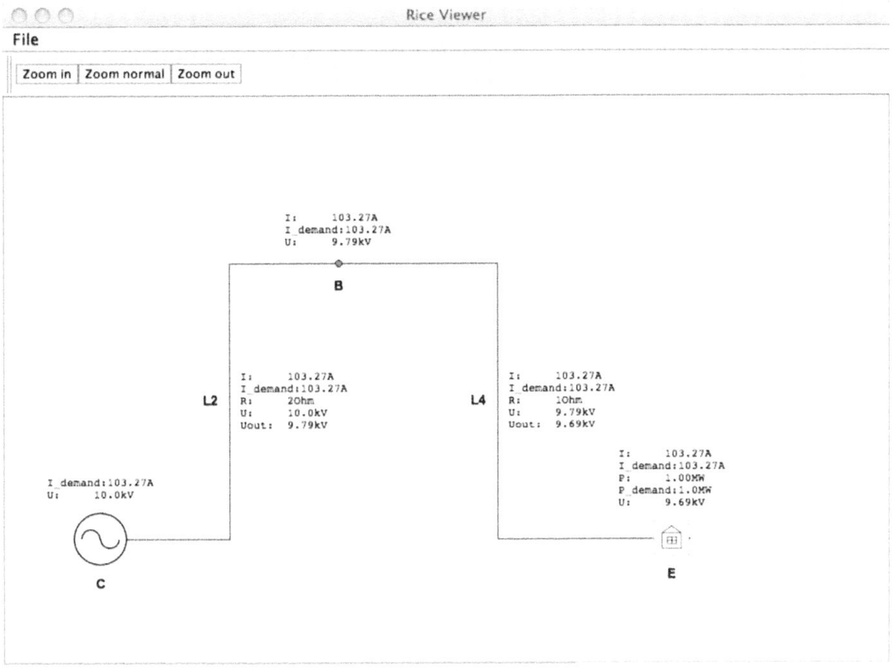

Fig. 2 Rice network viewer

The consumer agent's behavioral primitives are:

1. generate *P_demand* according to the time of the day and the given consumption table
2. the power consumption *P_demand* has changed $\longrightarrow$ change the corresponding current *I_demand* according to the relation (1)
3. I has changed $\longrightarrow$ consider if the power demand is saturated
4. the consumption is not saturated $\longrightarrow$ increase *I_demand*
5. the predefined consumption table/function

As can be seen in Table 1, the consumer node uses the network with established topology to initiate the actual energy flow throughout the network. When *P_demand* is set to non-zero value, *I_demand* is computed and dispatched to the subscriber (input line). The request then propagates through the whole network to the power source which decides if the request can be satisfied. If yes, it sets the I knowledge which then propagates through the network back to the consumer. All power losses on lines and the current splitting is computed automatically by the agents representing the appropriate network elements.

The power flow granted to the consumer is diminished by the losses. The agent then increases the *I_demand* and the whole process repeats until the difference between demanded and granted power flow reaches the given threshold which effectively defines the desired accuracy of the energy flow computation.

Throughout the whole process, the consumer agent simulates the power demand variation in time. The Rice system contains the timer routines which trigger the event in the pre-set time periods. As an reaction to this event the agent computes (or retrieves from a table) a power consumption value corresponding to the actual time of the day and sets the requested actual power consumption *P_demand*. All the power flow indicators get recomputed automatically.

4 Conclusions

We have presented the network power flow model implementation in the Rice multi-agent simulation environment. The system is based completely on local interactions between autonomous agents and on-demand knowledge propagation. No information is communicated without being requested.

Since each part of the simulation process is based on changes of agents' knowledge and the reactive nature of the agents, it is possible to manually change any of the involved indicators interactively during the simulation, including the network topology.

The Rice system also offers tools to visualise the network,[2] and to store the computed values in any moment for off-line post-processing or statistical analysis. The concept of freely-combinable little pieces of code defining the reactive behavior of the agents is suitable for machine generation of the simulation definitions.

In the time of exploitation of electric appliances in all kinds of human activity, any larger power outage, or blackout, may have a significant social and economical effects. Although it is impossible to prevent the blackout completely or to predict it for a certainty, a computer simulation may help to detect weak parts of the power network which should be replaced or back-uped.

The future work will be focused on the implementation of complex models involving intelligent decision making agents in highly dynamical environment.

References

1. Abel, E., et al.: A multiagent approach to analyze disturbances in electrical networks. In: Proceedings of the Fourth ESAP Conference, Melbourne, Australia, pp. 606–611 (1993)
2. Finin, T., Fritzson, R., McKay, D., McEntire, R.: KQML as an agent communication language. In: Proceedings of the third international conference on Information and knowledge management, pp. 456–463. ACM, New York (1994)

[2] See the Rice viewer in Figure 2.

3. Group, O.M.: Common object request broker architecture core specification (2004), http://www.omg.org/technology/documents/formal/corba_iiop.htm
4. Hcbson, E.: Network constrained reactive power control using linear programming. IEEE Transactions on Power Apparatus and Systems PAS-99(3), 868–877 (1980)
5. Kundur, P., Balu, N., Lauby, M.: Power system stability and control. McGraw-Hill Professional, New York (1994)
6. Prmek, M., Hork, A.: Multi-agent framework for power systems simulation and monitoring. In: Proceedings of ICICT 2005, pp. 525–538. Information Technology Institute, Giza (2005)
7. Prýmek, M., Horák, A.: Optimizing KQML for usage in power distribution network simulator. In: Proceedings of 3rd International Symposium on Communications, Control, and Signal Processing (ISCCSP 2008), IEEE Catalog Number: CFP08824, St. Julians, Malta, pp. 846–849. IEEE, Los Alamitos (2008)
8. Rehtanz, C.: Autonomous Systems and Intelligent Agents in Power System Control and Operation. Springer, Berlin (2003)
9. Russel, S., Norvig, P.: Artificial Intelligence, A Modern Approach, 2nd edn. Pearson Education, Inc., Upper Saddle River, New Jersey (2003)
10. Schweppe, F.: Power system static-state estimation, Part III: Implementation. IEEE Transactions on Power Apparatus and Systems, 130–135 (1970)
11. Wooldridge, M.: Introduction to MultiAgent Systems. John Wiley & Sons, Chichester (2002), http://www.amazon.com/exec/obidos/redirect?tag=citeulike07-20&path=ASIN/047149691X

Asking for Help through Adaptable Autonomy in Robotic Search and Rescue

Breelyn Kane, Prasanna Velagapudi, and Paul Scerri

Abstract. Robotic search and rescue teams of the future will consist of both robots and human operators. Operators are utilized for identifying victims, by means of camera feeds from the robot, and for helping with navigation when autonomy is insufficient. As the size of these robot teams increases, the mental workload on operators increases, and robots find themselves in precarious situations with no assistance for resolution. This paper presents an approach that utilizes multiple levels of autonomy to allow a robot to consider a range of options, including asking for operator assistance, for dealing with problematic situations to maximize efficient use of the operator's time. Individual robots use self monitoring to determine failures in task progression, a form of local autonomy. Upon this trigger, the robot evaluates decisions to properly route asking for help, consensus autonomy. A Call Center alerts the operator(s) to incoming requests for assistance. This results in a better use of operator time by focusing attention where it is needed. Experiments explore the effectiveness of agents' decisions, both local and team-level, with multiple simulators. A high fidelity simulator and user interface further evaluate how effective robot information is relayed to the operator, through human trials.

1 Introduction

Applications for multirobot systems (MrS) such as interplanetary construction [10], search and rescue [6], or cooperating aerial vehicles [1] will require close coordination and control between human operator(s) and teams of robots in uncertain environments. Some functions of a MrS, such as identifying victims among rubble, depend on human input. In general, humans are needed for situations where robotic sensing is inadequate or reasoning is

Breelyn Kane · Prasanna Velagapudi · Paul Scerri
Carnegie Mellon University, Pittsburgh, PA 15213
e-mail: breelynk@cmu.edu, pkv@cs.cmu.edu, pscerri@cs.cmu.edu

Q. Bai and N. Fukuta (Eds.): Advances in Practical Multi-Agent Systems, SCI 325, pp. 339–357.
springerlink.com

insufficient. Robot autonomy is needed because the aggregate demands of decision making and control of a MrS continually exceed the cognitive workload of a human operator.

As the size of the robot team becomes large, it is prohibitively difficult for an operator to constantly monitor and preempt or solve any difficulties the robots get into. Adding extra operators allows some scale up, but this can be inefficient if the robots are simply divided between the operators because there are times when some operators are overloaded while others are idle. Somewhat paradoxically, the more reliable robots become the more inefficiently an operator's time is used resulting in operators spending more time monitoring and less time doing useful work. Previous work has addressed this problem by giving an agent team the ability to proactively ask for input, including Schurr et al's team-level adjustable autonomy [9]. However, previous work has typically focused on team level problems and not looked carefully at balancing the value of getting operator input versus operator availability and the value of continuing to act autonomously in robot teams.

Previous papers [5] present ideas on the importance of robots knowing the context of the situation they are in and adapting their autonomy to make more informed decisions. Adaptable autonomy describes how robots may need to wear multiple hats depending on their current condition. This can happen if a robot is originally labelled with rank of a boss, then must transition to an underling that takes order as its levels of awareness become less clear. The ability to adjust autonomy allows the robot to be adaptable depending on the situation. This is important in applications where the environment changes dynamically.

This paper presents a new approach to the problem of operator interaction with a large team of autonomous robots. The approach relies on individual robots having some capability for understanding that they are in a situation where operator input or teleoperation can be helpful. This *self-awareness* can be very conservative, flagging more situations that require operator input than are actually difficult for the robot. For example, in this work, robots consider getting operator input when their pitch changes unexpectedly or when they are making no progress toward their goal for some period of time. If a robot decides it needs operator assistance, a request is sent to a queue system which can pass the request for assistance on to an available operator. The queue system prioritizes tasks according to the relative value of the task associated with the robot requesting help. If the wait for an operator's attention is too long, the robot can reason about other courses of action that may resolve its problem. The combination of self-reflection by the agents and the call center to route requests maximizes the efficient use of the operator's time, and improves the overall team utility.

When a robot decides that operator input may be useful, it has a range of possible options, some of which require it to collect information from other agents or from the call center to decide what action to take. For example, before choosing to wait for operator input, the robot needs to know the length of the queue of requests at the call center. The robot also attempts to estimate the length of time the operator will need to help it out of the current situation. Options for the robot include simply giving up its task, getting another robot to take over its task, or getting another robot to provide context to the current situation leading to a decrease in the amount of time the operator will take to resolve the robot's problem. The outcomes of all these actions are uncertain and the aim is to maximize overall performance, hence Markov Decision Processes are a natural choice for representing the reasoning.

This approach was extensively evaluated in two ways. First, the approach was evaluated in a series of pilot tests with real operators in a high fidelity robot simulation[1]. This determined a baseline for the need of human intervention in searching highly cluttered environments versus robots searching completely autonomously. Second, to understand the properties of the approach under a more diverse set of circumstances, a simple simulation of the approach was implemented and tested under a range of parameters. This included varying the probability of robots getting stuck, the number of operators, and the number of robots.

2 Problem

We intend to address the following questions: What is the best way to go about providing an operator with more information, and focusing their efforts to where they can be utilized most effectively? Additionally, how can robots in teams simultaneously make smart decisions, based on current resources, to autonomously search zones while continuing to progress in their task?

Defining the problem more abstractly, consider a team consisting of a set of robots $R = \{r_1, r_2, r_3, \ldots\}$ and operators $O = \{o_1, o_2, o_3, \ldots\}$. The team objective is to maximize the number of identified victims and total area searched with respect to time. This objective can be represented as the completion of a discrete set of exploration and discovery tasks $G = \{g_1, g_2, g_3, \ldots\}$ over the map. These tasks are of varying importance to the team goal, as determined by the joint reward function $u(g)$.

Robots operate in one of several states, depending on model fidelity, but they are primarily *fixed*, *stuck*, or *dead*. In order to complete a task g and gain reward $u(g)$, a robot must spend some number of timesteps in the *fixed* state. However, environmental circumstances can cause robots to transition to the *stuck* or *dead* states at each timestep. Operators have the ability to allocate a number of their time steps to restoring robots in the *stuck* state to the *fixed* state. However, they can only do this for one robot at a time. For time t, this is represented by the assignment A^t_{or} of operator o to robot

[1] While future evaluation will involve real robots, it is not feasible at this time to conduct extensive user evaluation with sixteen autonomous robots in a large, complex environment.

r at time t. Over a given time horizon T, the team must allocate operator attention in a way that maximizes completed tasks:

$$U_{team} = \arg \max_{A^0_{or},\ldots,A^T_{or}} \sum_{i=1}^{n} u(g_i)$$

It is important for an operator to not just be useful in assisting robots in need, but also to assist robots that generate a higher utility. The following illustrates a concrete example of the problem. Two operators are controlling twenty four robots. They must assist robots as necessary to maximize coverage of the search area, while watching live camera feeds in order to mark the location of potential victims. Each operator must constantly switch context between surveying the environment for victims, assisting robots that look like they are not being useful, and communicating with the other operator. Robots may often get stuck and stop searching, but with so many robots to control, this may go unseen to an operator. In another case, an

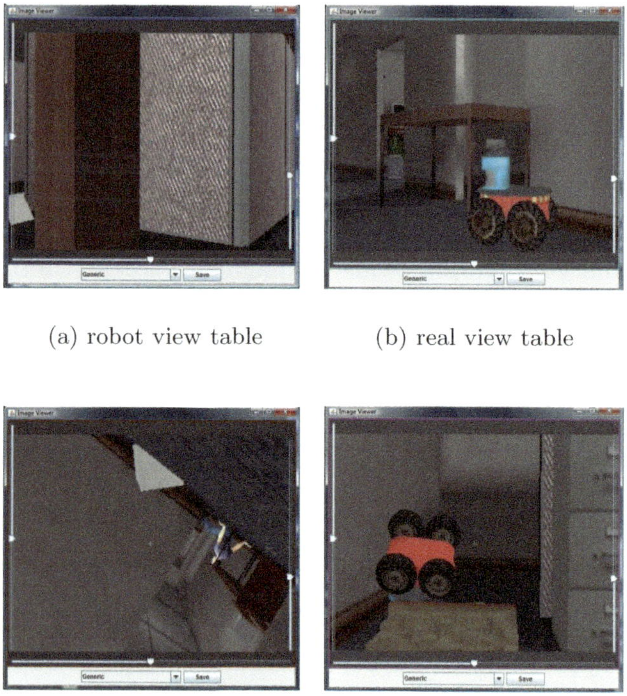

(a) robot view table (b) real view table

(c) robot view flip (d) real view flip

Fig. 1 Various stuck positions are displayed: The views on the left, (a) and (c), show what the operator sees from the robot camera vs. the actual position of the robot in the real world, corresponding views (b) and (d).

operator might try to assist a robot that is stuck, and spend an arbitrarily long time concentrating on how to get it out of a small, completely mapped room rather than marking victims. Or, the operator may not realize that getting this particular robot unstuck, instead of a different robot that is stuck in an area that is not yet mapped, hurts their ability to search a larger environment. Furthermore, the degree to which a robot is stuck may also vary. All of these instances present challenges in how the operator must go about balancing their current tasks with knowing which robot is critical and the time it may take to free the robot. Illustrating this point, Figure 1 shows a side by side display where a robot is stuck, and what an operator sees through the robot's forward camera versus an exocentric view of the current state of the robot. In robotic search and rescue domains, robots are often stuck in situations where even the most skilled operators have trouble understanding the severity or simplicity of the situation. This is because, provided only with egocentric camera displays, human operators have difficulty establishing complete situation awareness [12]. This leaves an opening for potential contributions from individual robots to provide more context to the user. The robot can assist the operator by autonomously making decisions, both individually and through the team, to accomplish the overall goal. Robots can delegate focused requests to the operator with a sort of scalable adjustable autonomy. Thus, the degree to which a robot should make decisions for the operator on where to concentrate is also an important question.

Figure 1 shows side by side displays illustrating the differences between what an operator sees through the robot's forward camera, versus the exocentric view of the current state of the robot.

3 Concept

Having self-awareness at the robot level can be beneficial to operators in multi-robot domains in dealing with these issues. The ability for robots to internally initiate decisions and act as a low pass filter over noisy information allows operators to focus on assisting higher priority robots, allowing the control of larger teams with less operators. In this work, we propose a multiagent approach that divides the problem of managing operator assistance into several subtasks that are simultaneously processed both on the robot and at the operator interface level. An additional actor, the queue meta-level manager, is introduced as a coordinator for interaction between operators and robots.

The motivation for this architecture comes from the structure of a call center. A call center is a system that allows multiple operators to service many requests and is robust to varying call loads. In this domain, robots would represent customers, and human tele-operators their servers. Rather than providing advice the operator is responsible for assisting robots in need. Robots, operators, and the queue meta-level manager all act as independent agents. Each agent is a separate process. This modularity allows for

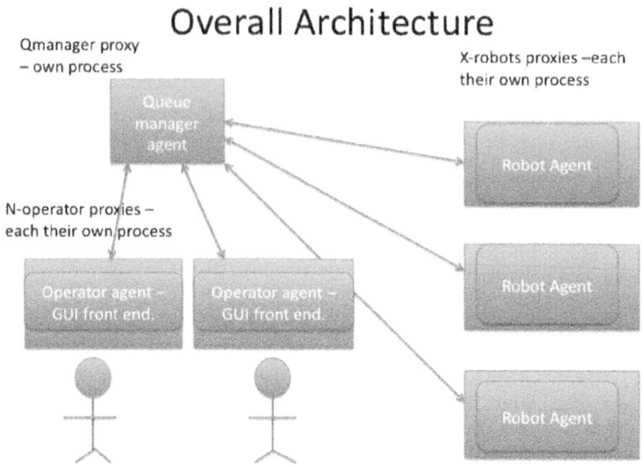

Fig. 2 Communication between multiple processes

independent interaction. For example, robots assess their situation and make informed decisions, while operators service requests, and a queue manager acts as a switching board for incoming requests.

As robots travel through the environment the evaluation of their stuck state triggers the creation of an internal reasoning proxy. This occurs when the percentage that a robot is stuck exceeds some threshold ω. The reasoner proxy acts as a self-contained piece of software that combines team level information to reason about its next action. This reasoning produces a policy, Π, which returns the optimal matching action u of the current state x. The series of possible actions are described in the MDP formulation given in 4.2. As described in Algorithm 1, if the robot decides to place a request for the operators help, it sends the request to the queue manager. This manager

---

Algorithm 1

---

if $Pr(stuck) > \omega$ **then**
  Π = generateReasoner(RobotID)
  $u = \Pi(x)$
  if u == ask operator for help **then**
    sendMessage(QMessage, RobotID)
    waitForUpdates()
    u_{t+1} = getNextAction()
  else
    execute alternate u
  end if
else
  executeTask()
end if

---

extracts the incoming request message, and places it on a literal queue. Statistic information is flagged for this request, and calculated as necessary. The queue manager can also send updated changes for values such as predicted service times, and the varying length of the queue as they change dynamically. The queue is upper bounded by the number of robot agents multiplied by their number of request types. It is useful for the robot to take updated queue statistics into account when deciding whether or not to get additional help from a team member, or try other behaviors while waiting for help from the operator. The communication between these entities is illustrated in the system level diagram, Figure 2.

4 Algorithm

The approach has two major components that we will describe in detail in the following sections. In one piece, the robot locally decides through monitoring if it is unable to progress in its current task. This can be thought of as the classification of a full or partial failure. Another section gives an analysis of how the robot takes information from itself, the team, and properties of the call center queue to decide how to go about resolving its current situation.

4.1 Self-monitoring

There are many instances where something unexpected occurs and prevents a robot from progression or even completion within their prescribed task. As has been stated above, if robots could provide feedback on their current state through a sense of self monitoring, it is easier to anticipate when things are not going as planned. If a seemingly naive robot is able to determine a fail or partial fail scenario this can act as a trigger for making decisions to help resolve the current situation. The current implementation is the beginning steps towards robot introspection. As described in [2] full introspection incorporates both a monitoring phase and a meta-reasoning phase. In this model, sensor level information is processed and then sent to a metareasoner which in turn 'thinks about thinking'. Control algorithms are replayed back to be translated into actions for the robot to execute. It is stated that the current implementation emulates the beginning steps, because our approach incorporates just the monitoring phase. Even though the metareasoning is employed at a team level, the individual agent does not really evaluate if an action is good or bad in determining its initial stuck state.

The robot determines if it's 'stuck' with different sensory information obtained while obstacle avoiding. The z-axis of the root reference frame is parallel to gravity but points up. The y-axis of the root reference frame points laterally and is chosen according to the right hand rule. The x-axis of the root reference frame points behind the robot. The rotation matrix is utilized to know if a robot is flipped completely over, facing the ceiling, or facing the ground. This is explicitly determined by looking at the z-axis, represented as the last row in the rotation matrix. If it is positive for the forward vector then the robot is flipped facing the ceiling, if it is negative for the forward vector the robot is flipped facing the floor, and if it is negative for the up vector the robot is flipped over. Laser scan readings reveal if the robot is detecting an obstacle. If progression is not achieved over a large threshold of time $> \beta$, even after trying to turn in obstacle-free directions, the robot is labelled as stuck in a corner. When a robot has not moved or turned any distance in a large number of iterations $> \delta$ it can also be said that it's wedged or blocked in some manner. The robot continues to explore unknown areas and keeps track of sensor information, pre-empting when it thinks it can't continue. Upon this realization the robot will stop and begin evaluating its next action.

(a) Driven up a wall

(b) High-centered on a step

(c) Trapped by victim and wall

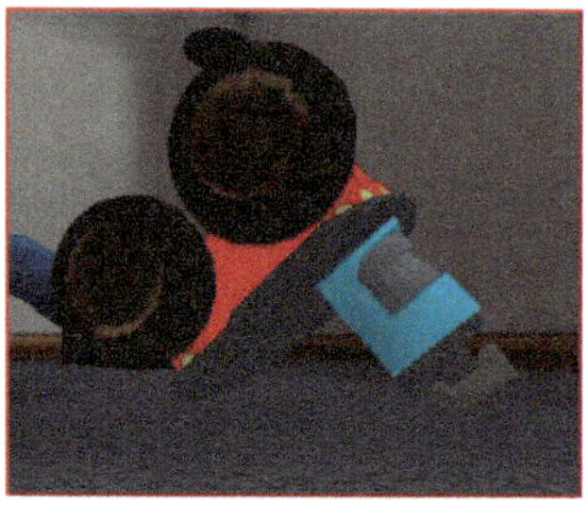

(d) Completely flipped

Fig. 3 Robots can become stuck in a variety of ways, with different amounts of effort required to extricate them.

4.2 Decision Model

The ability of a robot to identify that it may be stuck, use this information and that of the team to decide what to do to proceed, and either directly or indirectly assist the operator is important for accomplishing the goals of the team.

Once the robot has determined it is in need of assistance, a Markov Decision Process is used to optimally evaluate the next best action. The decision model uses internal robot states, and evaluates how the robot should proceed in requesting help. A Markov Decision Process is a way of mathematically formulating a problem that can be described in a four-tuple $\{S, A, C(S,A), T(S,A)\}$ The common general mathematical formulation for MDPs is described below:

- a set of actions $A = \{a_1, a_2, a_3 ... a_n\}$
- a representation of states $S = \{s_1, s_2, s_3 ... s_n\}$
- the one step cost of each associated action state pairing $C(S, A)$
- a transition function that denotes the next state given a state and action $T(S, A)$

The Markovian property is exploited where an action state pair relies only on the current state and does not account for all previous states. The system is fully observable as all states resulting from current actions are known. This MDP is inherently stochastic as the probability of many states are returned through the transition function rather than one discrete state for each state action pair. The value and optimal policies emerge when the algorithm converges by the value iteration form of dynamic programming. Here a policy is the mapping $\Pi : S \to A$ of optimal state and action pairings which minimizes the value $V : S \to C$, or total accumulated cost, for being in a state S, with discount factor γ. This is also known as the Bellman equation.

$$V(s) = \arg\min_{a \in A}\; C(s,a) + \gamma V(T(s,a)) \tag{1}$$

In this case, the MDP has four thousand states, choosing from five possible actions. It converges in approximately fifty iterations. Each state of our MDP has five dimensions represented with a variable for queue time, stuck state, distance of helping robot, distance of a robot providing more information, and a boolean indicator of whether or not this robot is currently waiting on the queue. The stuck states were determined after experimentation with P2AT robots in the USARSim domain, USARSim is described in 5.1. See Table 1 for the explicit state listings.

There are five possible actions a robot can take in each state: {*AskHelp, HelpSelf, AskSelf, AskRForInfo, AskRForHelp*}. Each of these corresponds to a potential action the robot can take in response to its current stuck state. The robot can ask for operator help, which places a request on the queue. The robot can try and help itself, meaning that it will attempt to move on its own

Table 1 The possible states of the MDP model.

States				
Queue Length	Stuck State	Distance Helper Robot	Distance Robot providing info	On Queue
Secshort	facedown	there	there	on
SecMedium	faceup	short	short	off
SeLong	wedged	medium	medium	
MinOne	goingoverobj	long	long	
MineTwo	cornerstuck	unknown	unknown	
MinFour	flippeddead			
MinSix	notstuck			
MinEight	unknownstuck			
MinTen				
MinLong				

to unstick itself. The robot can reassess its current state. This means that the robot takes movements to help clarify its stuck state if it is uncertain of this state previously. Also, another robot can be called in to help the current robot, this is with the intention of the helper robot getting the calling robot unstuck. Lastly, the robot may ask another teammate to provide information. This is where another robot can come and provide camera imagery for the current robot, to help give the operator more situational awareness for better assisting the robot.

Transition probabilities are generated by doing trials in the physics simulation USARSim, described below, to gauge an accurate approximation of how the environment affects the dynamics of a robot's stuck state. These probabilities are static heuristics, but as the state and action space increases these transition probabilities could be learned through an algorithm such as linear regression which will help maintain weights on the feature space.

Getting in a terminal stuck state of flipped over, or a not stuck state, act as the absorbing states for this model. The single step cost for a given state is dependent on the current action. For example, robots that ask for operator help incur a higher cost when helper robots are near. This happens because close robots may act as additional obstacles for getting stuck. Robots that decide to ask for other robots to assist them have a lower cost when the queue length is less and a helper robot is near, this is because helper robots are better at providing a different viewpoint for the operator and assisting the robot in need. It is not assumed that once assistance is employed to a stuck robot that it will become unstuck. The model tries to account for the percentages of times when the robot actually may get more stuck. The amount of assistance a helper, operator, or the current robot can provide at a given time is a direct result of the current state and action.

Trends emerge based on how the robot chooses actions from an optimal policy. When queue lengths are short, and the robot is not on the queue, the optimal action for the majority of states is to ask the operator for help. Also, when the queue length tends to be long, and the robot is on the queue, the best action is for the robot to help itself. If the distance of a helper robot is right near the current stuck robot, the robot tends to ask for robot help more often when the queue is of a higher length. A condensed version of this MDP is compared to other policies in the Simple Sim described below in 5.2.

5 Results

In order to explore the complex potential interactions between robots and operators as well as the gross effects of various system parameters, experiments were conducted in both a high-fidelity simulation environment with human participants and a low-fidelity discrete event simulation with simulated human operators.

5.1 USARSim

Pilot experiments were conducted in a high fidelity robotic environment developed as a simulation of urban search and rescue (USAR) robots and environments, and intended as a research tool for the study of human-robot interaction (HRI) and multi-robot coordination. USARSim is open source and freely available[2]. It uses Epic Games' UnrealEngine2 to provide a high fidelity simulator at low cost. USARSim supports HRI by accurately rendering user interface elements (particularly camera video), accurately representing robot automation and behavior, and accurately representing the remote environment that links the operator's awareness with the robot's behaviors. MrCS (Multi-robot Control System), a multirobot communications and control infrastructure with accompanying user interface developed for experiments in multirobot control and RoboCup competition [11] was used with appropriate modifications in both experimental conditions. MrCS provides facilities for starting and controlling robots in the simulation, displaying camera and laser output, and supporting inter-robot communication through Machinetta [8], a distributed multiagent system. This distributed control enables us to scale robot teams from small to large. The operator interface for this experiment can be seen in Figure 4.

Initial experiments comparing robots autonomously searching with the assistance of a human operator did not provide conclusive evidence of needing operator assistance. The cubicle map was then customized to include many more obstacles including desks, ramps, and chairs. This increased the difficulty of the task. In this environment we ran a series of ten experiments for

[2] *http://www.sourceforge.net/projects/usarsim*

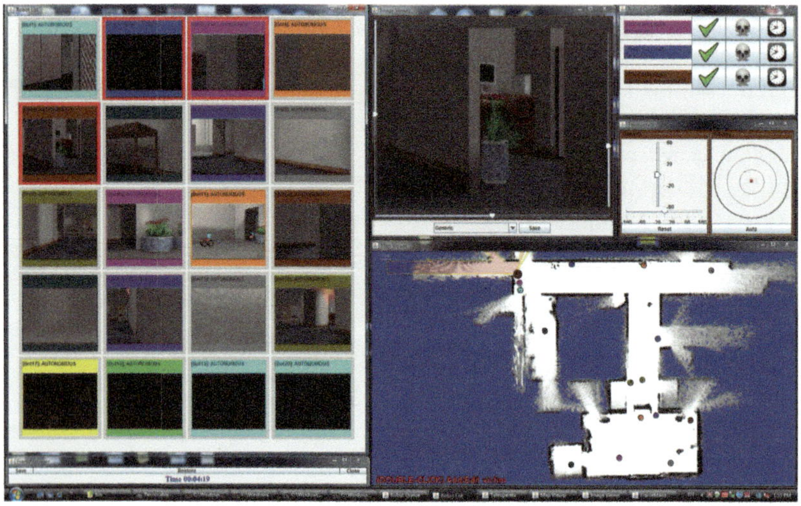

Fig. 4 Setup and user interface for USARSim experiments. Users are presented with streaming video feeds from each robot (left), a global map panel (lower right), teleoperation controls (middle right), and notification queue (upper right).

eight different single operators controlling sixteen robots in a highly cluttered cubicle environment. Two trials were run for each variant. The performance metric was the total area searched by the robot team. The first variant involved sixteen robots searching autonomously; using their laser scans to probabilistically map uncertain terrain populating an occupancy grid. Unexplored areas contained more entropy or uncertainty so robots tended to branch into unmapped regions automatically by maximizing their information gain. The second variant included the addition of a supervisory operator. The operator was told the goal was to maximize search space and could teleoperate robots of their choosing to explore areas of the map or assist as necessary. The third variant included the use of a queue system. In this experiment robots still traverse autonomously while an operator supervises. The difference is that now the robot can detect it is in need of assistance, pre-emptively stop, generate a request to the operator, and wait indefinitely for help. The underlying queue is a priority queue sorted based on the robot's priority. The robot's task priority is assigned based on how much utility they can generate by searching the environment, or the amount of information gain when searching a particular area. The queue system user interface employs a standard alarm scheme to alert the user to new help requests. The video feed of the respective robot needing help blinks until user acknowledgement at which point the queue message is displayed to the operator. The operator is given the option of clicking next to the message to mark the request resolved, dead, or ignore the request. Marking the request dead results in marking the robot as unusable. This symbolizes that an operator has done all they can to free

Table 2 Results of USARSim simulation trials. Fully autonomous robots are compared to robots being assisted by human operations in a direct supervisory and call center role.

Round	Area (m^2)
Autonomously Searching Robots 1	482.42
Autonomous Searching Robots 2	504.13
Operator Assistance / Autonomous Search 1	583.99
Operator Assistance / Autonomous Search 2	665.92
All Above plus Call Center Interface 1	626.96
All Above plus Call Center Interface 2	645.56

the requesting robot, but is unable to do so. The ignore button allows the user to multi-task and not address current request when they may be overloaded with other tasks.The results are displayed in Table 2.

The total search area achieved in conditions in which an operator assisted the robots showed an increase in more than 25% over robots simply searching autonomously with no assistance at all, proving that operator assistance is helpful. The results between the call center and operator conditions did not yield signficant differences in performance.

The results from the USARSim experiments do not conclusively show that the call center approach improves performance over that of an operator which decides what robots to assist independently. There may be several reasons for this result. The current experiments had the robots functioning very unintelligently. Once the robot determined it was stuck it waited for the operator's help possibly indefinitely, a naive policy. This significantly decreases the efficiency of the robots and may not be the optimal thing to do when the operator is not available to address a particular request. The reason robots are left idle in their down time is to allow an incremental approach to experiments. Bringing in too many additional variables may skew results. The results also did not take into account metrics that may be specific to the operator. For example, operators that take a very long time to address each robot request will tend to have other requests queue up in the meantime. Also operators that may not be very good at un-sticking robots could possibly not benefit from a call center scheme, as all of the operators used in these pilot experiments were inexperienced. It does seem that with more operators and robots, where an operator can not possibly address every situation in an optimal way, the call center setup is beneficial. Since a good baseline has been established without negative results, the next logical step is to test variations among human subjects which include varying the operators, number of robots, and the complexity of decision autonomy the robots can employ. To gain insight into how these variants effect the system a simpler simulator was designed that allowed for exploring all these variations to provide validity for future wide-scale human experiments.

5.2 *SimpleSim*

A simple discrete event simulation was used to explore a large number of parameters when evaluating the policies generated by the MDP, as well as different dynamics related to how a call center functions. The simple simulator is made up of Unmanned Ground Vehicles (UGVs), and Unmanned Aerial Vehicles(UAVs) that travel the span of a grid world at each timestep. All robots create paths to goals through an A* planner. The UGVs travel around planning a path to a goal marked with a 'T'. The ability of the UGV to travel to that 'T', without getting stuck, is a completion of their task. UAVs also travel through the environment and generate paths, but they do not get stuck, they are purely used as helper robots in the context of these experiments. Squares in the grid environment are marked with different colors to represent different stuck states such as the robot being terminally stuck versus stuck and needing help. An underlying queue collects incoming help requests for the operator. A simple operator model based on a Gaussian distribution is used for varying queue resolution times.

In generating our results, the MDP model is constrained to the state space of the simple simulator. An example of how initial states might transition is shown in Figure 5. There are only six hundred states with three possible actions. All state values remain the same accept for the stuck state and distance of a helper robot. The stuck state changes to include $\{dead, stuck, unstuck\}$, and the distance of helper robots no longer is important, using only $\{long, unknown\}$ states. The actions now only allow the robot to ask an operator for help, ask itself for help, or call in an additional robot to provide more information. The optimal policy gets fed into the simulator that maps states and actions.

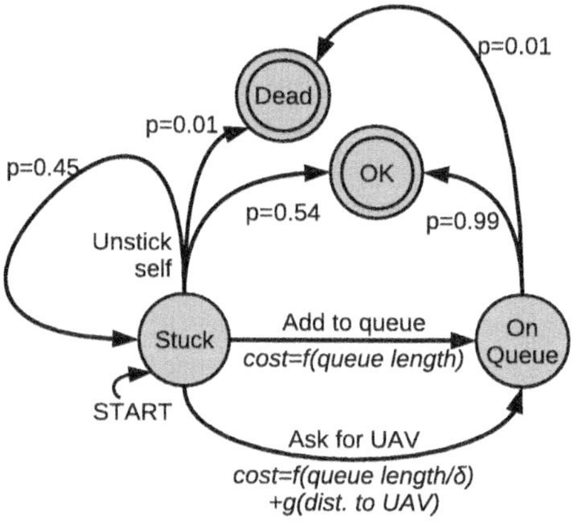

Fig. 5 An example of a partial MDP sequence.

This MDP policy is compared over varying parameters to two other simple policies. These policies are, a policy that puts every stuck robot on the queue (Is it good to always ask an operator for help?), and a policy where the robot always tries to help itself before placing itself on the queue. The 'operator policy', where every stuck robot places itself on the queue, was chosen as it is similar to the policy used in the human trials that were described in 5.1.

A series of experiments studied the effects of varying the number of operators, the number of robots, the proportion of dangerous map cells (affecting how often robots were stuck), and the average time for an operator to help a robot. In each experiment, one parameter was varied, while the others were held at the nominal values given in Table 3.

Table 3 Nominal values for SimpleSim experiments.

Parameter	Value
# of robots	256
# of operators	2
% dangerous cells	4
Average assist time	30

At each setting 100 simulation runs were executed, each lasting 1000 time steps. Robots and tasks were initialized uniformly random for each run, with tasks varying randomly in value. The performance of each run was measured by a sum of complete task value, additionally recording average queue length over the run. More specifically, as a task is completed the value for that task was accumulated in the total utility. The results of the experiments can be seen in Figure 6.

Queue length increases and value decreases as conditions become adverse for the robots, as the number of dangerous cells increases in Figures 6(e) and 6(f).

The behavior of the MDP is apparent in its effect on queue length across all conditions. While other policies show an increase in the queue length as conditions become increasingly adverse, the MDP policy becomes increasingly frugal with operator requests, efficiently keeping the operator workload down as the queue length plateaus to a constant length, as operators take longer to assist 6(b), and when the number of operators are low 6(d), and when amount of dangerous cells increase 6(f), and as the number of robots increase while operators are held constant 6(h). This is a good result and supports the need for autonomous robot decision making using the environment's state information.

Since the MDP is additionally aware of task utility, it is also capable of prioritizing requests to improve performance, yielding higher overall value than the other policies. Figure 6(c) shows an interesting trend where the MDP policy's value decreases over time but still finishes at a higher value

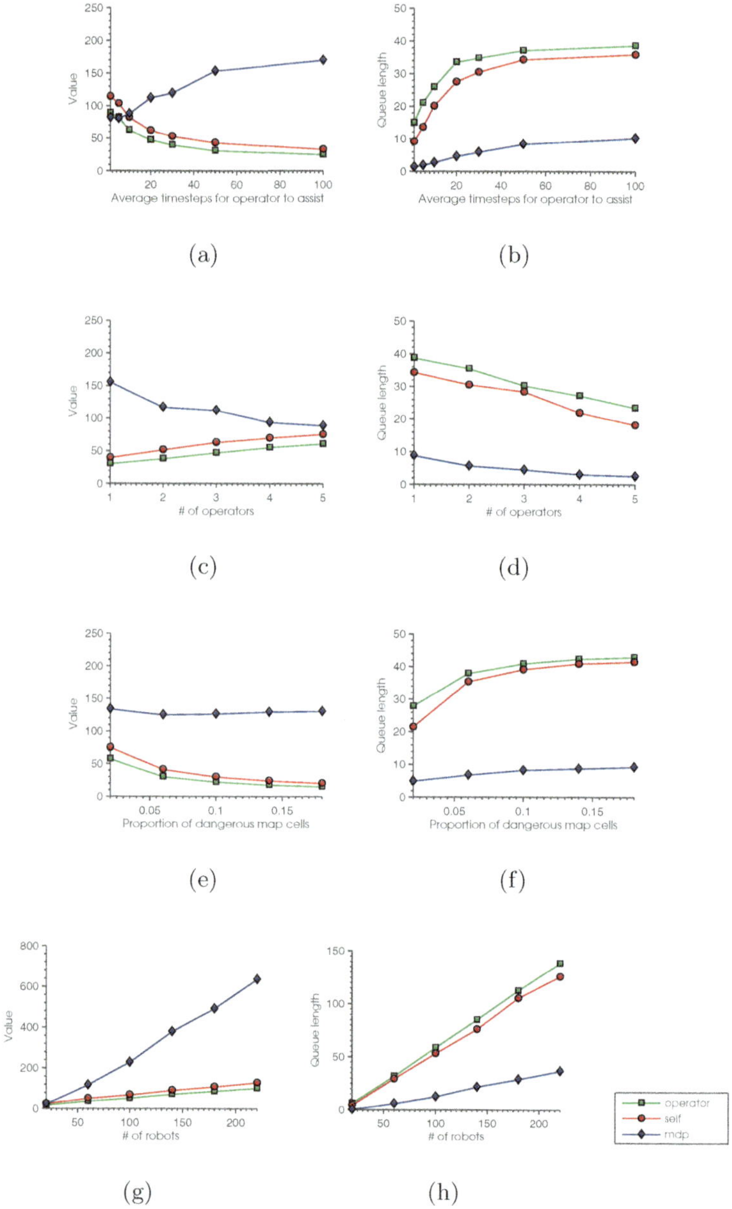

Fig. 6 Results of the simple simulator experiments over a range of parameters.

than the other policies. This may occur because increasing the number of operators does not necessarily lead to the assistance of more robots, especially when the MDP does not always ask for operator help. Also interesting is the trend as operators take longer to assist in Figures 6(a), this shows that the accumulation of task value increases without the need of an operator, for the optimal policy.

6 Related Work

For human robot teams in uncertain environments, it is necessary to balance between robot autonomy and the use of teleoperation for robot assistance. Others [4] explore the effectiveness of a robot that navigates through the environment by teleoperation and asks an operator for input. This is referred to as collaborative control. While the framework described above also employs a line of communication between the operator and robot agents, it additionally utilizes team autonomy information to make more informed decisions. Also, the robots proactively take actions when an operator may be busy. This allows an efficient use of an operator's time by balancing when it is necessary for an operator to interact with a robot versus the robot acting autonomously. This balance is illustrated in models between neglect time and interaction time [3]. The model describes a point in robot autonomy that reaches degradation caused by neglect tolerance. This suggests that overall tasks would improve with an operator causing the operator to devote interaction time to the robot. The architecture just presented addresses this balance without the need for dedicated operators.

To be clear the focus of this work was not to evaluate the graphical user interface between the operator and the system. Many previous work addresses interface design. Some of this work [7] touches on how questions should be relayed with the support of robot state information to users to help their understanding. The work presented here does provide context to the user, but focuses more on the underlying intelligent decision making of the system. Additionally, previous work [13] conducted tests for interface designs specific to the task of robotic search and rescue. Users often felt more comfortable with interfaces that provide more information about the robot's orientation. A preference for fixed windows was noted based on user feedback explaining how this allowed them to better judge the current situation. Although, an evaluation of the interface design is not the purpose of this work, the system architecture supports the need for presenting more situational awareness to the user, thus increasing their comfort level, and improving their ability to complete the task.

7 Conclusion and Future Work

This paper presents a new way to assist a multi human robot team in accomplishing its goal through a local autonomy trigger, and reason provided the context of a team. The analysis of algorithm presented shows the importance of the concept of robots sharing responsibility in making smarter decisions.

This represents a baseline for future experiments and directions in teams with higher scale and more dynamic environments.

The major contributions of this work include the presentation of a system architecture that addresses neglect and interaction time profiles. Also, a robots adaptable autonomy in this architecture efficiently allocates operator time, provides situational awareness to the operator, improves team performance, and is robust to partial system failure.

Future work includes extending experimentation to real robot platforms, and delving deeper into investigating each subset of the problem. Extensions on self monitoring might incorporate the ability of a robot to evaluate how good their decisions are, and if they can learn, or set up a prediction of areas they may want to avoid based on knowing histories of getting stuck. The robot's current decision model can be expanded with additional states and actions. This may include the use of heterogeneous agents, and the ability for an agent to give help in addition to asking for help. A call center formulation does not have to be constrained to just agents requesting help. The queue can also be utilized in other applications such as UAVs for holding important video playback data that the operator may need for better completing their task.

References

1. Chaimowicz, L., Kumar, V.: Aerial shepherds: Coordination among uavs and swarms of robots (2004)
2. Cox, M., Raja, A.: Metareasoning: A manifesto. BBN Technical (2007)
3. Crandall, J., Goodrich, M., Olsen, D., Nielsen, C.: Validating human-robot interaction schemes in multitasking environments. IEEE Transactions on Systems, Man, and Cybernetics, Part A: Systems and Humans 35(4), 438–449 (2005)
4. Fong, T., Thorpe, C., Baur, C.: Multi-robot remote driving with collaborative control. IEEE Transactions on Industrial Electronics 50(4), 699–704 (2003)
5. Macfadzean, R.H., Barber, K.S., Interrante, L., Martin, C.E.: Supporting dynamic adaptive autonomy for agent-based systems (1996)
6. Murphy, R., Stover, S.: Rescue robots for mudslides: A descriptive study of the 2005 La Conchita mudslide response. Journal of Field Robotics 25 (2008)
7. Rosenthal, S., Dey, A., Veloso, M.: How Robots Questions Affect the Accuracy of the Human Responses. In: Proc. International Symposium on Robot-Human Interactive Communication, pp. 1137–1142 (2009)
8. Scerri, P., Liao, E., Lai, J., Sycara, K., Xu, Y., Lewis, M.: Coordinating very large groups of wide area search munitions. Theory and Algorithms for Cooperative Systems (2005)
9. Schurr, N., Marecki, J., Lewis, J., Tambe, M., Scerri, P.: The defacto system: Training tool for incident commanders. In: Proc of the AAAI, vol. 20 (2005)
10. Sellner, B., Heger, F., Hiatt, L., Simmons, R., Singh, S.: Coordinated Multiagent Teams and Sliding Autonomy for Large-Scale Assembly. Proc of the IEEE 94(7), 1425–1444 (2006)

11. Wang, J., Lewis, M.: Human control for cooperating robot teams. In: Proc of the ACM/IEEE international conference on Human-robot interaction, p. 16. ACM, New York (2007)
12. Wickens, C.D., Lee, J., Liu, Y.D., Gordon-Becker, S.: Introduction to Human Factors Engineering, 2nd edn. Prentice-Hall, Inc., Upper Saddle River (2003)
13. Yanco, H.A., Keyes, B., Drury, J.L., Nielsen, C.W., Few, D.A., Bruemmer, D.J.: Evolving interface design for robot search tasks: Research articles. J. Field Robot. 24(8-9), 779–799 (2007)

Conceptual Framework for Design of Service Negotiation in Disaster Management Applications

Costin Bădică and Mihnea Scafeş

Abstract. Interactions between stakeholders providing their services in disaster management might involve negotiations for optimal selection of service providers. In this paper we propose a conceptual framework for designing cooperative multi-issue one-to-many service negotiations. The framework allows us to define: negotiation protocols, negotiation subject, properties of negotiation subject issues, deal spaces, and utility functions of participant agents. We also consider a simple example of how this framework can be instantiated to particular negotiation in a disaster management information system.

1 Introduction

The increased complexity of socio-economic systems in combination with natural hazards might create conditions that can evolve into dangerous chemical incidents in or nearby urban areas. These are complex processes that face involved stakeholders with handling of chemical substances threatening themselves, the population, as well as the natural environment. Moreover, as we deal with increasingly complex chemical substances, the risks of using them in a wrong way are higher. Places such as harbors, as well as industrial areas (e.g. the port of Rotterdam) are constantly at risk because many companies are using, transporting and/or processing dangerous substances in those areas. Agencies specialized in environmental protection and emergency management are continuously monitoring the situation with the goal of taking corrective actions when things tend to go wrong. For example, DCMR[1] in

Costin Bădică · Mihnea Scafeş
University of Craiova, Software Engineering Department,
Bvd.Decebal 107, Craiova, 200440, Romania
e-mail: {scafes_mihnea,badica_costin}@software.ucv.ro

[1] DCMR Milieudienst Rijnmond.

Q. Bai and N. Fukuta (Eds.): Advances in Practical Multi-Agent Systems, SCI 325, pp. 359–375.
springerlink.com

the Netherlands and DEMA[2] in Denmark are two organizations that protect the people and the environment against chemical hazards. Their goal is to monitor the situation and, when needed, to discover the source of the disaster, handle the chemical substances, secure the dangerous area and eventually decide to evacuate the population or take other protection measures.

However, in responding to a threat caused by an incident, these agencies cannot and usually do not act in isolation. Efficient response to a chemical incident requires collaborative effort by exchange of information and expertise between several stakeholders that can belong to different organizations. For example, information from the Police (the traffic volume, congestions that might be present near the incident location) and the City Hall (what is the population in the area of the incident and how is it distributed?) help agencies to take decisions whether to evacuate or shelter affected people. Additionally, the agencies might also use volunteers that are present in the incident area and are able and willing to give precious information about the source of the incident (do they smell something unusual? do they see weird smoke in the area?) apart from their own specially trained staff.

Among the difficult challenges that environmental and emergency organizations have to face during their responses to incidents, we mention:

- Difficulty in finding the source of an incident, or at least obtaining a list of potential sources.
- Inability to get an accurate on-site situational overview.
- Difficulty in finding useful resources and experts and in establishing communication lines between them for information exchange.
- Inability to log the incident handling.
- Inability to try "What if" scenarios that help the decision makers.

See [3] for other problems and typical scenarios that are faced by these organizations.

In order to tackle these kinds of problems faced by organizations such as DEMA and DCMR, the DIADEM project[3] has been started. DIADEM aims to develop a distributed information system that supports experts from various geographically distributed organizations (e.g. environmental agencies, police, fire fighters) with heterogeneous techniques from various domains including software engineering, decision making, human-computer interaction, chemical engineering, environmental management, and crisis management.

The core platform of the DIADEM distributed information system – the DIADEM Framework – is modeled using multi-agent systems techniques [17]. Software agents support human experts to achieve their tasks. They can provide user interfaces to various tools that run on different devices that experts might utilize or carry with them (e.g. PCs, laptops, PDAs, mobile phones, etc.).

---

[2] Danish Emergency Management Agency.

[3] DIADEM: Distributed information acquisition and decision-making for environmental management. See project website at `http://www.ist-diadem.eu/`

An important feature of DIADEM is the communication system. Software agents help to automatically route and exchange information between experts. Experts' tasks are wrapped in agent processes. Thus, agents run processes that contribute to the global goal of the system: mitigation of a chemical disaster. This wrapping of tasks into agent processes helps the system to connect the experts through agents' reasoning and interaction. Agents advertise their processing capabilities as services by using a simple Yellow Pages mechanism. Each time an agent is not able to fulfill his task (for example because of lack of knowledge to gather or derive necessary data or of the incapacity of the expert to move from his current location), he asks another agent for help, that is, he asks another agent to provide him with a certain service. This is the basic mechanism that enables connection of the agents in the system (for more details about the DIADEM Framework, see [12,11]). In this context, agents are faced with the problem of selecting the optimal service provider taking into account a set of parameters specific to each service. Optimal selection of service providers is accomplished using *service negotiation*, which is the subject of our work and of this paper. Note that results presented in this paper are also a continuation of our preliminary work on service negotiation mechanisms for disaster management reported in [14].

For better understanding, in what follows we will consider a simple example of service negotiation in the system. The example was adapted from [3] and it is also described in [14] (for more complex examples, see also [3]).

Assuming that an incident emerged and that the fire brigade already arrived at the location of the incident, it is necessary that the *Fire Fighter Commander* (FFC) agent marks a safety distance to keep people away from the incident (we assume that: a spill of a chemical substance has leaked from a tank; the substance has already been identified; the substance turns into a gas). The FFC agent cannot compute the safety distance by himself because there are parameters that he doesn't know: (i) how does this gas react to the atmosphere?; (ii) in what measure is this gas harmful to people?; (iii) what are the atmospheric conditions at the location of the incident? Therefore, the FFC agent requires the *Safety distance* service and therefore he starts a negotiation for this service with all the agents that are able to provide it.

The location of the incident is important to the negotiation as the safety distance should be computed taking into account the weather and environmental conditions that are obviously location dependent. In many cases weather conditions (for example wind speed and direction) can only be gathered for a place or area that is close to the required location. The FFC agent must carefully filter the proposals from weather provider agents taking into account the location parameter. Another important aspect of this negotiation is the precision of the data that will be provided. The FFC agent will surely take this aspect into account when selecting the service provider agent. Therefore, the FFC agent negotiates about the group of items *(service name, location, precision)*, that is known as negotiation subject. Note that other important negotiable parameters that take into account the context of the incident and of the service providers, like for example *up-to-dateness* or *completeness* of data, can be included in the negotiation subject. See [9] for a list of quality of context parameters and methods to compute them.

This paper is structured as follows. In Section 2 we give an introduction to negotiation in multi-agent systems, with a focus on service negotiation. Section 3 follows with a discussion about the framework we propose, including examples of utility functions, and worked out example. In section 4 we briefly compare our framework with other existing negotiation frameworks. Finally, in Section 5 we draw some conclusions and point to future work.

2 Service Negotiation

Negotiation is a process that describes the interaction between one or more participants that want to agree on a subject by exchanging deals or proposals about this subject. Negotiation about a service that one or more participants agrees to provide to other participants is called *service negotiation*.

There are at least two interpretations for services in software systems: *Software as a service* and *Service as a software*. In the context of the DIADEM system, we assume both interpretations. The first interpretation is more suited as a software delivery model by application service providers for component-based distributed applications. In DIADEM, we develop specialized components that are wrapped by agents and have their functionalities provided as software services. Some examples are decision making components and prognosis components. They can provide an overview of the situation, predict how the situation will evolve, rank the predictions according to certain criteria, etc. These capabilities of software data processing are provided to requester agents (representing humans as well as other software components) as software services.

We use the second interpretation by viewing a service performed by humans or organizational entities as being delivered through software and IT infrastructure, i.e. software agents in our particular situation. For example, human experts or entire organizations can advertise their *physical* capabilities as services. Therefore, they can provide their services by agreeing to perform physical tasks included in their set of capabilities and returning results via software agents assigned to them. Additionally, on-site experts and trained staff can use computing devices (mobile phones, laptops, etc.) to interact with their personal software agents that act as an external interface of the DIADEM system.

There are three important aspects that need to be addressed in the definition of a negotiation problem: *negotiation protocols*, *negotiation subject* and *negotiation strategies* [8].

The *protocol* describes the public rules of the agents' interaction process. Agents must comply to the protocol of the negotiation in order to be able to communicate. The protocol specifies the steps of negotiations, the types of messages that can be exchanged as well as when messages can be send and/or received. Among well known negotiation protocols, we mention Contract Net (CNET) [15], Monotonic Concession Protocol (MCP) [2], and different types of auctions [17].

A *negotiation subject* describes what is being negotiated between the partners. In service negotiation, the negotiation subject is the provided service with its associated parameters. One of the preliminary steps of designing a negotiation for a specific problem is to identify the type of the subject: whether it is single-issue (e.g. only the service name) or multi-issue (e.g. service name, time, quality measures, etc). The operations that can be performed over the range of issues must be also identified: whether the set of issues can be modified or not (e.g. participants can add or remove issues) and whether the values of issues can be changed or not and in what way.

An agent's *strategy* represents the private decisions that the agent makes during the negotiation in order to reach the desired agreement that maximizes some form of welfare. The chosen strategy must be defined in accordance with the negotiation protocol.

An initial evaluation of different negotiation mechanisms with respect to their usefulness for the DIADEM system is presented in [14] and [2].

3 Conceptual Framework

Negotiations in DIADEM are managed by a specialized software component – *negotiation engine* that is instantiated inside each agent. The negotiation engine supports the addition of new negotiation protocols by offering an interface with support for negotiation demands in DIADEM. The negotiation engine is *generic*, i.e. it can run negotiations described by various negotiation protocols that were adapted for the DIADEM domain.

Negotiations in DIADEM are one-to-many and multi-issue. In a one-to-many negotiation, a single agent is negotiating simultaneously with a group of one or more agents. Therefore, one-to-many negotiations assume a special agent that has a different role than his opponents. For example, in CNET, there is one manager and possibly multiple contractors. The manager part of CNET protocol is different from the contractor part. The manager sends and eventually re-sends calls for proposals, selects the winner(s) and waits for task results, while the contractor(s) make proposals or refusals and provide task results.

Another important characteristic of service negotiations in DIADEM is that they are cooperative. Cooperativity stems from the fact that the overall goal of negotiation participants is the optimization of the response to the chemical incident. Basically this means that during a negotiation: (i) the manager is the leading decision factor that will try to optimize the assignment of the negotiation task to the best available contractor; (ii) the contractor will try to make the best proposal for serving the manager, taking into account his current duties and availability.

In DIADEM, a typical situation occurs when one agent requests some data that can be made available through services provided by other agents in the system. We found useful to describe such situations by borrowing the role names from CNET. So in DIADEM there are only two negotiation roles: (i) *manager* associated to the requester agent and (ii) *contractor* associated to the provider agents. Protocol descriptions will be divided into multiple sections to describe each role, in particular manager and contractor roles in DIADEM. Agents should be able to identify their roles from the protocol description and the context where the negotiation is applied.

3.1 Negotiation Steps

Negotiations in DIADEM share a common set of generic steps (see table 1). All implementations of negotiation protocols in DIADEM are supposed to follow this set of steps. These steps are generic in the sense that:

i) Each step must be specialized for a given type of negotiation by indicating the information exchanged between agents about proposals, acceptance/rejection and agreements made.
ii) The control flow of executing these steps must be specialized for a given type of negotiation. Mainly, the flow should follow the sequence in table 1. Note however that some steps are optional, while other steps can be iterated, as it is for example the step of proposal submission.

3.2 Negotiation Subject

In our service-oriented application, agents provide services taking various input parameters that are needed to produce their outputs. Some service parameters can be important decision factors during negotiation, i.e. they are taken into consideration during the process of selecting the proper service providers (e.g. time to start the service, estimated time to provide the service, distance to incident location etc). Therefore, such service parameters are in fact negotiation issues. Then, after the negotiation has taken place, all the service parameters become inputs to the service and are used or consumed during service invocation to compute their output. In conclusion, during negotiation some of the service parameters, i.e. those that designers consider useful for the selection of the service providers, will become negotiation issues. These issues together with the service description form the negotiation subject. Moreover, each issue has a set of properties.

The description of a multi-issue protocol introduces the subject issues and their associated properties. The issues can then be referred in the protocol description, both on the manager and contractor sides. Additionally, the negotiation engine supports querying of the issue properties to

Table 1 Common steps of DIADEM negotiations

Step No.	**Step Name**	**Description**
1	Identification of subject, role	Setting up the subject and taking roles. Usually only the manager takes this step into account, as he needs a certain provided service.
2	Start negotiation	At this point, the negotiation starts. Depending on implementation, this step might include registering the negotiation at a negotiation server.
3	Registering to negotiation	Whoever is interested in negotiation, is registering to it. Sometimes this step might not be taken into account.
4	Proposals	Receiving and sending proposals. The utility functions and deal space are used in this step.
5	Agreement/Conflict	Deciding whether an agreement or a conflict is reached, selecting the service provider. This step also uses utility functions.
6	Termination	The negotiation process really ends in this step, but an extra step might be necessary to receive the results. A data connection between the manager and the contractor is created.
7	Service provision/results (Cleanup)	The results are sent/received in this step. The connection between the agents is broken after this step.

determine their values. A subject issue that can be negotiated has the following properties:

- *Name*: a sequence of characters that uniquely identifies an issue of a given subject.
- *Data type*: the type of the value the issue is allowed to take. For example, possible values of this property are: NUMBER that describes an integer or real number and STRING that describes a sequence of characters.
- *Value type*: specifies if the issue can be modified in proposals or it should keep its original value set by the manager. Possible values are: FIXED (this describes a non-negotiable issue, i.e. and agents cannot propose values for this issue) and DYNAMIC (agents can propose values for this issue).
- *Value*: in case of a fixed issue, the manager sets this property when he defines the subject.

Depending on their role – manager or contractor, participant agents can privately configure the negotiation issues using additional properties:

- *Monotony* (for both manager and contractor): specifies whether the manager or contractor prefers high values or low values of this issue. Possible values are: INCREASING (the agent prefers high utility values of the issue) and DECREASING (the agent prefers low utility values of this issue).

For example, in many negotiations the manager agent usually prefers to get some result in shorter rather that longer time. In this case *time* is a decreasing issue. On the other side, the contractor would have to spent more effort by allocating additional computational or other resources in order to provide the result in shorter, rather than longer time. So for the contractor *time* is an increasing issue.

- *Reference value* (for manager): represents the ideal value of the issue from the point of view of the manager. The manager compares the actual issue value from a proposed deal with this value in order to take a decision about how to proceed in the negotiation. Usually a distance-like function is applied to the actual issue value and the reference value in order to estimate its contribution to the overall utility of the proposal for the manager. For example, 0, meaning "a very short time duration", might be the reference value of the *time* issue discussed above.
- *Weight* (for manager): represents the relative importance of the issue in the subject, compared to the other issues.

During a negotiation, agents propose deals from their *private deal spaces*. Only deals that match the negotiation subject are valid for proposals. A deal issue has the same properties as the corresponding subject issue, with the requirement that its value is always set (i.e. it cannot be unknown or left blank). We assume that issues are independent. That is, a change in the value of an issue does not have any influence on the other issues. In other words, an issue does not compensate other issue. We are aware that this is a restriction and we plan to address it in the future (for example using ideas inspired by [7], [6]).

Monotony of subject issues is useful for agents to make concessions to their opponents during negotiation. For example, if an agent proposal is refused during a negotiation round then the agent will know based on the opponent's monotonicity of the subject issues how to update its proposal during the next round.

3.3 Agent Preferences

Negotiation participants – either managers or contractors, use utility functions to measure their preferences over deals. Therefore in multi-agent systems, agent preferences are expressed by *utility* functions. The utility function of agent a can be defined as:

$$u_a : \mathcal{X} \to [0, 1]$$

where $\mathcal{X}$ is the set of potential deals of agent a. An agent's utility function is a mapping from his set of potential deals to a real number between 0 and 1.

In what follows we show how utilities of manager and contractor agents can be specified in our proposed framework.

3.3.1 Manager Preferences

Taking into account multi-issue negotiations, let $\mathcal{I}$ be the set of issues partitioned into sets $\mathcal{I}^{\uparrow}$ and $\mathcal{I}^{\downarrow}$ of monotonically increasing and decreasing issues. Let x_i be the value of an issue $i \in \mathcal{I}$ of a deal $x \in \mathcal{X}$. Each issue $i \in \mathcal{I}$ has a weight $w_i \in [0, 1]$ and a partial utility function f_i. Note that weights are normalized i.e. $\sum_{i \in \mathcal{I}} w_i = 1$. Intuitively, the weight of an issue represents the relative importance for the manager of that issue in the set of all issues associated to a negotiation subject.

The manager uses a weighted additive utility function to evaluate proposals and to select the service provider. Therefore his utility function takes the following form:

$$u_m(x) = \sum_{i \in \mathcal{I}^{\uparrow}} w_i * f_i(x_i) + \sum_{i \in \mathcal{I}^{\downarrow}} w_i * (1 - f_i(x_i))$$

In other words, the utility of a proposal (i.e. a deal proposed by a contractor agent) is evaluated as a weighted sum of issue utilities. Depending on the issue monotonicity, the term $(1 - f_i(x_i))$ for monotonically decreasing issues or the term $f_i(x_i)$ for monotonically increasing issues is used for evaluating the contribution of a single negotiation issue to the utility of the proposal. The partial utility of an issue maps the issue domain to a value in the interval $[0, 1]$:

$$f_i : \mathcal{D}_i \rightarrow [0, 1]$$

where $\mathcal{D}_i$ is the value domain for issue i. An issue utility represents the manager's preference over an issue value as a real number between 0 and 1.

The definition of function f_i depends on the domain of the issue. For example, a possibility to define the partial utility function of a real valued issue when $\mathcal{D}_i = [x_i^{min}, x_i^{max}]$ is as follows:

$$f_i(x_i) = \frac{|x_i - x_i^*|}{x_i^{max} - x_i^{min}}$$

where x_i^* is the reference value assigned by the manager to the issue i.

For text valued issues, the partial utility function might measure the semantic difference between two texts or simply their lexical difference.

As issue utilities have values in the interval $[0, 1]$ and the issues weights are normalized, the utility function of the manager evaluates a proposal to a real number in the interval $[0, 1]$. Value 0 means that the proposal is the worst for the manager, while value 1 means the ideal proposal, i.e. the best for the manager.

3.3.2 Contractor Preferences

Typically in bargaining problems agents use tactics to make proposals [5]. In our task allocation problem, the contractor agent uses his utility function to select the deals that he will propose to the manager. The utility function of the contractor agent is a mapping from his deal space to a real number in the interval [0, 1] as follows:

$$u_c : \mathcal{X} \rightarrow [0, 1]$$

The utility function of the contractor agent can be defined as shown below:

$$u_c(x) = 1 - effort(x)$$

Here $effort(x)$ represents the total effort that needs to be deployed by contractor agent c to be able to provide the terms and conditions specified by proposal x of the manager. The contractor will obviously choose to propose a deal that maximizes his utility, i.e. minimizes the required effort for achieving the task required by the manager.

In practice function $effort(x)$ can take into account several factors, including for example the cost of providing service x (i.e. $cost(x)$), the existing commitments of agent c that were previously contracted and are not yet finalized, and/or the amount of resources that are required to provide the service.

Obviously, a higher number of commitments not finalized yet of agent c or a higher cost of providing x will result in a higher value of $effort(x)$ and consequently to a lower value of the utility function $u_c(x)$ of contractor agent. The $effort$ function can be defined in a similar way to the optimizer in [13]. For example, the agent might require a smaller effort if the task he is negotiating depends on another task that the agent is currently working on. For example, an agent located in the port of Rotterdam working on a previous task would have to invest a smaller effort for a new task requiring measurements of gas concentrations in the port of Rotterdam than for a new task requiring the same type of measurements but in another location, because in order to accomplish the former task the agent would not need to move to another location.

Moreover, in the domain of disaster management, participation of contractor agents in negotiations might be constrained by the availability of required resources. For example, an agent should not be able to carry out in parallel two tasks that both require the associated expert to be in two different places at the same time. Additionally, agents should not be able to perform in parallel too many tasks as they would be overwhelmed. To model such situations we can endow agents with resources and constrain their participation in negotiations for service provisioning by the availability of these resources. The *load factor* of an agent represents the degree of unavailability of the agent, i.e. the higher is the agent load factor, the less available is the agent to provide services. This factor can be estimated by the percent of resources from the total amount of resources available to the agent that are currently in use by the agent to provide already contracted services. The agent load factor is represented by a real number between 0 and 1. Using the load factor, the cost

for providing the service can be determined as the percent of resources that will be added to the agent load factor if the agent will agree to provide the services under the terms and conditions of the deal. If the agent has a load factor of 0 then all his resources are available (this means he is not performing any task at the moment), while if the load factor is 1 the agent cannot engage in new negotiations until he finishes some of the currently performed tasks. The load factor is very important to define the agent strategy for making proposals, as the agent cannot propose a deal requiring more resources than the available resources.

3.4 Levels in Negotiation Specification

It is important to observe that, with the introduction of our conceptual negotiation framework, there are actually two levels of specifying negotiations that we support in DIADEM.

i) *Negotiation type* level. This corresponds to the definition of a negotiation protocol by specifying negotiation steps and their control flow. Usually at this level negotiation subjects, associated issues and some of their properties need also to be specified. Specifically, at this level the following items will be configured:

   - the negotiation protocol.
   - the subset of the service parameters that will be considered negotiation issues and default values for their associated properties, as defined in Section 3.2.
   - a local utility function for each issue.

   A *negotiation type* is created by humans at design time using the conceptual artifacts provided by the negotiation framework.

ii) *Negotiation instance* level. This corresponds to the definition of a particular negotiation by setting custom values for properties of the negotiation issues. A *negotiation instance* is created by agents at run time by instantiating and configuring a given negotiation type.

Note that if we supported only the first configuration level then an expert would have no possibility to tune the issue properties for a certain situation. If we supported only the second configuration level then an expert would probably be overwhelmed by the set of properties he would have to check (because the properties would not have appropriate default values). This is the reason for having two levels of configuration. A set of default parameter values will be configured at the first level and the expert will be able to tune the properties at the second level, before the start of each negotiation. For example, an expert may want to set the ideal value for issue "location" to his current location each time a negotiation starts. Note that this process of negotiation configuration is linked to the description and configuration of

DIADEM services. However, a discussion of service description and configuration in DIADEM are outside the scope of this paper.

3.5 Example

In this section we consider a simple practical example of how the framework supports negotiation in a disaster management scenario. Let us suppose that a manager agent needs to find a provider for the service "Measure gas concentration". For simplicity, we will not take into account the precision or location of the measurement, so *time* is the only negotiable parameter of the service. In this context, *time* means the time duration in which the measurement of the gas concentration will be available. Therefore, the manager agent starts a negotiation concerning the negotiation subject (*service*: "Measure gas concentration", *time*). *service* is a fixed issue with the meaning that agents cannot propose deals in which this issue has a value different than the initial value specified by the manager. *time* is a dynamic issue and agents can propose any value from their deal space for this issue. Let's suppose there are three other contractor agents in the system, as described in table 2.

Table 2 Possible contractor agents in the system

Agent name	**Deal space**	**Load factor**
C_1	(*service*: "Measure gas concentration", *time*: 1) with *cost* = 1	0
C_2	(*service*: "Measure gas concentration", *time*: 1) with *cost* = 0.7 (*service*: "Measure gas concentration", *time*: 5) with *cost* = 0.5	0.4
C_3	(*service*: "Get weather data", *time*: 2) with *cost* = 0.3	0

The entire negotiation process structured according to the negotiation states in Table 1 is depicted in Table 3.

The manager agent sends calls for proposals to all contractor agents in the system. Agent C_1 can propose the deal in his deal space, as this deal matches the subject (the value for *service* issue matches the value of the fixed issue *service* set by the manager and the deal also contains an issue named *time*). Additionally agent C_1 can afford spending the associated effort (we assume the effort is simply computed by adding the values of the cost and the load factor, which is 1 in this particular case). Agent C_2 has two options in his deal space and both match the negotiation subject. However, he proposes the second option as he cannot handle the effort for providing the first option. Agent C_3 does not have deals matching the subject and consequently he refuses to send a proposal. After the manager had received the proposed deals from all the participants, he computes the utility of each deal. Resulted utility values are shown in table 4.

Table 3 Negotiation process

Neg. step	Manager	C1	C2	C3
1	Define subject (service: Measure gas concentration, time) and role (manager)	-	-	-
2	send CFP to C1, C2, C3	receive CFP	receive CFP	receive CFP
3	-	search the deal space and decide	search the deal space and decide	search the deal space and decide
4	receive proposals from C1, C2, C3	propose deal 1	propose deal 2	refuse
5	Compute utilities, send agreement to C1, conflict to C2 and C3	Agreement	Conflict	Conflict
6	End negotiation, wait for results	Start processing	End	End
7	Receive task information from C1, End	Send task information, End	-	-

Table 4 Utilities of the deals received by the manager. As agent C3 did not send a proposal, it is not mentioned in the table

Sender	Utility
C1	$0.5 + 0.5 * 0.99 = 0.995$
C2	$0.5 + 0.5 * 0.95 = 0.975$

When computing the utilities in table 4 the manager takes into account the following:

- the *service* issue takes text values;
- the *time* issue takes number values;
- the domain of the *time* issue is $[0, 100]$;
- the *time* issue is descending;
- the reference value of the *time* issue is 0;
- all issues have equal weights of value 0.5.

Note that the issues (*service* and *time*) as well as the default values of their properties (reference values, weights, monotony, data types, value types) are defined at the first level of negotiation configuration.

The manager selects the sender of the deal with the highest utility as the service provider. According to table 4, it is not difficult to observe that the manager agent would pick the option proposed by agent C_1.

Note that in this example we have included among the possible contractors agents that are not able to provide the required service – agent C_3 in our example. However, in practice the manager will send the call for proposal

only to agents that are (at least in theory) able to provide the required service by first querying a Yellow Pages server. Note that this does not mean that contractors receiving the call for proposal will necessary propose something, because they might be too busy or they would not be able to propose an appropriate deal that matches the negotiation subject (for example, the manager might add to subject an extra issue that some agents might not be aware of).

4 Related Work

Recently, a service negotiation framework has been proposed in paper [10]. That framework was applied to model Web service negotiations between insurance companies and car repair companies. The negotiation is implemented as an interaction between web services and the participants' preferences are represented using an ontology. The framework is utilizing the CNET protocol for negotiation, implemented as Web service method invocations. Note that in our proposal we do not constrain the negotiation protocols to a particular interaction protocol. Rather, we define a set of generic negotiation steps that are likely to be followed by many negotiation protocols. Our framework is generic in the sense that it allows creation and integration of different negotiation protocols. The approach of combining generic negotiation with components for representation of negotiation participants' deal space, utilities, and resources allows us to design and evaluate different negotiation protocols useful for service negotiation in disaster management. Also, different from [10] where the authors utilize a fixed set of subject issues specially tailored for their scenario, in our approach we give the possibility to define new subject issues with their properties that suit best the application in hand. Additionally, contractor utilities have a different semantics inspired by task allocation problems ([13]) that is more appropriate for the collaborative rather than competitive context for which our system is designed. However, similarly to [10], we use additive weighted utility functions for managers that take into account partial utility functions defined for each negotiation issue.

In [1] the authors discuss SAMIN – System for Analysis of Multi-Issue Negotiation. SAMIN is an advanced Prolog-based software for analysis of one-to-one, multi-issue, both automated and human negotiations. In our system we are dealing with one-to-many service negotiations. We would like to develop a formalization of negotiation processes similar to the work in [1], by generalizing the results to one-to-many and many-to-many negotiations.

In [16] the authors discuss applications and improvements of the max-sum algorithm for the coordination of mobile sensors acting in crisis situations. Our negotiation system differs from the coordination system in [16]. While the agents in our systems also need to coordinate, they are doing it by establishing workflows of services. Agents that require services negotiate with agents that are capable of providing those services using one-to-many

negotiation protocols, meaning that one of the agents participating in a negotiation has a special role and specifies the subject of the negotiation. In our problem agents can negotiate without being neighbors; agents present at the location of incidents can negotiate with agents located at the headquarters of environmental organizations (here *location* can be a negotiable issue). Moreover, as the problem domain can be very complex, human experts play an important role in our system. They can be involved in negotiations (e.g. propose conditions for service provisioning, select optimal service providers). Moreover, in our system, human agents can negotiate with fully automated agents (e.g. sensors) for configuring their services.

In [4] it is proposed a generic framework for auctions in electronic commerce. The main difference from our approach is the centralized way in which auction services are thought to be provided in an electronic commerce environment. Rather than encapsulating auction capabilities in each agent, authors of [4] propose to factor out and in some sense standardize the auction services such that each agent will be able to use them for buying and/or selling goods whenever they decide to do so.

5 Conclusions and Future Work

In this paper we have proposed and discussed a generic framework for defining service negotiations in agent-based collaborative processes for disaster management. The framework addresses negotiation protocols, negotiation subject, properties of negotiation subject issues, deal spaces, and utility functions of participant agents. Examples covering utility functions and sample negotiations were discussed. The framework is currently under development and will be incorporated into the DIADEM framework. We plan to address the following issues in the near future:

- We have made the assumption that all agents define the same issue over the same domain. However, this is a non-realistic constraint because agents usually have different knowledge about the world. The domain of the same issue of different agents can be different. We must study this in more detail and see how it can incorporated into our conceptual framework.
- Our utility functions are currently linear and do not take into account inter-dependent issues. While currently this is not a problem, we might encounter situations where more complex utility functions with inter-dependent issues must be used to better model agents' preferences for more complex disaster management scenarios.
- We plan to develop methods and tools for experimental evaluation of negotiation processes defined using our framework. These methods will be applied on negotiation scenarios derived from the DIADEM project. We are currently building a negotiation workbench that will be used to test and evaluate service negotiations in DIADEM.

- We plan to develop principled approaches for integrating service negotiations developed and validated using the proposed framework, into the DIADEM multi-agent framework.

Acknowledgements. Work of Mihnea Scafeş was partly supported by the IOSUD-AMPOSDRU contract 109/25.09.2008 and partly supported by the DIADEM project. Work of Costin Bădică was supported by the DIADEM project. The Diadem project is funded by the European Union under the Information and Communication Technologies (ICT) theme of the 7th Framework Programme for R&D, ref. no: 224318.

References

1. Bosse, T., Jonker, C.M., Meij, L., Robu, V., Treur, J.: A system for analysis of multiissue negotiation. In: Software Agent-Based Applications, Platforms and Development Kits, pp. 253–280. Birkhaeuser Publishing Company (2005)
2. Bădică, C., Scafeş, M.: One-to-many monotonic concession negotiation protocol. In: Proc. of 4th Balkan Conference in Informatics – BCI 2009, pp. 8–13. IEEE Computer Society Press, Los Alamitos (2009)
3. Damen, D., Pavlin, G., Kooij, V.D., Bădică, C., Comes, T., Lilienthal, A., Fontaine, B., Schou-Jensen, L., Jensen, J.: Diadem environmental management requirements document, issue 1.13.1 (2009)
4. Dobriceanu, A., Biscu, L., Bădică, A., Bădică, C.: The design and implementation of an agent-based auction service. International Journal of Agent-Oriented Software Engineering (IJAOSE) 3(2/3), 116–134 (2009), `http://dx.doi.org/10.1504/IJAOSE.2009.023633`
5. Faratin, P., Sierra, C., Jennings, N.R.: Negotiation decision functions for autonomous agents. Int. Journal of Robotics and Autonomous Systems 24(3-4), 159–182 (1998), `http://eprints.ecs.soton.ac.uk/2117/`
6. Hattori, H., Klein, M., Ito, T.: A multi-phase protocol for negotiation with interdependent issues. In: IAT 2007: Proceedings of the 2007 IEEE/WIC/ACM International Conference on Intelligent Agent Technology, pp. 153–159. IEEE Computer Society, Washington (2007)
7. Hemaissia, M., El Fallah Seghrouchni, A., Labreuche, C., Mattioli, J.: A multilateral multi-issue negotiation protocol. In: AAMAS 2007: Proceedings of the 6th international joint conference on Autonomous agents and multiagent systems, pp. 1–8. ACM Press, New York (2007)
8. Jennings, N.R., Faratin, P., Lomuscio, A.R., Parsons, S., Sierra, C., Wooldridge, M.: Automated negotiation: Prospects, methods and challenges. International Journal of Group Decision and Negotiation 10(2), 199–215 (2001), `http://eprints.ecs.soton.ac.uk/4231/`
9. Manzoor, A., Truong, H.L., Dustdar, S.: On the evaluation of quality of context. In: EuroSSC 2008: Proceedings of the 3rd European Conference on Smart Sensing and Context. LNCS, vol. 4405, pp. 140–153. Springer, Heidelberg (2008)
10. Paurobally, S., Tamma, V., Wooldrdige, M.: A framework for web service negotiation. ACM Trans. Auton. Adapt. Syst. 2(4), 14 (2007), `http://doi.acm.org/10.1145/1293731.1293734`

11. Pavlin, G., Kamermans, M., Scafeş, M.: Dynamic process integration framework: Toward efcient information processing in complex distributed systems. In: Papadopoulos, G., Bădică, C. (eds.) Intelligent Distributed Computing III. SCI, vol. 237. Springer, Berlin (2009)
12. Pavlin, G., Wijngaards, N., Nieuwenhuis, K.: Towards a single information space for environmental management through self- configuration of distributed information processing systems. In: Hrebicek, J., Hradec, J., Pelikan, E., Mirovsky, O., Pillmann, W., Holoubek, I., Bandholtz, T. (eds.) Proceedings of the TOWARDS eENVIRONMENT Opportunities of SEIS and SISE: Integrating Environmental Knowledge in Europe, Masaryk University (2009)
13. Sandholm, T.W.: An implementation of the contract net protocol based on marginal cost calculations. In: Proceedings of the 12th International Workshop on Distributed Artificial Intelligence, pp. 295–308. Hidden Valley, Pennsylvania (1993), citeseer.ist.psu.edu/sandholm93implementation.html
14. Scafeş, M., Bădică, C.: Service negotiation mechanisms in collaborative processes for disaster management. In: Eleftherakis, G., Hannam, S., Kalyva, E., Psychogios, A. (eds.) Proceedings of the 4th South East European Doctoral Student Conference (DSC 2009), Research Track 2: Information and Communication Technologies, pp. 407–417. SEERC, Thessaloniki (2009)
15. Smith, R.G.: The contract net protocol: High-level communication and control in a distributed problem solver. IEEE Trans. Comput. 29(12), 1104–1113 (1980), http://dx.doi.org/10.1109/TC.1980.1675516
16. Stranders, R., Farinelli, A., Rogers, A., Jennings, N.: Decentralised coordination of mobile sensors using the max-sum algorithm. In: Proc 21st Int. Joint Conf on AI, IJCAI (2009), http://eprints.ecs.soton.ac.uk/17268/
17. Vidal, J.M.: Fundamentals of Multiagent Systems: Using NetLogo Models (2006) (unpublished), http://www.multiagent.com/fmas

An Optimized Solution for Multi-agent Coordination Using Integrated GA-Fuzzy Approach in Rescue Simulation Environment

Mohammad Goodarzi, Ashkan Radmand, and Eslam Nazemi

Abstract. Agents' coordination, communication and information sharing have been always open problems in multi-agent research fields. In complex rescue simulation environment, each agent observes a large amount of data which exponentially increases through the time while the capacity of messages in which agents' information is shared with others and also the time needed to process the data is limited. Apparently, having a suitable coordination strategy and data sharing system will lead to better overall performance. Therefore, agents should select and spread out the most useful data among their observed information in order to achieve better coordination. In this paper, we propose an unique method based on integration of Genetic Algorithms and Fuzzy Logic theory to decide which part of data is more important to share in different situations. We also advise a new iterative method in order to obtain admissible experimental results in rescue simulation environment which is a good measurement for our research.

1 Introduction

The main objective in rescue simulation environment is to decrease the amount of damages imposed to civilians and city by rescuing buried and injured civilians and extinguishing the burning buildings as much as possible while a simulated earthquake happens. Three types of agents including Fire brigades, Ambulances and Police forces act in this environment. In order to reach high-performance coordination, agents should effectively share a large amount of information which they percept from observed objects in surrounding environment. Due to lack of time to process and limited capacity

Mohammad Goodarzi · Ashkan Radmand · Eslam Nazemi
Faculty of Electrical and Computer Engineering, Shahid Beheshti University, Iran
e-mail: {mo.goodarzi,a.radmand}@mail.sbu.ac.ir,nazemi@sbu.ac.ir

Q. Bai and N. Fukuta (Eds.): Advances in Practical Multi-Agent Systems, SCI 325, pp. 377–388.
springerlink.com

of messages used to send mentioned information, agents need to optimally manage these messages.

Information observed by agents in rescue simulation environment are so various that messages carrying them are needed to be divided into a through set of message types [9]. Therefore, we propose a method to produce set of message types which covers all the data in the environment. Afterwards, considering introduced message types, we describe some parameters by which infinite possible situations of disaster space are categorized to a limited range of parameters. Then, using a GA-based method, we approximate some optimum solutions for specific training situations. Each solution is an array of values which shows the weight of different message types to send. The adequate amount of these specific solutions is obtained from an iterative procedure which explained in section 5. Finally, since possible situations in rescue simulation environment are not limited to training ones, a generalization method is required to obtain solutions for all situations. A fuzzy inference system based on Takagi-Sugeno modeling [10] is proposed that generalizes gained solutions to all situations.

In section 2, the environment categorizing method and definition of parameters which classify the situations are presented. Section 3 describes the GA-based method used to obtain solutions for training situations. The generalization fuzzy system is explained in section 4. In section 5, experimental results of presented method are given. Finally, the conclusion and future works are presented in sections 6 and 7 respectively.

2 Environment Categorizing

2.1 Data Categorizing

Figure 1 demonstrates a categorization of entities in rescue simulation environment [7].

According to figure 1, each entity has a set of states that describes its condition and agents can do some limited actions based on their type. At the other hand, every change in rescue environment is the result of state-changing or agents' actions. Therefore, the set of all message types can be defined as follows:

$$\{MessageTypes\} = \cup_{i=1}^{m} \cup_{j=1}^{n} State(i,j) + \cup_{k=1}^{o} \cup_{l=1}^{p} Action(i,j) \quad (1)$$

Where,
$State(i,j)$ is j^{th} state of the ith entity,
$Action(k,l)$ is l^{th} action of the kth agent,
m is number of the entities,
n is number of the ith entity's states,
o is number of the actions
and p is number of the kth agent's actions.

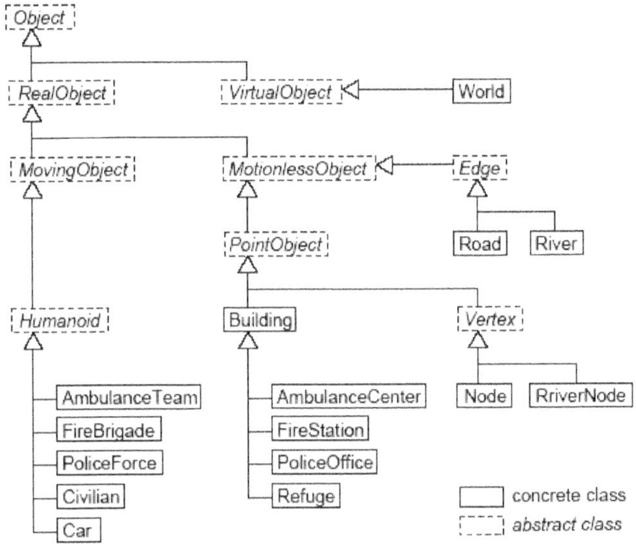

Fig. 1 Categorization of RoboCup rescue simulation environment entities.

The above equation covers all the events in the environment according to the entities and agents' actions. Therefore, by using equation (1) for the entities which demonstrated in figure 1 and compounding the messages which express similar information, we can obtain the entire message types. Table 1 shows the most important message types generated through (1).

Table 1 Important message types in RoboCup rescue simulation environment

Message Name	Related Entity	Action/State
Civilian Dead Message	Civilian	State
Civilian Critical Message	Civilian	State
Civilian Average Message	Civilian	State
Civilian Healthy Message	Civilian	State
Buried Civilian/Agent Message	Civilian/Agent	State
Building Burning Message	Building	State
Building Burned Message	Building	State
Building Semi Burned Message	Building	State
Road Blocked Message	Road	State
Position Message	All Entities	State
Agent Buried Message	Agent	State
Road Clear Message	Police Force Agent	Action
Extinguish Message	Fire Brigade Agent	Action
Rescue Message	Ambulance Agent	Action
Victim Find Message	Agent	Action
Move To Refuge Message	Ambulance Agent	Action

2.2 Parameters Definition

Since the RoboCup rescue simulation is a highly stochastic multi-agent environment, it is almost impossible to define all situations in a discrete set. However, as we focus on agents' coordination and their communication management in this paper, it is reasonable to define some parameters to classify the situations.

It should be considered that these parameters must completely cover all the imaginable situations. In addition, they should be measurable parameters in order to be beneficial for our purpose. Based on our experience in RoboCup rescue simulation league, we consider the most important parameters to classify the simulations as follows:

- Number of the burning buildings
- Number of the blocked roads
- Number of not rescued victims (unseen or buried civilians or agents)

Although these parameters are not capable of classifying all the scenarios in rescue simulation system, but combination of the mentioned parameters can describe all the possible communication-based scenarios occur in RoboCup rescue simulation.

3 Extracting Solutions to Defined Situations

In section 2 we defined some parameters which covers all the situations concerned in this paper. Using these parameters and utilizing a Genetic Algorithmic [2,4,5] method, we are able to extract solutions to certain training situations. To determine the adequate number of training situations, we should first specify possible ranges of parameters and then calculate an initial number of needed solutions based on these ranges. Calculating the initial number situations will be explained in section 5. The minimum and maximum value of each parameter is shown in table 2. This information is obtained from the RoboCup rescue simulation league rules in 2008 competitions.

Table 2 Minimum and Maximum value of proposed parameters

Parameter	Minimum	Maximum
Number of Burning Buildings	0	1100
Number of Blocked Roads	0	1480
Number of Not Rescued Civilians/Agents	0	195

Based on ranges above, a definite number of training situations are generated and using our GA method, an optimum solution to each of them is obtained in an iterative procedure. This iteration continues until the solution becomes admissible.

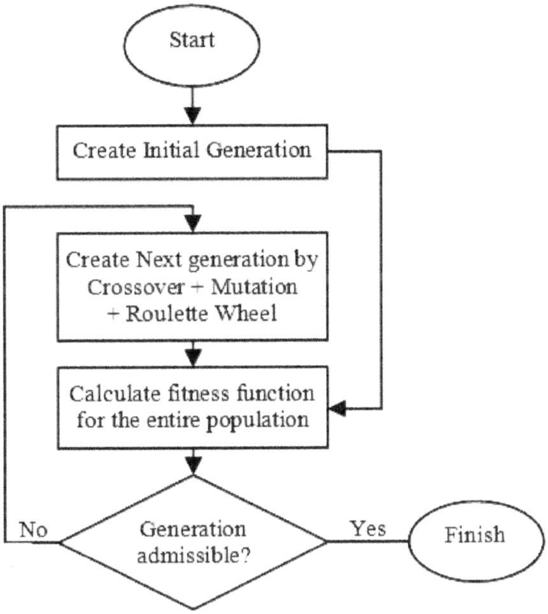

Fig. 2 GA system iterative procedure

To initialize the first generation of the GA system, we use approximate data based on our experience in rescue simulation environment which results to finding the optimal chromosome faster. Figure 2 depicts block diagram of GA training section.

Chromosome structure is defined in section 3.1; creating generations and its algorithms including crossover and mutation operators and selection algorithms are explained in sections 3.2 and 3.4 and fitness function calculation for each chromosome is explained in section 3.3.

3.1 Chromosome Structure

The structure of chromosomes is an array of values that indicates message types weights in the sending packet. The chromosome size is equal to the number of message types. For example, if we have a chromosome like the one in figure 3, 21% of the whole sending packet should be assigned to message type 1, 12% should be occupied by message type 2, 7% by message type 3 and so on.

Each chromosome has a constraint that the accumulative value of a its genes must be equal to one.

$$\sum_{i=1}^{n} gene(i) = 1 \qquad (2)$$

0.21	0.12	0.07	0.32	0.04	0.01	0.09	0.07	0.06	0.01

Fig. 3 An example of a chromosome structure

3.2 Fitness Function

As our experiment environment is very complex and the elements which affect the final result are in a wide range, finding a fitness function that can exactly measure the effectiveness of communication in agents' performance is infeasible [8]. However, We use the new scoring system developed for RoboCup Rescue Simulation league, called *Score Vector* [9] as our fitness function which has a dynamic and well defined structure that allow us to add or remove parameters and exactly measure the chromosome's fitness value for each situation.

3.3 Crossover and Mutation

In proposed GA method, there are limitations in both crossover and mutation caused by satisfying the constraint explained through equation 2 in section 3.1.

While our chromosome structure is an array of values, it is a simple structure and does not need any complex crossover algorithms. Two parents are selected and two new children are generated from them. Break points in each parent chromosome are selected randomly and segments between these break points are substituted by parents to generate new individual children. In order to satisfy the mentioned constraint a proportional value should be added or subtracted to or from all genes.

For the mutation process, a parameter called *mutation constant* defined which is the maximum valid value of mutation for each gene. In every mutation process, one random gene is selected and its value changes according to the mutation constant. Afterward, a method similar to the one in crossover operation is used to satisfy the mentioned constraint.

3.4 Selection Algorithm

The method's selection algorithm is based on the fitness values of each chromosome. Roulette wheel method [1] is used for the selection part of the algorithm. The main idea of this model is to select individuals stochastically from one generation to create the next generation. In this process, the more appropriate individuals have more chance to survive and go forward to the next generation. However, the weaker individuals will also have a probability to select.

Some chromosomes that have the most fitness value will be selected to make the next generation automatically. Rest of the needed chromosomes will be extracted from the crossover and mutation operations.

4 Generalizing Using Fuzzy Logic

By finding some solutions to the whole rescue system using proposed GA method in section 3, a set of message types weights gained which are proper in specific learned situations. In order to evaluate the answer in the case of unknown situations, generalizing the extracted solutions is inevitable. The generalization method in this paper is based on Fuzzy Logic theory [3,6,12]. The procedure of using fuzzy logic system is shown in figure 4.

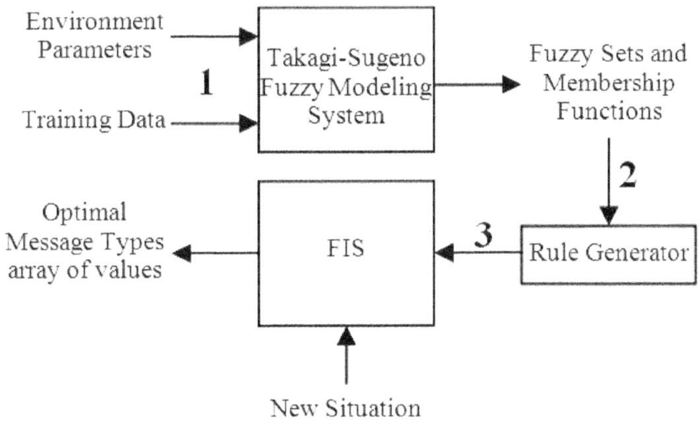

Fig. 4 Procedure of using fuzzy logic system

As it is shown above, the fuzzy system gets a set of environment parameters and training data as the input. The parameters are which described in section 2 and the training data is the output of GA system. Our method uses Takagi-Sugeno fuzzy modeling system [10] to obtain each parameter's membership functions which are used to generate fuzzy rules. Obtained fuzzy rules are used in FIS (Fuzzy Inference System) to evaluate the generalized solution for each unknown condition.

4.1 Fuzzy Sets and Membership Functions

One of the outputs of fuzzy modeling system is fuzzy sets on mentioned classifier parameters and each one has different membership functions. Figure 5 shows achieved fuzzy sets and membership functions of parameters considering their valid definition range after the system is trained.

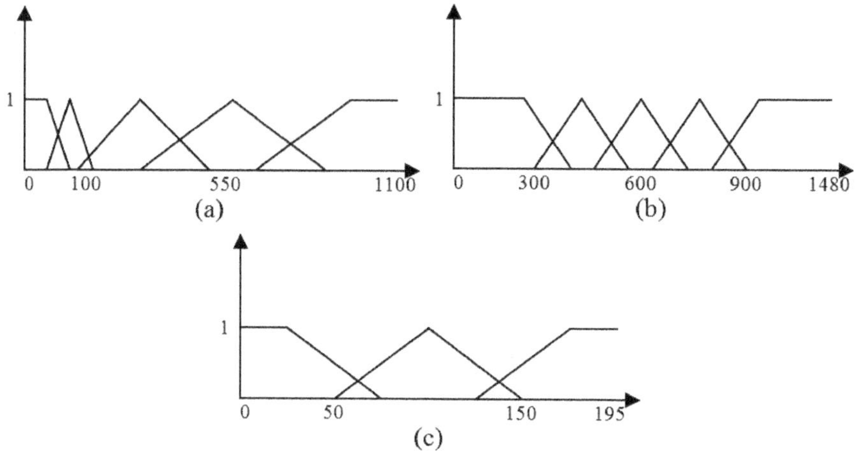

Fig. 5 (a) Burning buildings parameter, (b) Blocked roads parameter, (c) Not rescued civilians/agents parameter Membership Functions

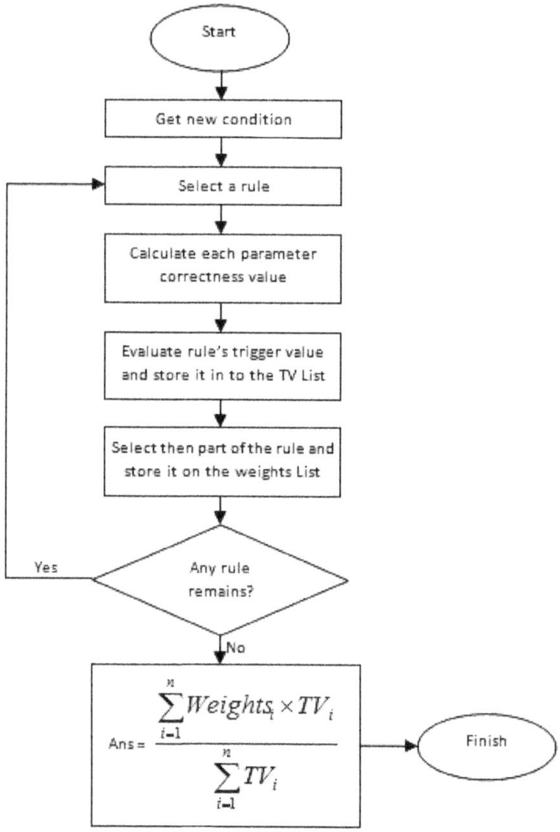

Fig. 6 Defuzzification procedure

4.2 *Fuzzy If-Then Rules*

By extracting fuzzy sets and membership functions, fuzzy If-Then rules are generated. Fuzzy If-Then rules have just *AND* connector and rules' parametric structure is as follows:

IF (x_1 is A_1 and x_2 is A_2 ... and x_k is A_k) THEN {R} is $\{r_1, r_2, \ldots, r_n\}$

Where $x_1, x_2, \ldots, x_k$ are environment's parameter, $A_1, A_2, \ldots, A_k$ are members of fuzzy sets with linear membership functions, $r_1, r_2, \ldots, r_n$ are weights of message types and R is a set which contains $r_1, r_2, \ldots, r_n$.

4.3 *Defuzzification*

With defining fuzzy sets, membership functions and fuzzy if-then rules, the system is ready to evaluate weight of message types for any given condition. Figure 6 indicates the fuzzy system's behaviour.

In defuzzification step, to achieve final weights for the given situation, *then parts* of all rules should be combined considering their trigger value. The method used for defuzzification step in this paper is *Weighted Average* method.

5 Experimental Results

The performance of proposed method is directly related to GA's output. An iterative procedure is devised in order to reach the desirable performance. The iteration is based on extracting training situations, trying to generalize them and evaluate the final result and it continues until the given result reaches the expected performance and become admissible.

Table 3 Detailed average scores on tested maps

Number of Trained Data	Average Score on 20 selected maps
64	1.75869
84	1.75993
104	1.80003
124	1.84353
144	1.87980
164	1.90342
184	2.00001
204	2.00123
224	2.00987
244	2.00991

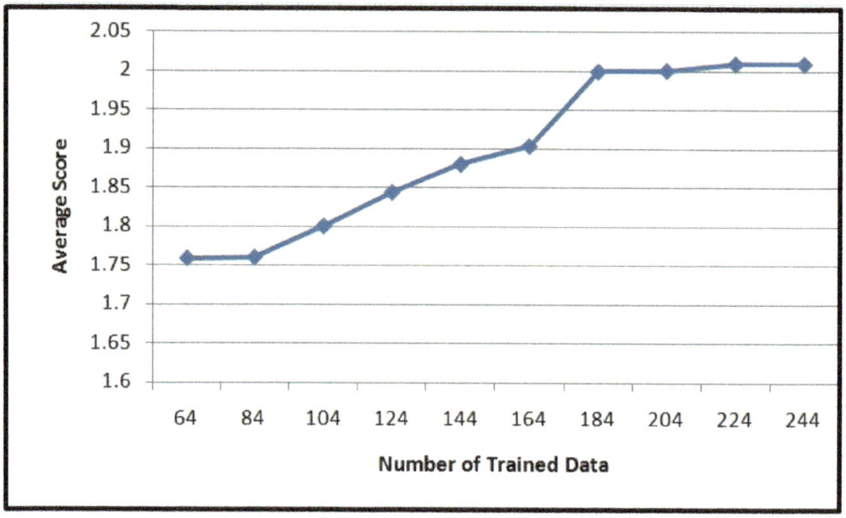

Fig. 7 Average scores on 20 selected maps while training data is increasing

Our goal in this paper is to reach the top score of about 20 maps which had been used as communication-based maps in China2008 and IranOpen2009 RoboCup Rescue Agent Simulation competitions.

Based on our experience in RoboCup rescue simulation system, we assume four partitions for each parameter in the fuzzy logic system as a start point (by partition we mean the definition of partition introduced by Takagi-Sugeno fuzzy modeling system). Thus, the first iteration of obtaining results has $4^3 = 64$ training situations. Table 3 and figure 7 show the detailed average scores and the final score progression respectively while the amount of training data

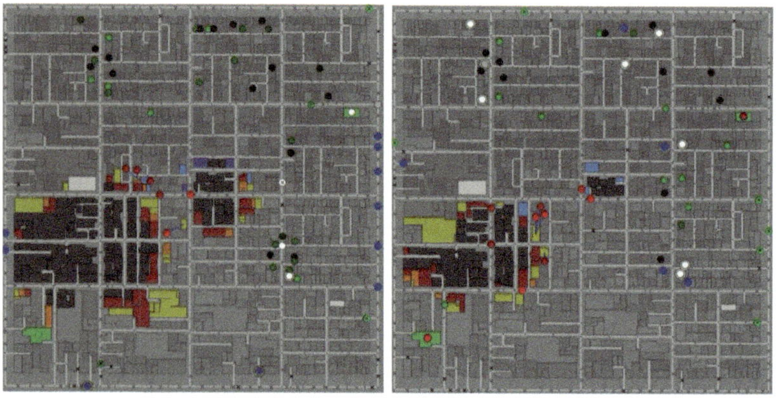

Fig. 8 Difference of agents' performance before and after implementation of the method on IranOpen2009 VC-6 (final day) map, Left: Before (score:1.4708), Right: After (score:1.6576)

is increasing. The GA part of the described method is very time demanding, Therefore because of time limitations and supervision on the detailed results we added 20 new training situations to our training data in each iteration. This number is just matched with our time and devices constraints and it directly affects the number of iterations to reach the desired performance.

Figure 8 shows the difference between agents' performance before and after implementation of the proposed method on SBCe_Saviour team source code which released after IranOpen2009 competitions.

6 Conclusion

In this paper, we proposed an integrated solution based on GA and Fuzzy logic in order to improve the agents' coordination in multi-agent environments specifically in RoboCup rescue simulation system. We categorized all the observed data in the environment to a set called *Message Types*. Using these types, we produced a set of parameters that can fully describe the environment on any possible communication-based situation. Afterward, we utilized a GA-Fuzzy method to first generate a set of specific solutions to some certain scenarios and then generalize them to the entire simulation system. In an iterative procedure, we reached a satisfiable result that improved agents' coordination in any unknown scenarios.

7 Future Works

In all multi-agent environments with large number of effective parameters finding an approximate optimal answer for a specific situation is a serious problem. We used a method based on Genetic Algorithms in this paper and obtained desirable results. However, Long time of execution for evaluating the fitness values in RoboCup rescue simulation system can not be ignored. We are looking to devise new faster method that can obtain suitable result in much less time. We have also looking to improve the proposed method in order to handle the scenarios which does not have any radio communications between agents.

One of the research topics in information sharing methods is to select a subset from the entire agents whom to send the messages according to data characteristics [11]. We are also looking to research in this aspect of information sharing methods and algorithms.

References

1. Al jaddan, O., Rajamani, L.: Improved Selection Operator for GA. Journal of Theoretical and Applied Information Technology, 269–277 (2005)
2. Banzhaf, W., Nordin, P., Keller, R., Francone, F.: Genetic Programming - An Introduction. Morgan Kaufmann, San Francisco (1998)

3. Dubois, D., Prade, H.: Fuzzy Sets and Systems: Theory and Applications. Academic Press, New York (1980)
4. Goldberf, David, E.: Genetic Algorithms in Search, Optimization and Machine Learning. Kluwer Academic Publishers, Boston (1989)
5. Holland, J.H.: Adaptation in Natural and Artificial Systems. University of Michigan Press, Ann Arbor (1975)
6. Kruse, R., Gebhardt, J., Kalwonn, F.: Foundations of Fuzzy Systems. John Wiley & Sons, New York (1994)
7. Morimoto, T.: How to Develop a RoboCup Rescue Agent. Rescue Technical Committee (2002)
8. Radmand, A., Nazemi, E., Goodarzi, M.: Integrated Genetic Algorithmic and Fuzzy Logic Approach for Decision Making of Police Force Agents in Rescue Simulation Environment. In: Baltes, J., Lagoudakis, M.G., Naruse, T., Ghidary, S.S. (eds.) RoboCup 2009: Robot Soccer World Cup XIII. LNCS, vol. 5949, pp. 288–295. Springer, Heidelberg (2010)
9. Siddhartha, H., Sarika, R., Karlapalem, K.: Score Vector: A New Evaluation Scheme for RoboCup Rescue Simulation Competition 2009, Rescue Technical Committee (2009)
10. Takagi, T., Sugeno, M.: Fuzzy Identification of Systems and Its Applications to Modeling and Control. IEEE Transactions on systems, man and cybernetics SMC-15(1) (January/February 1985)
11. Xu, Y., Lewis, M., Sycara, K., Scerri, P.: Information Sharing in Large Scale Teams. In: AAMAS 2004 Workshop on Challenges in Coordination of Large Scale MultiAgent Systems (2004)
12. Zimmermann, H.-J.: Fuzzy Set Theory and its Applications, 2nd edn. Kluwer Academic Publishers, Boston (1991)

A Study of Map Data Influence on Disaster and Rescue Simulation's Results

Kei Sato and Tomoichi Takahashi

Abstract. Applying the agent-based simulation system to real cases requires the fidelity of the simulations. It is desired to simulate disasters situations under environments that are close to real ones qualitatively and quantitatively. Maps are a key component of the disaster and rescue simulation system. We propose a method to create more real maps using open GIS data and discuss the simulation results between two maps. Buildings data of one map is generated by programs and the other is created from real data. The experiments show clear difference between the simulation results, and the causes of the difference are discussed.

1 Introduction

In disasters, human behaviors range from rescue operations to evacuation activities. We cannot do experiments on human evacuation or rescue behaviors physically at the level of real scale. Agent-based simulation makes possible to simulate the human behaviors and interactions between their rescue operations and disasters. The simulation results can be used to predict damages to civil life, and to make plans to decrease the damages. The RoboCup Rescue Simulation (RCRS) is an agent based simulation system and was designed to simulate the rescue measures taken during the Hanshin-Awaji earthquake disaster [5].

RCRS has been used in RoboCup competition since 2000, and has been used in various situations [8]. Most of the usages come from the requirement to decrease the damage of disasters. Applying the simulation systems to practical usage requires that the results of simulation are explainable to

Kei Sato · Tomoichi Takahashi
Meijo University
e-mail: e0428299@ccalumni.meijo-u.ac.jp,
ttaka@ccmfs.meijo-u.ac.jp

Q. Bai and N. Fukuta (Eds.): Advances in Practical Multi-Agent Systems, SCI 325, pp. 389–402.
springerlink.com

users from viewpoint of theories, experiments in laboratories, or past experience. Simulation using real parameters or data is one of ways to increase the fidelity of simulations.

Simulation using real size requires a huge number of computer resources. For example, Takeuchi et al, simulated evacuation behavior of thousands of agents at $4 \times 4km^2$ on distributed computing system [11]. The simulation takes a lot of time and the simulation clock is slower than wall time clock. The slow simulation becomes a serious problem when the simulation is used during disaster. Lightweight simulations are alternative when the results are qualitatively guaranteed.

In this paper, we propose a method to create maps with road networks and buildings from open GIS data, and check how the reality of maps influences the simulation results. Section 2 describes of agent based disaster and rescue simulation in practical usages. Section 3 discussed creation of maps from open sources. Experiments using RCRS and the summary are presented in Section 4, Section 5 respectively.

2 Agent Based Disaster and Rescue Simulation for Practical Usage

2.1 Role of Disaster and Rescue Simulation

When a disaster occurs, rescue agents rush to the disaster sites and the local government will play a role of rescue headquarters [13]. They will make plans that will decrease damages for civil lives, or estimate the effect of plans. Figure 1 shows an image of the disaster-rescue simulation that will be used by local governments. Before disaster, they estimate damages of their town and make disaster prevention plans to expected the damages. During disasters, emergency center will use them to evacuate people, deliver relief goods, etc effectively.

The following scheme is assumed to make effective plans.

$$\underbrace{plan \rightarrow \underbrace{compute\ damages \rightarrow compare\ past\ cases}_{as\ a\ estimation\ of\ models} \rightarrow estimate\ plans}_{usage\ as\ disaster\ management\ tool}$$

In order to use simulation results as one of disaster management tools, it is important to make followings clear,

1. what targets of the simulations are,
2. under what conditions the simulations are done,
3. what computational models of simulations are employed,
4. what criteria are used to assure the simulation results.

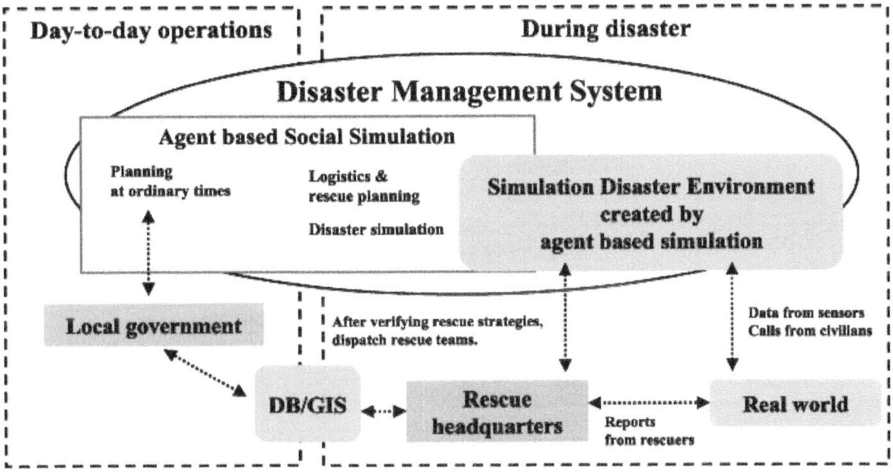

Fig. 1 An image of a disaster management system used by local governments. The system and data used for day-to-day operation will be utilized at rescue headquarters. The left part shows the day-to-day operations of the local governments. Agent based social simulations are used for their planning at ordinary times such as traffic flow estimations.

2.2 RoboCup Rescue Simulation System

The RoboCup Rescue simulation is an agent based system and is assumed to play a role of a component of *compute damages* in the above scheme. The RCRS is built with a number of modules [10]. Figure 2 shows the architecture of the RCRS and following are brief explanations of the modules.

Agent: Civilians and fire fighters are examples of agents. These individuals are virtual entities in the simulated world. Their decision processes or action-selection rules are programmed in the corresponding agents, and their intentions are sent as commands to the kernel.

Component simulators: Disaster simulators compute how disasters affect the world. These correspond to disasters such as building collapses and fires. A traffic simulator changes the positions of the agents according to their intentions. It resolves conflicts of the intentions, for example, many people want to go the same place.

GIS: This module provides the configuration of the world defining where roads and buildings are, and their properties such as width of street, floor area, number of stories, fire-resistant building or not. The roads are represented in the form of network and the data buildings are presented as polygons.[1]

[1] The numbers of agents and fires, and their positions are also specified as parameters of the simulation. The parameters are set in a file, `gisini`, in Table 4.

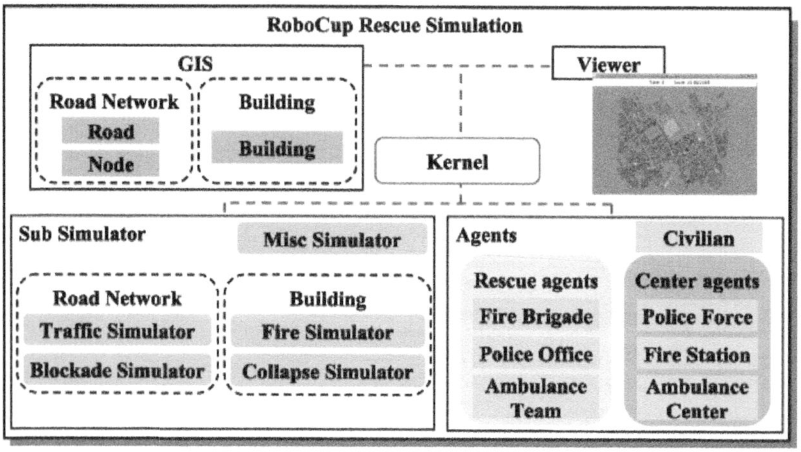

Fig. 2 Architecture overview of RCRS. Components are plugged into kernel via specified protocol.

Viewers: Viewers visualize the disaster situations and agent activities on a 2D or 3D viewer and display predefined matrices that are obtained from the simulation results.

Kernel: The kernel works as a server program to the clients and controls the simulation process and facilitates information sharing among the modules. The kernel exchanges protocols among agents and simulations, and checks the actions whether they are valid or not.

3 Map Generation from Public GIS Data

3.1 Open Source Maps

It is desirable to simulate disasters and rescue operations under environments as realistic as possible at any places where disaster occurs. Maps are key components to satisfy the above requirement. GIS data are available from commercial mapping agencies. They are expensive and the copyrighted data are restricted to use them.[2]

Recently, GIS data can be obtained in a form of GML from governmental institutions and volunteer organizations. Academic institutions may use the GIS data provided from their governmental institutions. Table 1 shows features of these maps.

Geographical Survey Institute (GSI): Japanese GSI has provided digital geospatial information of road network data in a form of Japanese

[2] Maps available at the Internet such as Google Maps are sourced from the mapping agencies.

Standards for Geographic Information (JSGI) 2.0 since 2002 [2]. From 2007, they have updated the JSGI as ISO 19100 compatible one and have started GSI data available from the Internet [3][3] .

Ordnance Survey(OS): Ordnance Survey is Great Britain's national mapping agency. It has provided digital data of OS Master Map. The digital data are sold to business and individuals [6].

Open Street Map(OSM): OpenStreetMap creates and provides free geographic data. As the data are input by volunteer, the data do not cover all area. But they can be used without legal or technical restrictions [7].

Table 1 Features of geographical data provided by governmental institutions and volunteer organizations.

data	start year	coordinate system	road network	road property	building data
OSM	2004	WGS84	○	○	
OS Topo	2001	British National Grid		√	√
OS ITN	2001	British National Grid	√		
GSI	2002	Tokyo Datum	√	√	
GSI	2007	WGS84		○	○

○ indicates the dates are available free.
OS ITN:OS Integrate Transport Network Layer [4].
OS Topo:OS Topography Layer [12].

3.2 Creation Simulation Map from Open Source GIS Data

RCRS simulations can simulate disasters and human behaviors with rescue operations at any places by replacing GIS data. Road network and building data of the area are components of the GIS data. Takahashi et al. proposed approaches to generate realistic maps from free GIS data that are used without any restriction [9]. Moritz et al. presented a tool using OpenStreetMap [1]. OpenStreetMap contains data of roads, landmark buildings and their relevant data. All buildings of the area are not contained in OpenStreetMap, so they generate the data of buildings when there are no building data.

As Table 1 shows, OSM and GSI2007 are available free from viewpoints of money and copyright. We create real GIS data from OSM and GSI2007. Followings are processes of creating simulation maps (Figure 3).

1. Figure 3 (a) and (b) show the network from OSM and building from GSI of Tokorozawa, a neighboring town of Tokyo. OSM has road network data, the nodes and edges of network correspond to crossing points and roads, respectively. The middle figure shows the outlines of buildings, parks and other areas.[4]

[3] GIS(2007) does not contain the road network data. GIS has provided the road network data in a format of SGI2.0 as commercial goods since 2007.

[4] We can recognize roads from Figure 3 (b), however, GSI has no data on roads.

(a) road network(OSM) (b) building area(GSI) (c) composite map

Fig. 3 Road network map from OSM (left), building data from GSI (middle) and composite map (right) of Tokorozawa, Japan

2. Figure 3 (c) shows a composite map created by our method.
   a. Imports road network from OSM map, and building data from GSI map.
   b. Inserts building entrance roads and connect them to the buildings.

   The connections are edges between nodes of road network and entry points of buildings. They serve ways to buildings from outside and agents use them to enter or move out the house.

   Figure 4 shows maps generated by Moritz's program and our program. Möritz's method generates one big house at the left upper block, where there are ten houses at real maps. Dots on the roads are nodes of the road network. They involve entrance nodes of buildings, so the map created by our method contains more nodes than Moritz's map.

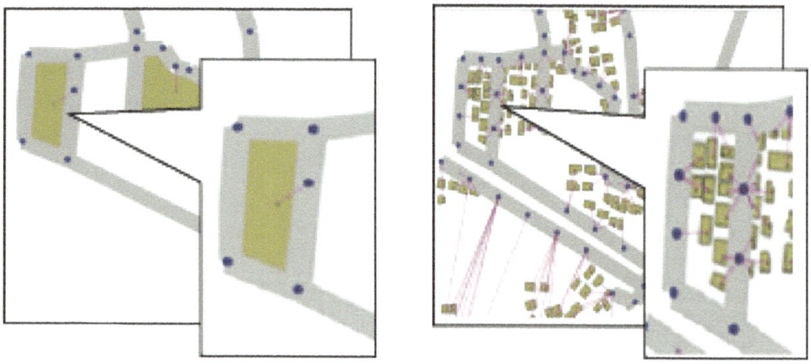

Fig. 4 Generated map by Moritz program(left) and by our program(right). Light gray parts are roads, and dark gray parts are buildings. Dots on the roads are entrance points to buildings.

4 Simulation Sensitivity of Environments to Simulations

It is conceivable that simulation with more realistic GIS data outputs better results, while the simulation with more realistic environments needs more computation resources. Simulations with simplified environments are one of ways to do with less resource under the conditions that the results are qualitatively guaranteed. We check whether results of simulation dependent on GIS data by comparing the correlation the results of simulation.

4.1 Created Maps from Open Source Data

Tokorozawa is one neighboring town of Tokyo and one of areas that maps are available. And Tokorozawa is the area where OSM and GSI(2007) provide data.[5] Three maps of Tokorozawa are created, Tokorozawa1, Tokorozawa2, and Tokorozawa3 from the smallest to the biggest one.(Figure 5) Followings are the features of three areas.

Torokozawa1. A railway track runs diagonally and only one road crosses the railway track.

Tokorozawa2. One railway track runs diagonally and two roads cross the railway track. There are more open areas where no buildings are than Torokozawa1.

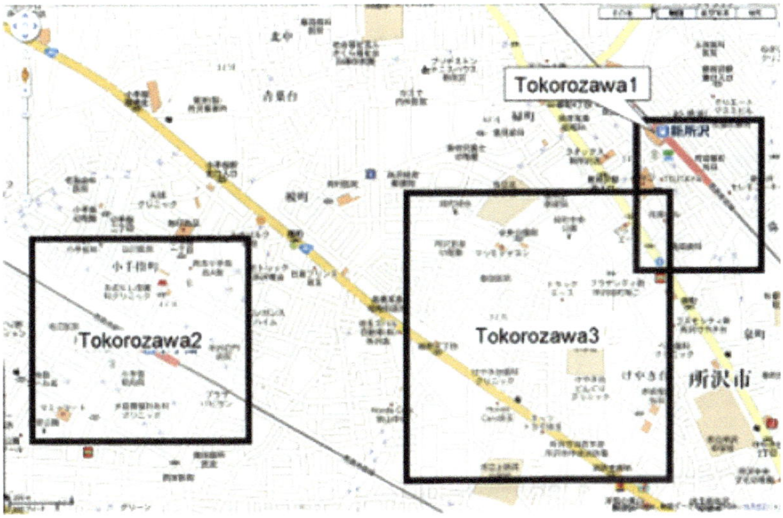

Fig. 5 Tokorozawa area displayed at Google Earth.

[5] OSM and GSI(2007) do not cover the whole of Japan.

Table 2 The futures of created maps

area		Tokorozawa1		Tokorozawa2		Tokorozawa3	
size ($m \times m$)		483×599		755×937		838×1052	
program		Moritz	our	Moritz	our	Moritz	our
network	node	150	228	278	461	400	563
	road	177	255	314	497	449	612
building	number	104	638	134	1146	321	1604
	floor area($m \times m$)	23,401	21,319	41,171	38,089	89,444	42,941
	average of area	225	33	307	34	279	27
	standard deviation	101	66	195	63	185	43
Simulation Time	S1	362.2	382.0	347.5	348.0	372.3	374.7
(CPU, second)*	S2	355.3	361.9	343.0	355.9	374.5	382.0
	S3	-	374.9	-	372.9	-	413.4
	S4	-	369.3	-	404.1	-	426.0

*: The simulations are done one computer. (CPU Intel 2Quad 3Ghz, Memory 3GB).
*: The time is the average of five runs.

Tokorozawa3. A big road diagonally runs instead of a railway track. There is a big open area in the center.

Table 2 shows the properties of Tokorozawa maps. Moritz program generates building data by dividing open areas first and placing houses sizes of that are almost equal to the size of the divided areas. So the sizes of buildings are larger and the number of building is smaller than the buildings created by our program. Figure 6, Figure 7, and Figure 8 show snapshots of rescue simulation every 100 step, at Tokorozawa1, Tokorozawa2, and Tokorozawa3, respectively. The features are shown in them.

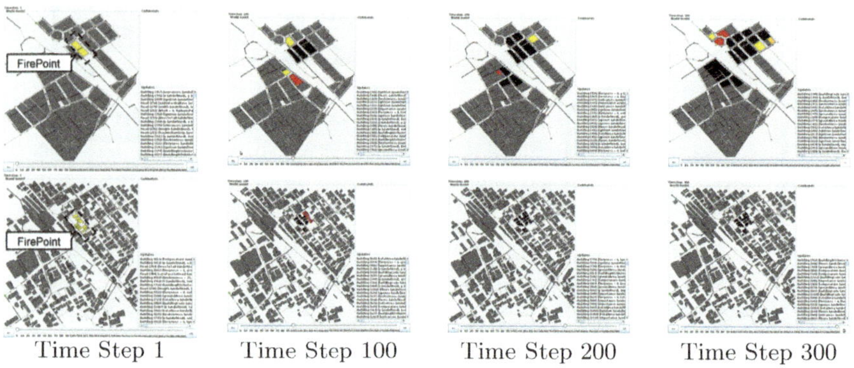

Fig. 6 Fire simulations at Tokorozawa1 map. Up figures are maps created by Moritz program and down ones are created by our program.

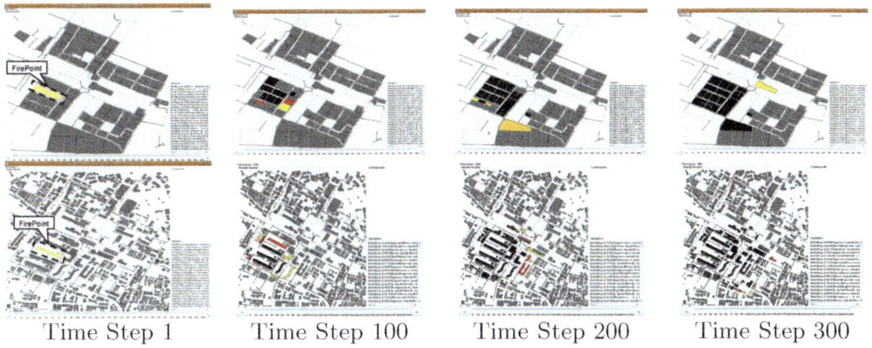

Fig. 7 Fire simulations at Tokorozawa2 map. Up figures are maps created by Moritz program and down ones are created by our program.

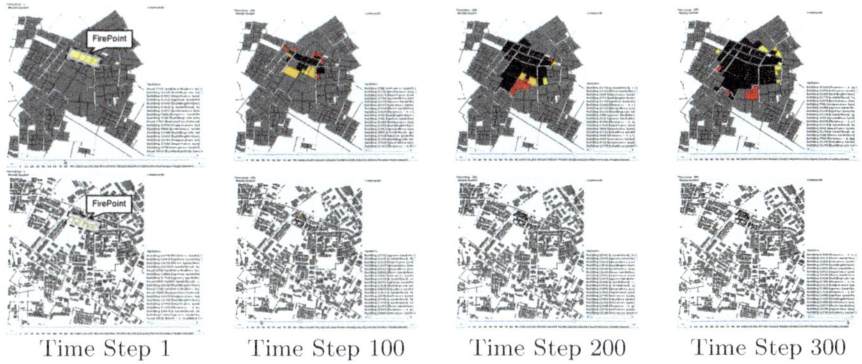

Fig. 8 Fire simulations at Tokorozawa3 map. Up figures are maps created by Moritz program and down ones are created by our program.

4.2 Sensitivity Analysis of Maps to Simulations

Fire simulation is one of disasters simulation that is influenced to the properties of maps.The down four rows in Table 2 show the run times of simulation using each map.The simulation time is the average of five runs for four disaster situations, S1, S2, S3 and S4; the situations are described in subsection 4.3. It takes longer time to simulate using the maps created by our program than maps created by Moritz program.

Table 3 shows the spread of fires and correlations between maps created by two programs. For Tokorozawa1 and 3, the correlations among the unburned building ratio from start to 300 steps are lower than Tokorozawa2. Figure 6,7 and 8 show fire spreads at maps created by Moritz method and our method. From the figures, it can be seen that fires do not spread after 100 and 200 time step at simulation the maps created by our program. Actually, from time step 100 and 160 for Tokorozawa1 and 3, respectively, the fire spread stopped.

Table 3 Spread of fire and correlation between two maps created by Moritz program and ours.

		Tokorozawa1		Tokorozawa2		Tokorozawa3	
elapsed time step \ map		Moritz	our	Moritz	our	Moritz	our
number of ignition point		2		6		6	
spread of fire by rate of unburned buildings(%)	100	96.1	98.0	92.7	90.5	95.7	98.5
	200	92.6	98.0	83.4	77.8	88.7	96.4
	300	80.7	98.0	77.9	71.4	80.1	96.4
correlation between maps	0-100	0.978		0.863		0.905	
	0-200	0.837		0.977		0.935	
	0-300	0.560		0.987		0.834	

Correlation factors are calculated using the ratio of unburned area of every step.

Table 4 Parameters of four scenarios

		Tokorozawa1				Tokorozawa2				Tokorozawa3			
scenario		S1	S3	S2	S4	S1	S3	S2	S4	S1	S3	S2	S4
the number of rescue agent	civilian	70		80		80		90		90		100	
	fire fighter	10		15		10		15		10		15	
	police	10		15		10		15		10		15	
	amblance	10		15		10		15		10		15	
the number of ignition points		5	35	5	35	8	72	8	72	10	100	10	100

The correlation of Tokorozawa1 is 0.56. The rates of unburned building are the same for Tokorozawa1 at our map. The fire doesn't spread in this case, and this causes the low correlation.

Followings are thought to cause the differences between simulations.

- Size of building: it takes longer time to burn a big building than a small one.
- Space between buildings: the space prevents fire from spreading.

4.3 Sensitivity Analysis of Rescue Agents to Simulation Results

Rescue agents extinguish fires. The agents and the initial condition of disaster the number of fire ignition points and their locations change the simulation results. The parameter setting of rescue agents and fires defines rescue scenarios of simulation.Table 4 shows four scenarios that are set by changing the number of rescue agents and the number of ignition points:

- The number of agents is different between (S1, S2) and (S3, S4),
- The number of ignition points is different between (S1, S3) and (S2, S4). The difference of ignition points comes from the size of generated buildings.

Table 5 Correlation between simulations under different environments

time step \ scenarios		Tokorozawa1				Tokorozawa2				Tokorozawa3			
		S1:S1	S2:S2	S1:S3	S2:S4	S1:S1	S2:S2	S1:S3	S2:S4	S1:S1	S2:S2	S1:S3	S2:S4
living persons	0-100	0.959	0.797	0.975	0.874	0.994	0.895	0.980	0.924	0.983	0.991	0.977	0.954
	0-200	0.997	0.990	0.989	0.982	0.997	0.984	0.998	0.993	0.992	0.996	0.996	0.981
	0-300	0.997	0.993	0.990	0.986	0.998	0.989	0.998	0.995	0.994	0.997	0.997	0.987
unburn-ed area	0-100	0.966	0.965	0.970	0.970	0.951	0.933	0.971	0.931	0.955	0.960	0.946	0.964
	0-200	0.879	0.955	0.894	0.942	0.703	0.927	0.948	0.968	0.939	0.945	0.928	0.872
	0-300	0.879	0.901	0.900	0.849	0.561	0.794	0.852	0.945	0.937	0.826	0.886	0.677

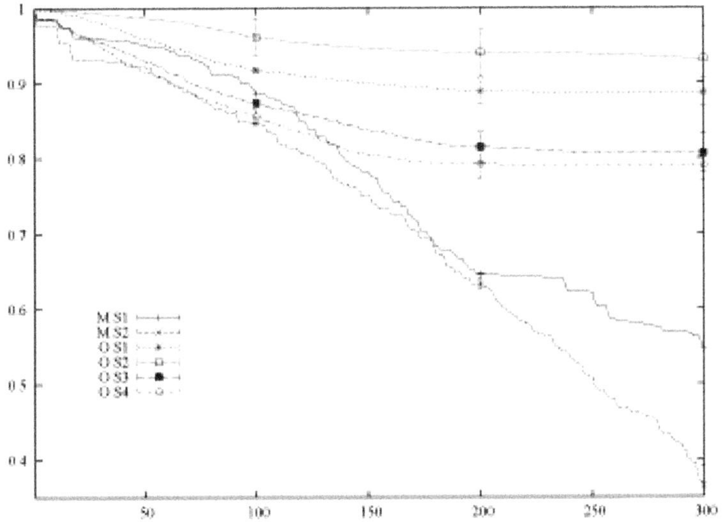

Fig. 9 Chronological change of unburned building area rate for map Tokorozawa1.

While the same numbers of ignition points are allocated in cases of S1 and S2, more ignition points are allocated to bigger houses in cases S3 and S4 on assumption that there are many origins of fire in a big building.

At Figure 6, 7 and 8, the ignition points are set within some area. Table 5 shows correlations between the simulations using maps created by Moritz and our programs. For example,

- Column S1:S1 shows the correlation between the result of scenario1 at Moritz map and ones of scenario1 at our program map.
- Column S2:S4 shows the correlation between the result of scenario2 at Moritz map and ones of scenario4 at our program map.

The values are different from ones in Table 3 because the disaster situations are different and rescue agents are included in the simulations.

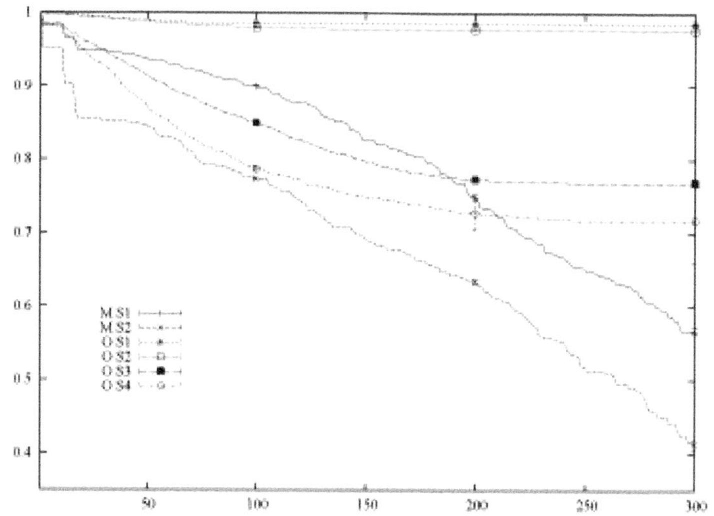

Fig. 10 Chronological change of unburned building area rate for map Tokorozawa2.

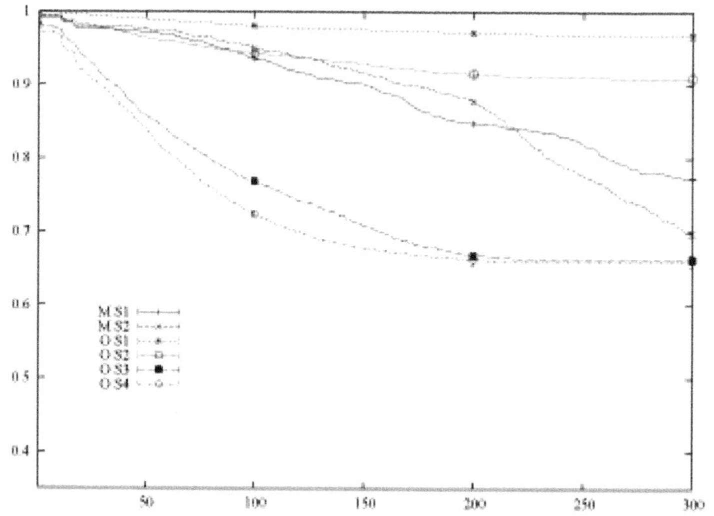

Fig. 11 Chronological change of unburned building area rate for map Tokorozawa3.

1. The values in Table 5 are strongly positive, except the correlation of unburned building between S1:S1 in Tokorozawa2 and S2:S4 in Tokorozawa3. Followings are thought to cause the two low values.

   a. low correlation for S1:S1 in Tokorozawa2
   Figure 9-11 show the chronological changes of unburned building area rate for three maps. Fire spread stops from 90 time step in a case of S1

using our program map in Tokorozawa2. (Figure 10) The fire spread is affected by size of building as described in subsection 4.2.

b. low correlation for S2:S4 in Tokorozawa3
Fire spread stops from 200 time step in a case of S2 using our program map in Tokorozawa3. (Figure 11) On the other hand, fire spreads wider from 200 time step in a case of S4 using Moritz map in Tokorozawa3, because of large building begin to burn from about 180 time step.

2. The deviations of simulation results are different among scenarios and maps. The difference is thought to come from fire fighting by fire brigade. The deviation of unburned building of S2 using our program in Tokorozawa1 (Figure 9)) is the biggest. The fire spread stopped until 300 time step for 3 cases.
3. The chronological change using our program maps looks like inverse proportion. While fire spreads within a block, it seems difficult that fire spreads over roads. This becomes noticeable when the sizes of buildings are small and it is thought to cause the unburned rate looks like inversely in our maps.

5 Discussion and Summary

Agent-based disaster and rescue simulations make possible to simulate situations that we cannot do experiments physically on real scale. Applying the agent-based simulation system to practical cases requires the fidelity of the simulations to satisfy the user's demands. It is desirable to simulate disasters and rescue operations under environments where they are close qualitatively and quantitatively to real one.

Maps are a key component of the disaster and rescue simulation system. Real maps contain a huge number of objects. Simulations using the real size map require computation resources and preparing related properties of the buildings. Simplifying environments is one way to reduce the computational resources under the conditions that the simulations results are qualitatively equal to the results with realistic environments.

In this paper, we propose a method to create maps with road networks and buildings from open GIS data and compare the results of simulations, one is simulated using the building data generated by programs and the other is created from the real data. The experiments show different aspects in correlations between simulations using two maps. The differences come from not only the properties of map but also the rescue agent actions.

Acknowledgements. Authors appreciate RoboCup Rescue community, GSI, OS and OSM. Without their software and GIS data, we cannot do this work in a short time and with reasonable resources.

References

1. Göbelbecker, M., Dornhege, C.: Realistic cities in simulated environments - an open street map to robocup rescue converter. In: Fourth International Workshop on Synthetic Simulation and Robotics to Mitigate Earthquake Disaster (SRMED 2009) (2009)
2. http://zgate.gsi.go.jp/ch/jmp20/jmp20_eng.html (September 5, 2009)
3. http://www.gsi.go.jp/ENGLISH/page_e30210.htm (September 5, 2009)
4. http://www.ordnancesurvey.co.uk/oswebsite/products/osmastermap/userguides/docs/OSMM_ITN_userguide_v1.0.pdf (September 5, 2009)
5. Kitano, H., Tadokoro, S., Noda, I., Matsubara, H., Takahashi, T., Shinjou, A., Shimada, S.: Robocup rescue: Search and rescue in large-scale disasters as a domain for autonomous agents research. In: IEEE International Conference on System, Man, and Cybernetics (1999)
6. http://www.ordnancesurvey.co.uk/oswebsite/ (September 5, 2009)
7. http://wiki.openstreetmap.org/wiki/Main_Page (September 5, 2009)
8. Takahashi, T.: RoboCup Rescue: Challenges and Lessons Learned, ch. 14, pp. 423–450. CRC Press, Boca Raton (2009)
9. Takahashi, T., Ito, N.: Preliminary study to use rescue simulation as check soft of urban's disasters. In: Workshop: Safety and Security in MAS (SASEMAS) at AAMAS 2005, pp. 102–106 (2005)
10. Takahashi, T., Takeuchi, I., Matsuno, F., Tadokoro, S.: Rescue simulation project and comprehensive disaster simulator architecture. In: IEEE/RSJ International Conference in Intelligent Robots and Systems(IROS 2000), pp. 1894–1899 (2000)
11. Takeuchi, I., Kakumoto, S., Goto, Y.: Towards an integrated earthquake disaster simulation system. In: First International Workshop on Synthetic Simulation and Robotics to Mitigate Earthquake Disaster (2003), http://www.dis.uniroma1.it/~rescue/events/padova03/papers/index.html
12. http://www.ordnancesurvey.co.uk/oswebsite/products/osmastermap/userguides/docs/OSMMTopoLayerUserGuide.pdf (September 5, 2009)
13. de Walle, B.V.(edited), Tyroff, M.: Emergency response information systems: Emerging trends and technologies. Communications of the ACM 50(3), 28–65 (2007)

Evaluation of Time Delay of Coping Behaviors with Evacuation Simulator

Tomohisa Yamashita, Shunsuke Soeda, and Itsuki Noda

Abstract. As described in this paper, we analyzed the influence of the time necessary to begin coping behaviors on the damage caused by chemical terrorism. To calculate the damage of a chemical attack in a major rail station, our network model-based pedestrian simulator was applied with systems designed to predict hazards of indoor gas diffusion. Our network model is designed to conduct simulations much faster, taking less than few minutes for simulation with ten thousands of evacuators. With our evacuation planning assist system, we investigated a simulated chemical attack on a major rail station. In our simulation, we showed a relation between the time necessary to begin coping behaviors of the managers and the damage to passengers. Results of our analyses were used for the instruction of rail station managers in a table-top exercise held by the Kitakyushu City Fire and Disaster Management Department.

1 Introduction

Terrorist acts using chemical and biological agents and radioactive (CBR) mate-rials have typically been nonselective attacks on crowds in urban areas.

Tomohisa Yamashita
National Institute of Advanced Industrial Science and Technology (AIST),
2-3-26 Aomi, Koto-ku, Tokyo 135-0064 Japan
e-mail: tomohisa.yamashita@aist.go.jp

Shunsuke Soeda
National Institute of Advanced Industrial Science and Technology (AIST),
1-1-1 Umezono Tsukuba, Ibaraki 305-8568 Japan
e-mail: shunsuke.soeda@aist.go.jp

Itsuki Noda
National Institute of Advanced Industrial Science and Technology (AIST),
1-1-1 Umezono Tsukuba, Ibaraki 305-8568 Japan
e-mail: I.Noda@aist.go.jp

Q. Bai and N. Fukuta (Eds.): Advances in Practical Multi-Agent Systems, SCI 325, pp. 403–414.
springerlink.com

These ha-zardous materials might be sprinkled, vaporized, or spread by an explosion. First responders to these accidents, such as fire protection and police agencies of muni-cipalities must prepare practical plans of coping behaviors against CBR terrorism, but they typically have insufficient experience and knowledge of CBR terrorism. Therefore, useful tools supporting planning and preparation are necessary to esti-mate and illustrate the damage done by CBR attacks. To meet those needs, the Ministry of Education, Culture, Sports, Science and Technology of Japan (MEXT) ordered a research consortium of Tokyo University, Advanced Industrial Science and Technology (AIST), Mitsubishi Heavy Industries (MHI), and Advancesoft to develop a new evacuation planning assistance system for use before CBR attacks.

With our evacuation planning assist system, a user can estimate the damage caused by CBR terrorism. These disasters, which are most likely to occur in urban areas, have numerous characteristics that differ from those of natural disasters. These disasters are caused intentionally, which means that responsible persons must prepare for the worst. There are still few cases in which CBR terrorism has actually been conducted, which means that we still know little about what damage might result from such terrorism.

As described in this paper, we have built a network model-based pedestrian simu-lator as a part of the evacuation planning assist system. Compared to previous grid-based and continuous space-based models that required hours to conduct si-mulations with fewer than thousands of evacuees, our network model is designed to conduct simulations much more rapidly, taking less than few minutes for simu-lations with tens of thousands of evacuees. Our pedestrian simulator is designed to function with hazard prediction systems of outdoor and indoor gas diffusion, which calculate how rapidly and at what concentrations harmful gases spread. Using data provided by hazard prediction systems, our pedestrian simulator is useful to estimate how much damage will be incurred under various evacuation scenarios. These results are expected to be useful to produce and evaluate evacuation plans as countermeasures against CBR terrorism.

As described in this paper, we explain our evacuation planning assist system, and share an example of practical use of our system. We dealt with coping behaviors against chemical attacks at a major rail station at the request of the Fire and Disaster Management Department of Kitakyushu City. In our simulation, we showed a relation between the time necessary to begin coping behaviors of the managers and the damage to passengers. Our analysis was used for enlightening the managers of the rail station in a tabletop exercise held by the Fire and Disaster Management Department of Kitakyushu City.

2 Evacuation Planning Assist System

Our evacuation planning assist system consists of three components: a pedestrian simulator constructed by AIST, a prediction system of outdoor gas diffusion by MHI, and a prediction system of indoor gas diffusion by

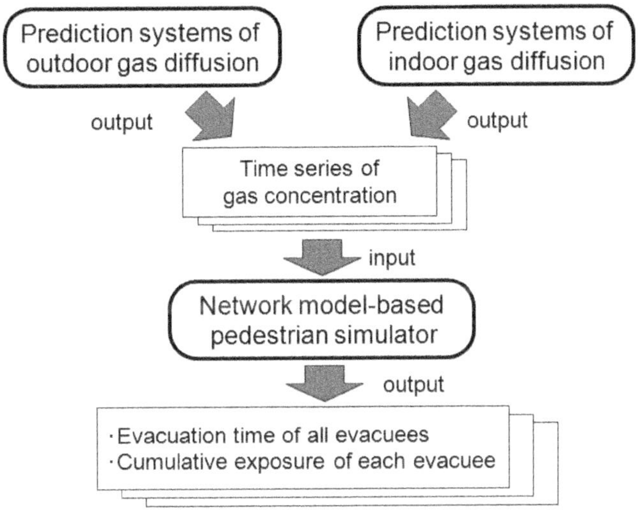

Fig. 1 Outline of dataflow of the evacuation planning assist system

Advancesoft. First, the prediction systems of indoor and outdoor gas diffusion calculate the concentration of hazardous gases. The output of these systems is a time series of gas concentration in designated areas. Then, the pedestrian simulator calculates the evacuation time of all evacuees and the cumulative exposure of each evacuee using a time series of gas concentration. An dataflow of our system is presented in Fig. 1.

2.1 Pedestrian Simulator

Pedestrian simulators of various kinds have been developed for various purposes. Pan roughly classified them into three categories: fluid and particle systems, ma-trix-based systems, and emergent systems [5]. However, all of

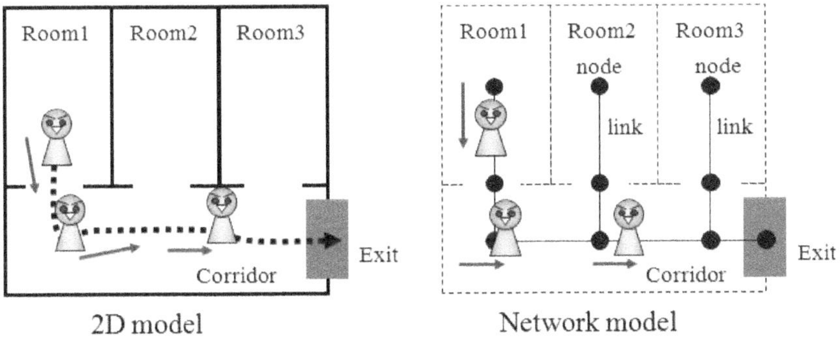

Fig. 2 Example of 2D model and network model

these systems are two-dimensional systems, which allow pedestrians to move around two-dimensionally. Unlike other pedestrian simulators, our simulator simplifies traffic lines by representing it with a graph model: a model with links and nodes (Fig. 2). The paths where the pedestrians move around are represented as links.

These links are connected at nodes. Because the pedestrians were able to move only along the links, our model is more one-dimensional than two-dimensional. This approach has often been used in traffic simulators [1], but not for pedestrian simulators. We chose a network based-model for our simulator because we need a high-speed simulator to examine the evacuation behaviors of many evacuees on the macroscopic side. A network based-model is unsuitable for simulating many pedestrians evacuating a large space precisely, but it might be useful to reveal bot-tlenecks and evacuation times quickly and thereby enable the comparison of many evacuation plans. The appearance of a network-based pedestrian simulator is shown in Fig. 2.

2.1.1 Pseudo-lane Model

The speed of a pedestrian is calculated using a model we call "pseudo-lane model". We assume that each path is made up of several pseudo-lanes, based on the width of the path. The speed of each pedestrian is calculated from the free flow speed (the speed of the pedestrian without any other pedestrians), and from the pedestrians immediately front of the pedestrian in the same pseudo-lane.

2.1.2 Speed of a Pedestrian

The speed of a pedestrian is calculated from the density of the crowd on the link. Each link has a width and a length, which is used to calculate the area of the link. Then, the density of the crowd on the link could be calculated from the number of the pedestrians on the link. Speed V_i of the pedestrian on link i is calculated from the following formula;

$$V_i = \begin{cases} V_f, & d_i < 1 \\ d_i^{-0.7945} V_f, & 1 \leq d_i \leq 4 \\ 0, & d_i > 4 \end{cases} \tag{1}$$

where V_f represents free flow speed of the pedestrians, which is the speed of the pedestrian when not in a crowd, and d_i represents the density of the pedestrians on link i.

Fig. 4 shows the relationship the speed of pedestrian and the density of the link. The exception to this formula is the pedestrian on the head of a crowd on the link. For this pedestrian, V_f is used regardless of how the link is crowded. Note that when the density of a link exceeds 4 pedestrians/m^2,

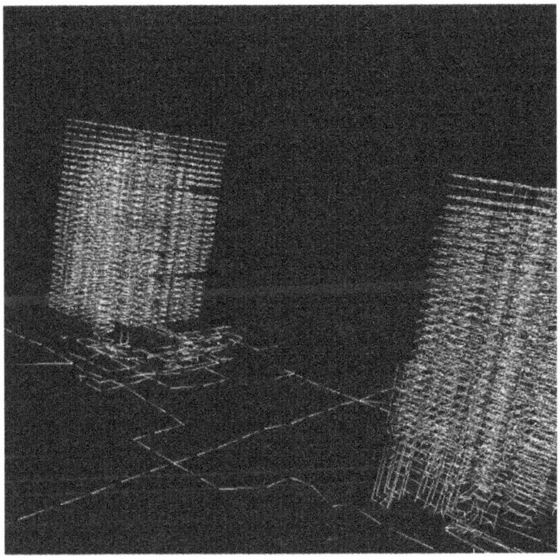

(a)High-rise building

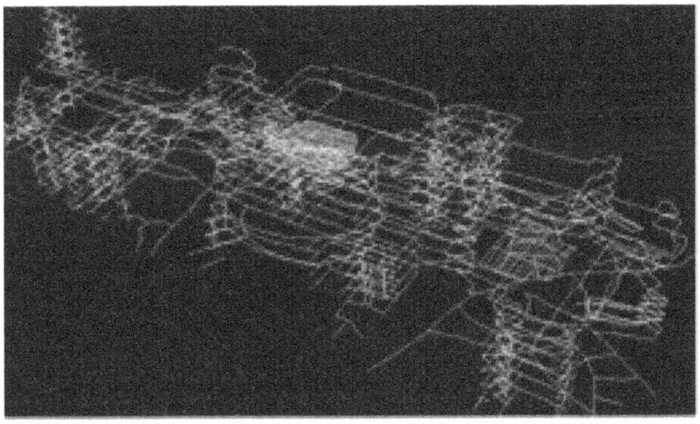

(b)Commercial complex

Fig. 3 3D view of our network-based model pedestrian simulator

all the pedestrian on the link cannot move except for the one who is on the head of the crowd.

2.1.3 Confluence

Confluences - where two or more paths meet together - slow down the speed of pedestrian. To illustrate slow-down by confluence, we used a simple model of limiting the number of the pedestrian who could enter a link. The maximum number of the pedestrian entering link l_{out} shown in Fig. 5 is determined from the width of the link. When there are pedestrians on l_{out} already, the number of the pedestrians that could enter l_{out} is decreased at some ratio.

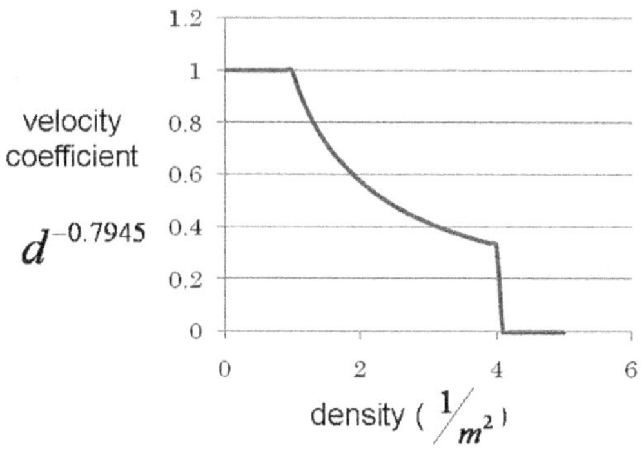

Fig. 4 Coefficient of the speed and the density of the pedestrians

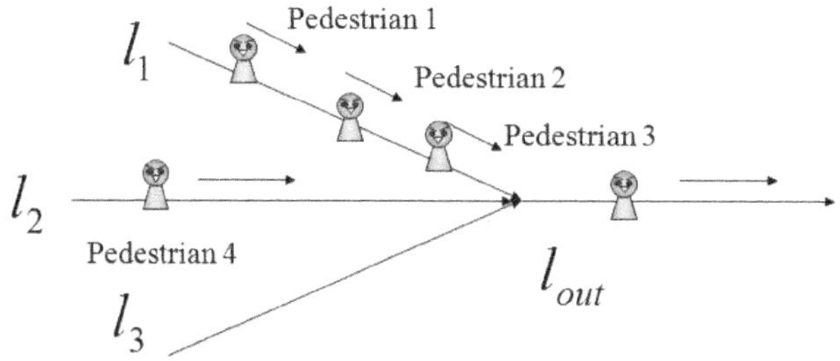

Fig. 5 Modeling confluence

When there are more than two links where the pedestrians are trying to enter l_{out}, this number is divided among the links depending on the number of the pedestrians trying to enter l_{out}. Also, in this case, the total number of the pedestrians able to enter l_{out} is also reduced.

2.2 Prediction System of Indoor Gas Diffusion

Recently, more subways, shopping malls, and high-rise buildings have large-scaled and intricate passages. Accordingly, the casualties on CBR attacks or fires there increase more disastrously. For prevention or reduction of these disasters, a hazard prediction systems of indoor gas diffusion "EVE SAYFA" (Enhanced Virtual Environment Simulator for Aimed and Yielded Fatal Accident) has been developed to aid their anticipation and the evaluation of the safety [2,7]. EVE SAYFA has two simulation models; one is EVE SAYFA 3D with highly accurate 3-dimensional model. The other is EVE SAYFA 1D with high-speed calculating 1-dimensional network model.

Using EVE SAYFA 3D, the simulation of diffusion of hazardous and noxious substances is carried out by Computational Fluid Dynamics (CFD)

software based on Large eddy simulation (LES), which is a numerical technique used to solve the partial differential equations governing turbulent fluid flow.

Using EVE SAYFA 1D, the macro-model is expressed in differential equations or algebraic equations in the whole of the large-scaled structures which consist of some elements corresponding to, e.g., rooms, corridors, stairs, walls, windows, etc. The total computing time can be reduced by saving the result of air movement simulation in database and reusing for diffusion simulation.

2.3 Prediction System of Outdoor Gas Diffusion

A hazard prediction system has been developed for CBR attacks in urban areas with the use of the mesoscale meteorological model, RAMS and its dispersion model HYPACT.

RAMS is equipped with an optional scheme to simulate airflow around buildings based on the volume fraction of the buildings within each grid cell. The HYPACT (HYbrid PArticle and Concentration Transport) code is an atmospheric diffusion code that can be coupled to RAMS. This code is based on a Lagrangian particle model that satisfies mass conservation in complex airflow and can adopt the finite difference method at large distances downwind to reduce computational time.

The developed simulation system, called MEASURES, consists of HYPACT, RAMS and an airflow database [3, 4]. Meteorological data can be loaded onto the system by the user directly or through the Internet. A time series of airflow is generated for the location of interest. In this procedure, the 3-dimensional wind data from the database of 48 atmospheric conditions are interpolated at each time step for the wind direction and the atmospheric stability observed at the location at that time step.

3 Simulation

With our evacuation planning assist system, we investigated a simulated chemical attack on a major rail station because of a request from the Fire and Disaster Management Department of Kitakyushu City. In our simulation, for enlightening the managers of the rail station, we showed a relation between the time necessary to begin coping behaviors of the managers and the damage to passengers.

3.1 Simulation Settings

In our simulations, a chemical attack with chloropicrin is set to take place in the station yard of a conventional line. Gas diffusion in the station yard is calculated using the systems to predict indoor gas diffusion. The movements

Table 1 The influence of exposure of chloropicrin

amount of exposure ($mg \cdot min/m^3$)	1	200	1,000	2,000	20,000
Influence in behavior	Pain in the eye & throat	Nausea & headache	Breathing trouble	Lethal dose 50%	Lethal dose 100%
Implementation in pedestrian simulation	Decrease in speed (-40%)	Decrease in speed (-90%)	Stop	Stop	Stop
Damage level	Mild	Moderate	Severe	Severe	Severe

Table 2 Coping behaviors of the managers and the times required for them

Manager		Coping behavior	Time required to begin(min) quick	slow
Conventional line	1	Detecting chemical attack	5	10
	2	Reporting to the fire department and the other managers	3	6
	3	Shutting down the trains	3	6
	4	Ordering an evacuation to the passengers	3	6
New bullet train line	5	Ordering an evacuation to the passengers	3	6
	6	Shutting down the trains	3	6
Monorail station	7	Ordering an evacuation to the passengers	5	6
	8	Shutting down the trains	3	6
Hotel	9	Ordering an evacuation to the guests	3	6
Fire and Disaster Management Department	10	Rescuing insured passengers	20	30

and damage to about 9000 passengers are calculated using our pedestrian simulator. The amount of exposure to chloropicrin of the passengers is calculated as the product of the chloropicrin concentration and the time spent in the area. The influence of chloro-picrin exposure [6, 8] on the passenger's behavior is described in the table. 1.

This rail station is a complex facility. It has facilities of four kinds: a conventional line, a new bullet train line, a monorail, and a hotel. Each facility has a manager. We assume the following 10 coping behaviors of the managers and the times required to begin these coping behaviors described in the table 2. The times required to begin these coping behaviors influence the damage because the damage to passengers increases the beginning of these coping behaviors are delayed. The sequence of the coping behaviors is portrayed in Fig. 6.

For example, the manager of the conventional line has four coping behaviors: (1) detecting the chemical attack, (2) reporting to the fire station and the other managers of the new bullet train line, monorail, and the hotel, (3) shutting down conventional trains, and (4) ordering an evacuation of the passengers of the conventional line. Each coping behavior is set to have the time of two kinds. For example, if coping behavior 4 is begun quickly, the

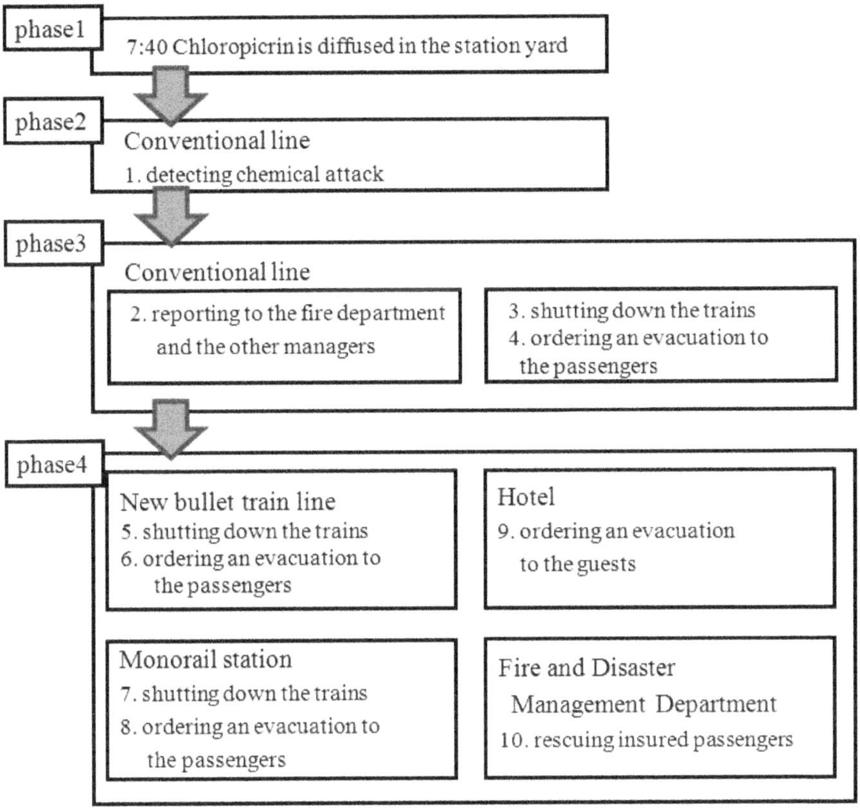

Fig. 6 Sequence of the coping behaviors

time necessary to begin is 3 min. Otherwise (begun slowly), the time required is 6 min.

The evacuation scenarios are 1,024 because the number of combinations of the times required to begin 10 coping behaviors is 2^{10} (=1,024). We calculate damage to approximately 9000 passengers in 1024 evacuation scenarios. In our pedestrian simulation, each passenger walks around normally, from and to outside areas of the station and from and to platforms, until the attack is detected and an alarm is given. After ordering an evacuation of the passengers, the passengers evacuate using a route directed by the station staff.

To assign a sequential serial number to each scenario according to whether 10 coping behaviors are begun quickly or slowly, we use a ten-digit number in the binary system. For example, if coping behavior 1 (detecting chemical attack) is begun quickly, then the first bit of the ten-digit number is set as 0. This scenario is represented as 0000000000 in the binary system if all coping behaviors are begun quickly. Then, the 10-digit number in the binary system is transferred to a serial number in the decimal system. The scenario represented as 0000000000 in the bi-nary system is assigned serial number

0 in the decimal system; the scenario represented as 1111111111 is assigned serial number 1023.

3.2 Simulation Result

The results of our simulation are shown in Figs. 7∼9.

In Fig. 7, the graph shows the number of the severely injured victims in 1,024 scenarios. Based on the number of severely injured victims, five characteristic clusters can be inferred. In each clus-ter, the scenarios have the same tendencies of coping behaviors 1 and 4. In scena-rios of cluster 1-1, both (1) detecting chemical attack and (4) ordering evacuation of passengers of the conventional line are begun quickly. In scenarios of cluster 1-2, (1) detection is begun quickly, and (4) ordering an evacuation is begun slowly. In scenarios of cluster 1-3a and 1-3b, (1) detection is begun slowly, and (4) order-ing an evacuation is begun quickly. The difference between cluster 1-3a and 1-3b is coping behavior 10. In the scenarios in cluster 1-3a, (10) rescuing injured passengers by the Fire and Disaster Management Department is begun slowly. On the other hand, in scenarios in cluster 1-3b, (10) rescuing is begun quickly. In scena-rios of clusters 1-4, both (1) detection and (4) ordering an evacuation are begun slowly. Therefore, it is confirmed that both (1) detecting, (4) ordering an evacua-tion, and (10) rescuing injured passengers are more important to decrease the number of severely injured victims.

In Fig. 8, the graph shows the number of the moderately injured victims in 1,024 scenarios. Based on the number of the severely injured victims, 4 characteristic clusters exist. The victims in cluster 2-1 are fewer than in other clusters. There is no great difference among the number of the victims of clusters 2-1, 2-2, and 2-3. Therefore, it is confirmed that (1) detection and (4) ordering an evacuation are more important to decrease the number of moderately injured victims.

In Fig. 9, the graph shows the number of the mildly injured victims in each of 1024 scenarios. Based on the number of the severely injured victims, four clusters can be inferred. The mildly injured victims in cluster 3-1 are more numerous than in cluster 3-2. However, the number of all victims in clusters 1-1, 2-1, and 3-1 is equal to that in cluster 1-2, 2-2, and 3-2. Therefore, it is confirmed that (1) detec-tion is extremely important to decrease the damage to passengers. Through com-parison of the damage incurred in each of the 1024 scenarios, we confirm that the most effective coping behaviors for decreasing the damage to passengers are i) detection of chemical attack and ii) ordering the evacuation of passengers of the conventional line. In a tabletop exercise held by Fire and Disaster Management Department of Kitakyushu City, our simulation result was shared to enlighten the rail station managers.

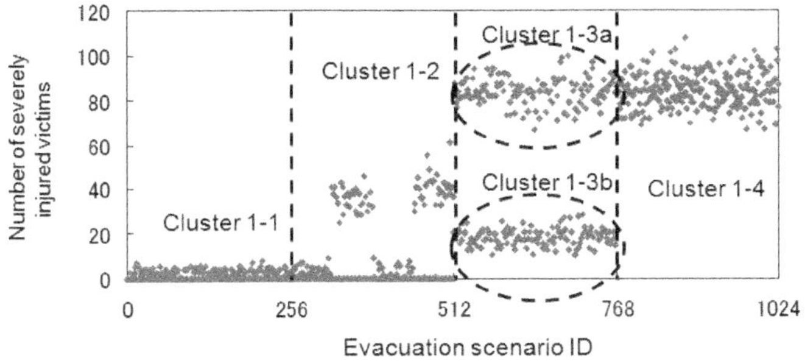

Fig. 7 Number of severely injured victims

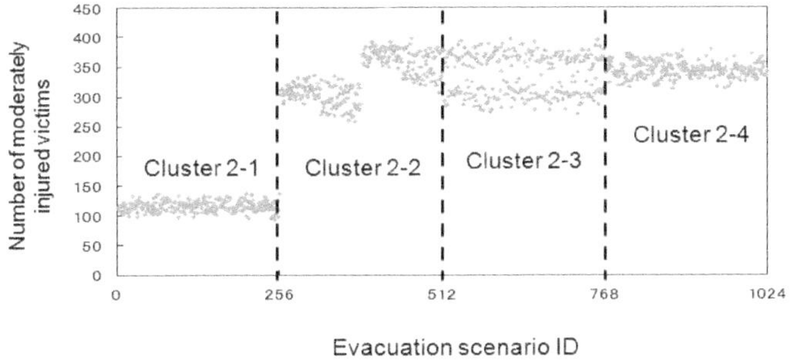

Fig. 8 Number of moderately injured victims

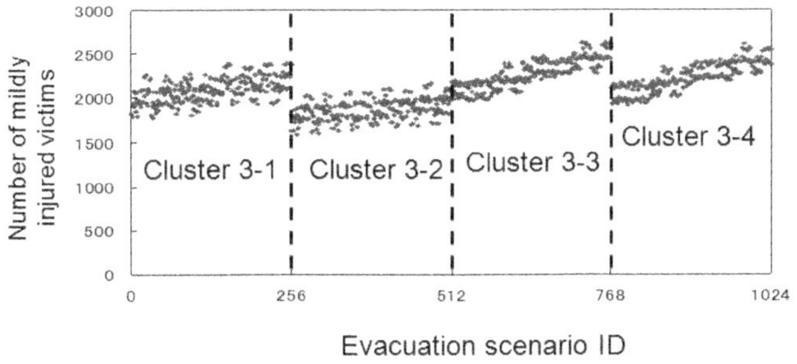

Fig. 9 Number of mildly injured victims

4 Conclusion

As described in this paper, we explained our evacuation planning assist system consisting of a pedestrian simulator, a prediction system of indoor gas diffusion, and a prediction system of outdoor gas diffusion. A chemical attack at a major rail station in Japan was taken as an example of the practical use of our system because of a request from the Fire and Disaster Management

Department of Kitakyushu City. In our simulation, we showed the relation between the time necessary to begin coping behaviors by managers and the damage to passengers. Results of comparisons of the damage incurred in 1,024 scenarios confirmed that, to decrease damage to the passengers, it is important to begin coping behaviors quickly by i) detecting the chemical attack and ii) ordering an evacuation of passengers.

Acknowledgements. This work was supported as a national project for urban safety commenced from 2007 by the Ministry of Education, Culture, Sports, Science and Technology of Japan (MEXT).

References

1. Burmeister, B., Haddadi, A., Matylis, G.: Application of Multi-Agent Systems in Traffic and Transportation. Software Engineering 144(1), 51–60 (2002)
2. Nara, M., Kato, S., Huang, H., Zhu, S.W.: Numerical analysis of fire on disasters with "EVE SAYFA", a fire simulator, and its validation – LES for thermal plume. In: Summaries of Technical Papers of 2006 Annual Meeting of Architectural Institute of Japan. pp. 317–318 (2006) (in Japanese)
3. Ohba, R., Kouchi, A., Hara, T.: Hazard Projection System of Intentional Attack in Urban Area. In: 11th Annual George Mason University, Conference on Atmospheric Transport and Dispersion Modeling (2007) (Poster)
4. Ohba, R., Yamashita, T., Ukai, O., Kato, S.: Development of Hazard Prediction System for Intentional Attacks in Urban Areas. In: Seventh International Symposium on New Technologies for Urban Safety of Mega Cities in Asia (USMCA 2008), pp. 687–696 (2008)
5. Pan, X.: Computational Modelling of Human and Social Behaviours for Emergency Egress Analysis. Ph.D. Dissertation, Department of Civil and Environmental Engineering, Stanford University (2006)
6. http://en.wikipedia.org/wiki/Chloropicrin
7. http://venus.iis.u-tokyo.ac.jp/english/Research/field-e/evesayfa/index.html
8. http://www.workcover.nsw.gov.au/Documents/Publications/AlertsGuidesHazards/DangerousGoodsExplosivesFireworksPyrotechnics/chloropicrin_fact_sheet_1371.pdf

Web-Based Sensor Network with Flexible Management by an Agent System

Tokihiro Fukatsu, Masayuki Hirafuji, and Takuji Kiura

Abstract. This paper presents the architecture and function of an agent system developed for sensor nodes equipped with a Web server such as Field Servers to manage a Web-based sensor network. The agent system accesses flexibly all Web components with a rule-based function and it performs various types of monitoring and complicated operations. By accessing useful Web applications such as image processing, the system also provides versatile data analysis. Moreover, the agent system has Web interfaces to control itself on the Web, so we can construct a multi-agent system in which one agent system controls not only Field Servers but also other agent systems to achieve high scalability and a robust system.

1 Introduction

To increase agricultural productivity and promote efficient management, it is important to monitor crop growth, the field environment, and farm operations at field sites. One of the key technologies in field monitoring is a sensor network constructed of many sensor nodes made up of small sensor units with radio data links [2] [10]. A sensor network enables large-scale environmental monitoring with high scalability and robustness, but it is difficult to perform complicated operations and to treat large amounts of data because of deficiencies in hardware performance. In the case of agriculture, a flexible monitoring system is required that collects various types of information, including image data, changes measurement procedure according to users' requests, and controls peripheral equipment from a remote site in response to the situation.

Tokihiro Fukatsu · Masayuki Hirafuji · Takuji Kiura
National Agricultural Research Center, 3-1-1, Kannondai, Tsukuba, Ibaraki 305-8666, Japan
e-mail: {fukatsu,hirafuji,kiura}@affrc.go.jp

Q. Bai and N. Fukuta (Eds.): Advances in Practical Multi-Agent Systems, SCI 325, pp. 415–424.
springerlink.com

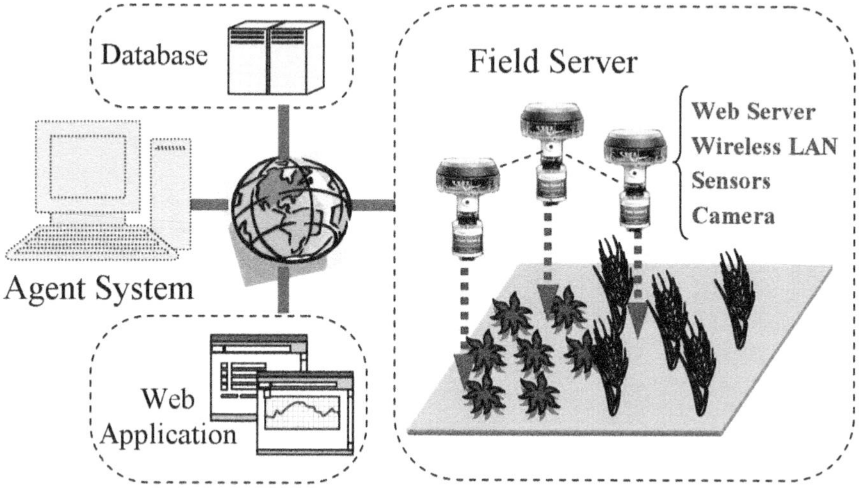

Fig. 1 System architecture of a Web-based sensor network consisting of Field Servers and the agent system. The agent system manages the Field Servers via the Internet at a remote site by using database and Web applications.

To meet such needs, we proposed a Web-based sensor network system in which useful sensor nodes (Field Servers [6]) equipped with Web servers are controlled by an autonomous management program (the agent system [7]) at a remote site (Fig. 1). Each Field Server has a wireless LAN to provide high-speed transmission and long-distance communication, various sensors including an Internet camera, and a monitoring unit with a Web server, which can be accessed by using a Web browser via the Internet. This system does not have intelligent processing on the inside like a conventional sensor network, but it allows various types of monitoring, complicated operations, and data analysis to be performed by the agent system. The agent system, which is an agent-oriented management system [14], is able to control Web servers flexibly and sophisticatedly with artificial intelligence functions on a remote computer, and it acts as an intelligent module of the sensor network. By using Field Servers and the agent system, the Web-based sensor network satisfies the requirements of an advanced monitoring system. In this paper, we describe the agent system architecture to realize a Web-based sensor network, and we evaluate its effectiveness and potential.

2 Architecture and Function of the Agent System

Research on agent systems capable of autonomous action [5] has been attempted in various fields, and there are many kinds of agent systems, such as search agents [4], conversational agents [13], mobile agents [11], and so

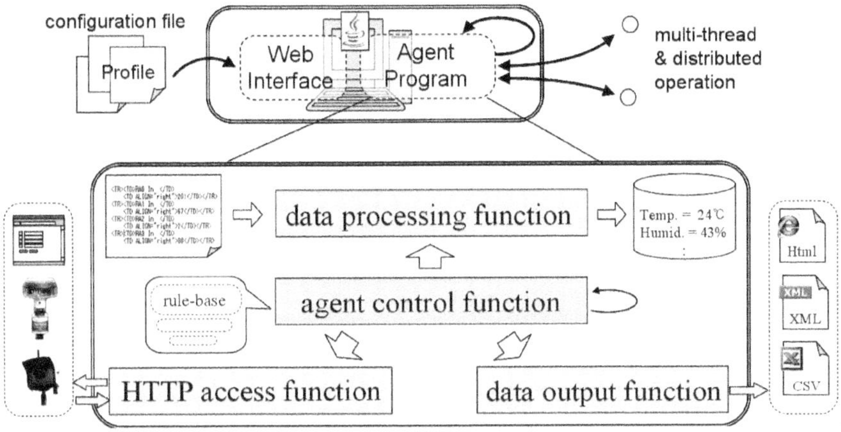

Fig. 2 System architecture of the agent system, consisting of the agent program, Profiles, and Web interface. The agent program has an HTTP access function, a data processing function, a data output function, and an agent control function.

on. In a Web-based sensor network, our agent system, which acts like a user based on an adaptive agent [12], accesses the Web pages of Field Servers, extracts sensor data from the Web pages, analyzes the data by using Web application tools, shows the resulting data on a Web page, and stores all data in a database. All modules of the Web-based sensor network are accessed via the Internet. The agent system seamlessly connects and uses them to manage Field Servers at a remote site, so users need not manage the system at a field site except when deploying Field Servers and connecting them to the Internet.

The agent system is constructed by an agent program written in Java, program configuration files ("Profiles") describing parameters of request operations in an XML format, and Web interfaces for easy management of the agent system (Fig. 2). The agent program can access all types of pages on the Web and operate them flexibly based on the Profiles. It autonomously executes the contents of Profiles with multi-thread and distributed functions. Users can utilize the agent program and manage Field Servers to their satisfaction by describing suitable Profiles.

The agent program is capable of not only basic functions such as accessing Web pages, extracting available data, and outputting calculation results, but also advanced rule-based control functions that enable the performance of complicated operations. By using Web applications that analyze input data, this program also provides versatile and easily expansible data processing activities and makes it possible to distribute calculation tasks. In this section, we show each function of the agent program.

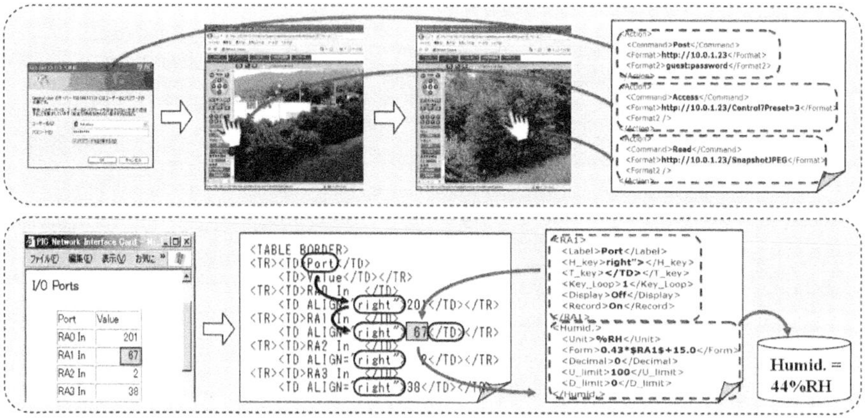

Fig. 3 Examples of agent actions with the basic function to obtain target image data according to a required procedure and to extract useful information from raw data with automatic calculation.

2.1 Basic Function

The agent program operates Field Servers based on Profiles, which consist of basic information about the Field Servers, definition of the associated sensors, and operating instructions combined with some tiny action commands. Each action command is mainly constructed of a simple operation that emulates user operations on a Web browser, such as Post, Get, Search, Copy, Save, and so on. The agent program executes these commands in the Profiles, and it can perform versatile and complicated operations in response to the user's requests. In the setting of sensors, there are important parameters to extract a desired value from raw html data and to calculate a meaningful value with calibration formulae. By describing these items based on target Web pages, the agent program can treat various kinds of Field Servers.

Fig. 3 shows examples of processing flow using the basic function. To obtain target image data from a certain Field Server, we perform a procedure such as inputting a password, selecting a camera position and downloading image data. To obtain measurement data, we also need to extract a target value and to calculate it with a certain relational expression, because Field Servers only show raw voltage data of each sensor on their Web pages. The agent program can perform these actions if adequate Profiles are provided.

The agent program interacts with not only Field Servers but also accessible Web pages on the Internet, and it stores collected data in a database with any format based on the Profiles. As the figure shows the agent program also performs as middleware for Web content data so that it can use any data with the same format and share various applications without data format problems.

2.2 Rule-Based Function

To realize intelligent control, the agent system has a rule-based function consisting of a number of IF-THEN rules [3]. We can establish simple conditional judgments and complicated rule-based algorithms by providing the Profiles described with these functions. The agent program evaluates a given rule-base with monitoring data, Profile parameters, and agent status, and then judges whether corresponding action commands are executed normally, skipped until conditions change, or executed with exception handling by other threads.

By using the rule-based function, we can design multi-mode applications according to the situation in a given environment. Fig. 4 shows the examples of a surveillance operation based on sensor data and a priority control operation based on battery level. In the case of the surveillance operation, the agent program periodically monitors the Field Servers in daytime and during the night. If some sensor data indicates an abnormal value by recognizing a suspicious person, the agent system changes its operation to warning mode with a gradual approach. The agent program can shift from one step to the next within the warning mode in response to sensor data based on the rule-based function.

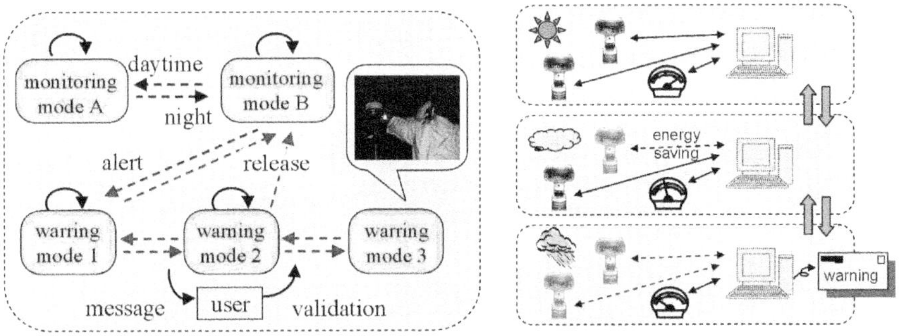

Fig. 4 Examples of agent action with rule-based function. The left figure shows a surveillance operation with gradual approach based on sensor and image data. The right figure shows a priority control of running device and access interval based on battery level.

In the case of priority control operations, the agent program suitably accesses Field Servers according to their condition and environmental situation. If the battery level is low, the agent program adjusts the power consumption of the Field Servers by switching the camera off and lengthening the access interval. It can also provide appropriate management based on an expected battery level from a Web weather forecast. A lot of IF-THEN rules act as autonomous decision functions, so the agent program deals with complicated and various requests.

2.3 Distributed Data Processing

The agent program can access all types of Web pages and Web applications. By preparing useful Web applications that perform image analysis and signal processing, the agent program can provide data analysis in combination with the calculated results from Web applications. It can realize versatile and easily expandable functions without changing or rebooting the agent program. This method makes it possible to distribute calculation tasks by means of Web applications, and it provides a scalable processing by increasing Web applications according to operation frequency and loading.

For example, the agent program itself does not have an image processing function, but it can obtain analyzed information by collecting image data from Field Servers through Web applications. Fig. 5 shows an example of event detection. By using some Web applications such as image masking of a target area, image subtraction to extract a difference, and counting image variations by binarization, the agent program automatically detects agricultural events from image data.

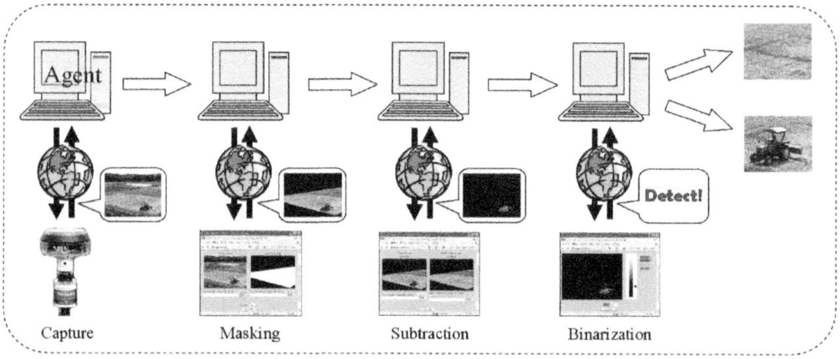

Fig. 5 Examples of data processing by using Web applications. After collecting image data from Field Servers, the agent system accesses image analysis Web applications and judges whether a certain agricultural operation is found based on the results from Web applications.

Via the distributed data processing, the Web-based sensor network can be used for various agricultural applications such as event detection in agricultural fields, automatic counting of insects, and monitoring of farming operations [9] [8]. Various types of data processing can be added simply by developing or finding the appropriate Web applications, and users' efforts can be reduced by sharing Web applications. Moreover, dynamic changing of data processing according to the situation can be realized by cooperating with the rule-based function.

3 Multi-Agent System

The agent system has Web interfaces that facilitate the control and monitoring of agent information such as checking operation status, creating and changing Profiles, and searching for and extracting necessary information from monitoring data. These functions allow the elements of the agent system to operate via an http protocol, so the agent system itself can be treated as just one of the nodes in our system. This also means that the operation of the agent program can be dynamically changed and modified by accessing the Web interfaces. By using a number of agent systems with Web interfaces, we can construct a multi-agent system in which one agent system controls not only Field Servers but also other agent systems. This structure is designed to achieve high scalability and a robust system.

Fig. 6 shows two types of multi-agent functions, a backup coordination for a robust system and task distribution for a scalable system. To establish a robust system, giving priority to Profiles, in which several agent systems have charge of several Field Servers redundantly with an access priority, is realized to provide strongly reliable performance. A subordinate agent system with a lower-priority Profile does not access the target Field Server, while a superior agent system with a higher priority works normally. The subordinate agent system periodically accesses the superior agent system to check its condition, and it takes over the task only when the superior agent system is down. With this method, an effective and robust backup function can be realized.

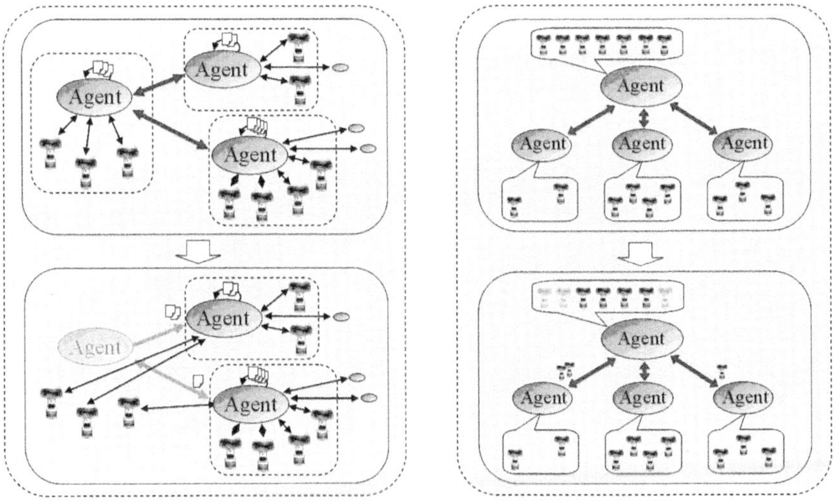

Fig. 6 Standard structure of a multi-agent system with robust and scalable management. The left figure shows a backup coordination of the system when one agent system is down. The right figure shows a task distribution of the system to smooth each task of the agent system.

To operate numerous Field Servers effectively, the multi-agent system performs task distribution in which each agent system compares its task with other tasks of the other member agent systems. A task is calculated for each agent system by considering its processing ability, the number of target Field Servers, the network condition, and the Profile contents. When a task of a certain agent system is heavier than other tasks of the other agent systems, the heavily loaded agent system deals out its excess task to other agent systems via the Web interfaces. By coordinating the agent systems, the multi-agent system can provide high scalability and efficient management.

4 Discussion and Future Work

To evaluate the agent system with regard to managing Field Servers effectively, we installed dozens of Field Servers in different locations all over the world, including severe-condition sites such as the Himalaya and India, and managed them with this system via the Internet. The agent programs were executed on PC clusters at a remote site with a VPN connection and on small computers at some field sites. This system has performed reliably since 2002, and the collected enormous amount of data, which is available on the Web [1], shows its capabilities and its reliability. Fig. 7 shows an example of archived data displayed on viewer Web applications.

Through field experiments, we have made improvements and discovered new requirements for the Web-based sensor network. In some field sites with

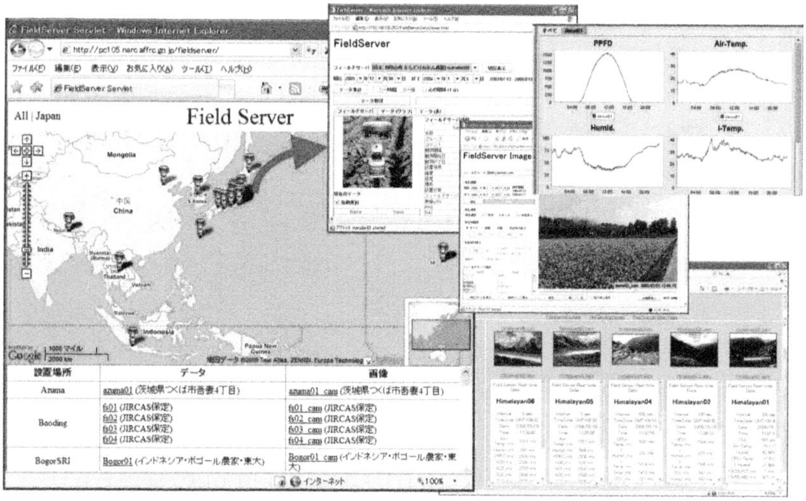

Fig. 7 Example of archived data displayed on viewer Web applications. By clicking a target icon of a Field Server on the main Web page (left figure), we can see the collected data displayed as graphs and animated images.

unstable network environments, a local agent system installed in a small computer is important to monitor Field Servers under the system. By coordinating the local agent system with a remote agent system while maintaining an Internet connection, we can manage the system successfully. In the project known as Geo-ICT and Sensor Network-based Decision Support Systems in Agriculture and Environment Assessment, we are trying to use modified Field Servers with a GPRS connection as another solution. In this project, we are also developing a hybrid sensor network system with Web-based sensor nodes such as Field Servers and conventional sensor nodes such as ZigBee units to realize an extensive and dense monitoring system.

The agent system with various functions can be used effectively in different situations. In this paper, we have described the fundamental application of the system, but the system can provide more intelligent agent algorithms by describing them in the Profiles. To develop the system more conveniently, there are tasks that need to be accomplished, such as the development of an easy operating procedure to treat enormous numbers of Field Servers, intelligent sensors that can perform automatic calibration, and effective management with advanced rule-based algorithms. By realizing a dynamical monitoring and analysis in response to users' requests regarding our agent system, we can construct an active database system to provide useful information for users autonomously. The agent system also has the potential to realize a more functional monitoring system by controlling mobile Field Servers and actuator units, and to realize a smart support system by providing suitable information and assistant operation to users in real-time.

References

1. Field Server Web Page, `http://model.job.affrc.go.jp/FieldServer/`
2. Akyildiz, I.F., Su, W., Sankarasubramaniam, Y., Cayirci, E.: Wireless Sensor Networks: A Survey. Computer Networks 38(4), 393–422 (2002)
3. Brownston, L., Farrell, R., Kant, E., Martin, N.: Programming Expert Systems in OPS5: an Introduction to Rule-based Programming. Addison-Wesley Longman Publishing Co., Inc., Boston (1985)
4. Etzioni, O., Weld, D.: A Softbot-based Interface to the Internet. Commun. ACM 37(7), 72–76 (1994)
5. Flores-Mendez, R.A.: Towards a Standardization of Multi-agent System Framework. Crossroads 5(4), 18–24 (1999)
6. Fukatsu, T., Hirafuji, M.: Field Monitoring Using Sensor-nodes with a Web Server. J. Rob. Mechatron. 17(2), 164–172 (2005)
7. Fukatsu, T., Hirafuji, M., Kiura, T.: An Agent System for Operating Web-based Sensor Nodes via the Internet. J. Rob. Mechatron. 18, 186–194 (2006)
8. Fukatsu, T., Nanseki, T.: Monitoring System for Farming Operations with Wearable Devices Utilized Sensor Networks. Sensors 9, 6171–6184 (2009)
9. Fukatsu, T., Saito, Y., Suzuki, T., Kobayashi, K., Hirafuji, M.: A Long-term Field Monitoring System with Field Servers at a Grape Farm. In: Proceedings of International Symposium on Application of Precision Agriculture for Fruits and Vegetables, pp. 183–190 (2008)

10. Kahn, J.M., Katz, R.H., Pister, K.S.J.: Next Century Challenges: Mobile Networking for "Smart Dust". In: Proceedings of the 5th Annual ACM/IEEE International Conference on Mobile Computing and Networking, pp. 271–278 (1999)
11. Lange, D., Oshima, M.: Programming and Deploying Java Mobile Agents Aglets. Addison-Wesley Longman Publishing, Massachusetts (1998)
12. Lesser, V., Oritz, C., Tambe, M. (eds.): Distributed Sensor Networks: a Multiagent Perspective. Kluwer Academic Publishers, Dordrecht (2003)
13. Maes, P.: Agents that Reduce Work and Information Overload. Commun. ACM 37(7), 30–40 (1994)
14. Russel, S., Norvig, P.: Artificial Intelligence: A Modern Approach. Prentice-Hall, New Jersey (1995)

Sensor Network Architecture Based on Web and Agent for Long-Term Sustainable Observation in Open Fields

Masayuki Hirafuji, Tokihiro Fukatsu, Takuji Kiura, Haoming Hu, Hideo Yoichi, Kei Tanaka, Yugo Miki, and Seishi Ninomiya

Abstract. Architecture of sustainable, robust and scalable sensor networks is proposed. To monitor environmental condition and ecosystems for many years in open fields, Web-based Sensor network can be employed. Agents control them by using purely HTTP. Functions of the sensor nodes can be reduced as much as the agent can assist the sensor nodes. Applications for the sensor networks are distributed Web services. All devices in the sensor nodes are Web servers such as data acquisition Web-server card and IP camera, and they are connected by Ethernet. That is, all devices are only Web servers. Power supplies for all devices in the sensor nodes are controlled by the agent, which can restart the freezing devices automatically. The Web-based sensor networks were tested at many sites, and rationality of the proposed architecture was examined.

1 Introduction

Long term observation data in open fields is very valuable for environmental science, agriculture and so on. Observation in the fields should be continued as long as possible, and data should be archived permanently. However lifetime of digital devices and data formats which depend on applications is not long, since digital technologies are rapidly advancing. For example, digital interfaces such as RS-232 had been popular, but currently RS-232C is treated as "legacy devices" and we are using USB instead. Data recorded, 30 years

Masayuki Hirafuji · Tokihiro Fukatsu · Takuji Kiura · Haoming Hu · Hideo Yoichi · Kei Tanaka · Yugo Miki · Seishi Ninomiya
National Agricultural Research Center, Tsukuba, 305-8666, Japan
e-mail: {hirafuji,fukatsu,kiura,haominhu,hyoichi}@affrc.go.jp, {tanaka.kei,mikiyugo,snino}@affrc.go.jp

Masayuki Hirafuji · Kei Tanaka · Seishi Ninomiya
University of Tsukuba, Tsukuba, 305-8666, Japan

Q. Bai and N. Fukuta (Eds.): Advances in Practical Multi-Agent Systems, SCI 325, pp. 425–433.
springerlink.com

ago, on 8-inch floppy diskettes were unreadable by recent PCs. Moreover data files for legacy software such as VISICALC (the first spread sheet program) also became to legacy.

Sustainable sensor network architecture as a kind of sustainable information systems is indispensable to monitor natural phenomena. Here we propose a sensor network architecture which can be maintained sustainably for long term.

2 Architecture

Objectives of data collection in agronomy, ecology, meteorology, environmental science and so on are to analyze complex phenomena of nature based on accumulated data for long term. A sustainable architecture for data format, protocol, hardware and software were chosen as followings:

Data format:	text(XML)	
Protocol:	TCP/IP	
Hardware:		
	Wired network:	Ethernet
	Wireless network:	WLAN(IEEE802.11b/g/n)
	Data storage:	NAS
	User interface:	Web
Services:		
	Protocol:	HTTP
	Control:	agent
	Integration:	agent
	Application:	servlet, applet

Here we predicted Ethernet, TCP/IP and HTTP will be sustainable for several decades. Standards for WLAN such as IEEE802.11a/b/g/n have been evolving rapidly so far, and lower performance WLAN devices were replaced into new one. Nevertheless, Ethernet and TCP/IP have been common features to all WLAN routers.

Sensor nodes based on above architecture must be Web servers, and also inside devices of the sensor nodes should be Web servers: at least all functions of the sensor nodes must be operable by using HTTP.

3 Web Devices for Pure Web

Operating systems such as MS Windows prohibit direct access to devices in PCs; device drivers are required to access its hardware, so programmers must develop device drivers. However it is not easy for ordinal programmers. On the contrary handling of TCP/IP and HTTP is quite easy, since there are rich APIs and class libraries concerning to TCP/IP and HTTP. Many

Fig. 1 iPic web-server (http://www-ccs.cs.umass.edu/~shri/iPic.html/). The one-chip CPU in the left picture worked as a Web server as shown in the right picture. This site is currently closed.

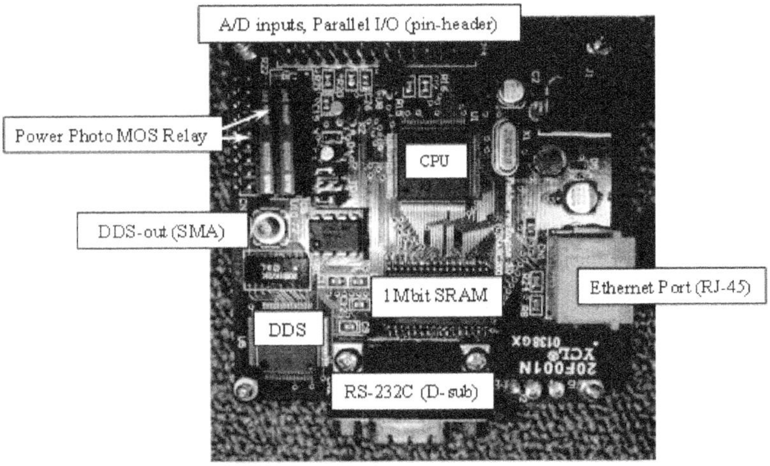

Fig. 2 Field Server Engine Ver. 3.1b.

ordinal programmers can develop easily applications for the Web-based sensor networks, and this advantage can sustain the Web-based sensor networks.

On 2001, the smallest Web server was a one-chip CPU (PIC microprocessor), which worked actually as ordinal Web servers (Fig. 1) [4]. This demonstration implied that all devices and appliances can be Web devices in low-cost, and Web devices will be more functional using advanced technologies such as more sensors and reconfigurable devices. We concluded, therefore, sensor networks should be constituted logically only by Web servers and we started a project to develop desirable functions and devices to realize the web-based sensor networks based on the proposed architecture (i.e. *pure web*).

The Web devices for Web-based sensor networks should be composed by:

1. A One-chip CPU for Web server,
2. A/D converters for sensors,
3. Solid state relays/Power photo MOS relays to control external devices,
4. DDS (Direct Digital Synthesizer) to make advanced low-cost sensors,
5. FPAA (Field Programmable Analog Array) for reconfigurable analog circuits,
6. Analog multipliers for frequency conversion and signal processing.

As a first step, we developed a simplified Web server card, Field Server Engine Ver.1 (FSE Ver.1), which was equipped with only 1-3 above. FSE Ver.3 was equipped with 1-4 (Fig. 2). The next version, FSE Ver.5, is equipped with all (1-6).

Hardware troubles may be caused under extreme environments, and we assume bugs such as memory leek and buffer overflow in firmware are unavoidable; sometime external IP devices may freeze by unknown bugs. So power supplies for all devices in the sensor nodes must be controlled manually or automatically by agents, which can restart the freezing devices automatically.

The digital parts above (1-3) can be integrated on FPGA (Field Programmable Gate Array) and the analog parts above (4-6) can be integrated on FPAA. After all, the Web device can be integrated on a set of FPGA and FPAA, and moreover this chip has many other advanced functions such as arbitrary function generator, encryption and neural networks [2]. We aim that this chip, in future, will take the place of current prototyping devices (FSE Ver.1-5) constructed by using discrete parts.

4 Experiments Using Field Server

A Field Server is composed of a main board (i.e. Field Server Engine), a rugged case, multiple sensors, and WLAN card. The sensors in a Field

(a) Prototype (2001) (b) 1st type (2001) (c) 2nd type (2003) (d) 3rd type (2008)

Fig. 3 Field Servers

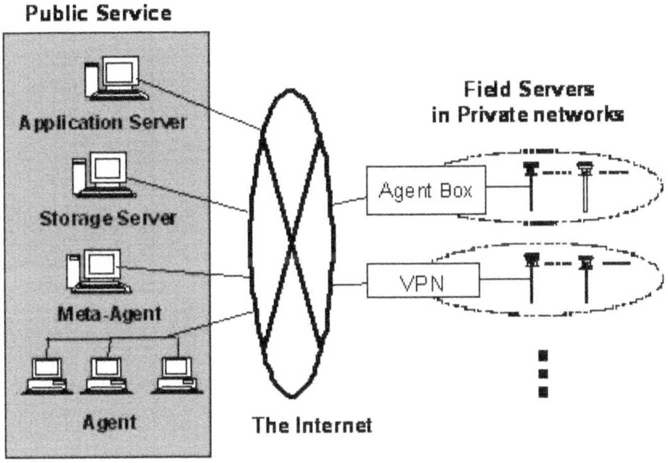

Fig. 4 High-level architecture of the Field Server network

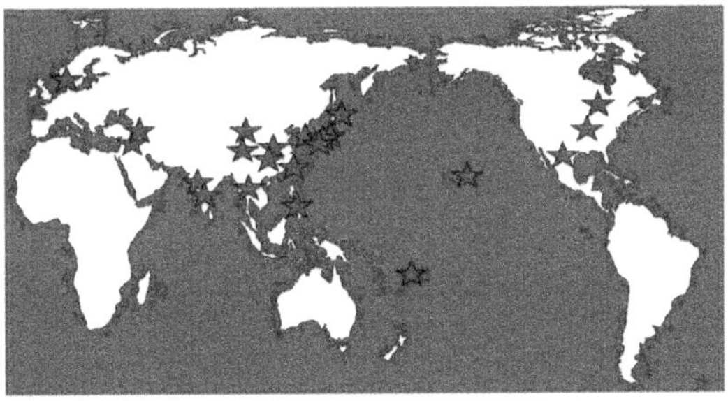

Fig. 5 Experiment sites

Server are a type of Asmann psychrometer which can accurately measure air-temperature and humidity. The air-temperature sensor and the humidity sensor are in air-flow as ventilated Asmann type psychrometer. Simultaneously, the air-flow cools the circuit boards in the case.

Other equipped functions of Field Servers can be added as PC clones. The first Field Server was equipped with four sensors, a one-board Web-server, an LED lighting system, and a wireless LAN access-point card. A network camera was added to second generation of Field Server (Fig. 3). Field Servers deployed in an area constitute private network, which are connected through a gateway using VPN by agents. The agents fetch data from the Field servers, and archive them on the storage server[1].

Sustainability of the proposed architecture has been tested actually by using the Field Servers. So far web-based sensor networks using Field Serves

[1] http://model.job.affrc.go.jp/FieldServer/FieldServerEn/Data-Storage.html

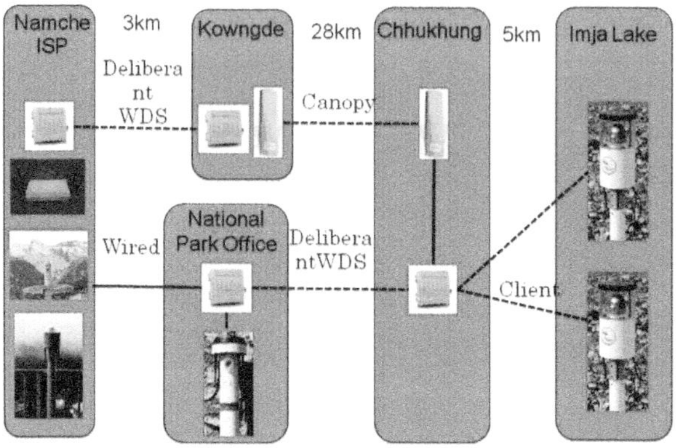

Fig. 6 The sensor network architecture for deployed in Himalaya

Fig. 7 Examples of image collected by the sensor network composed by 5 Field Servers deployed from Namche Bazaar to Imja glacial lake (totally about 35 kms)

were deployed at many sites in the world for testing of long-term operation and experiments of durability (Fig. 5) [3]. Recent deployment was Himalaya. Imja glacial lake and Namche bazaar have been monitored by 5 Field Servers totally (Fig. 6), which equip cameras (Fig. 7) and sensors (air temperature, relative humidity, solar radiation, CO_2 concentration and water level).

5 Results and Discussions

Sustainability of the proposed architecture was examined trough many accidental troubles. For example, the sensor network composed by Field Servers deployed in Himalaya is working for two years, although some devices were disconnected. At a site in Hawaii, a Field Sever was broken by wild pigs, and later two Field Servers under coconut trees were crushed by a falling coconut. The Internet connections to Field Servers were frequently disconnected by same reasons, such as human errors, freezing gateway routers and power failure, as in offices.

We could quickly find out the causes and improved the systems, since we could obtain rich information on webs of the devices respectively. Sometime we could restore the troubled sensor networks remotely from Japan only by accessing webs of the devices. This manner is similar to remote maintenance for space probes such as Mars surveyors.

Sensor networks which Field Servers constitute are "pull-type" sensor network; contrarily conventional sensor networks such as smart dust [5], Mote[2] and devices using ZigBee are "push-type". "pull-type" is advantageous to:

- Stability
  If enormous push-type sensor nodes send data onto a data storage server all simultaneously, their accesses may damage the network and the storage server like as DDoS (Distributed Denial of Service). On the contrary, pull-type sensor nodes (Field Servers) wait for access of the agent. The agent crawls over sensor nodes to collect data according to capacity of individual network connection.
- Scalability
  Even if too many pull-type sensor nodes are deployed for an agent, the agent can crawl on them in series up to its maximum capability. Or only by increasing number of PCs, capability of multiple agents can be reinforced.
- Qualia of connectivity
  Qualia is subjective quality of conscious experience [1]. Users can access pull-type sensor nodes manually using Web browser. Then the users can sense capacity of connection to a sensor node, and delay of its response as vivid feeling. Qualia in this sensation are very useful to estimate conditions such as network capacity to deploy sensor nodes in unknown fields.

These points were useful to deploy sensor networks in short time completely. Especially in extreme environment such as Himalaya, mission time for deployment in high altitude was limited for thin air and cold windy environment. We could check devices and connectivity at once by feeling response from sensor nodes. Later we could set up an agent to collect data from the deployed Field Servers and optimize data sampling rate in Japan at ease. Moreover we improved ventilation algorithm on the agent to protect inside devices against low temperature and condensation.

Qualia of connectivity were indispensible for diagnosis, trouble shooting and decision making in explorative deployments, where we must pay attention to enormous potential troubles. For example, only to choose the best antenna we must consider many related factors such as communication range, direction, antenna gain, tolerance against strong wind, depth of the snow and

[2] http://www.xbow.com

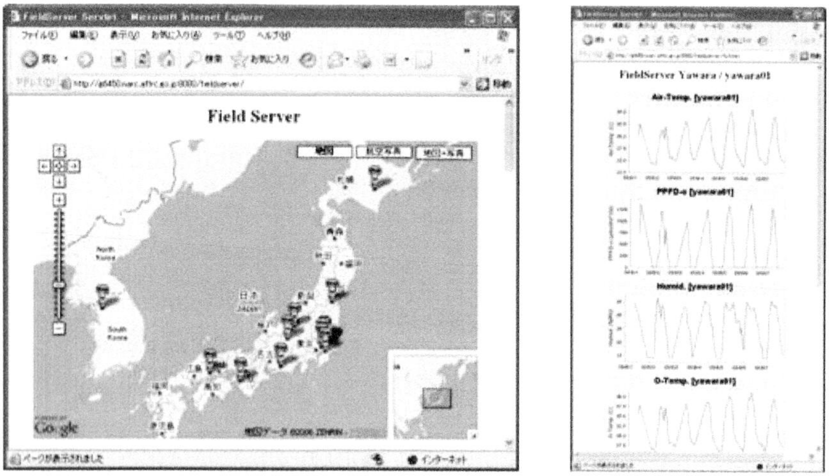

Fig. 8 An example of archived data

so on. Only a failure of these factors causes serious troubles. In a case that a device in a Field Server was disconnected at a site beside Imja Glacial Lake in Himalaya, we could estimate causes based on Qualia of connectivity to other live devices, and we repaired it there only by carrying tools prepared in advance; kinds of carrying tools and parts are limited.

Currently we are trying to deploy Field Servers at an experimental field in India, where air temperature is extremely high and almost other conditions are unknown for us. Only connection using cellular phone (GPRS) is available there, so agent box (agent installed in a Linux box) was deployed at the field instead. Currently we have a lot anxiety about operation of the Field Servers, since we cannot get Qualia of their operation in Japan.

All archived data collected by Field Servers are visualized by an applet viewer as shown in Fig. 8. However this service and the agent service also are not sustainable, since these services are only for scientific researches; these services must terminate after all projects for developing Field Servers are completed. Ideally such services should be operated in free as semi-permanent business like as Google and Yahoo! to realize sustainable sensor network.

The agent service may create a new network business; the agents can collect other numerical data provided on webs in the Internet by same manner as agents crawl on Field Servers. Users can find hidden relationships between time series data such as the price index of stocks and temperature in districts, which are collected by the agent service.

6 Conclusions

We proposed sustainable sensor network architecture based on philosophy of *pure web*: all sensor nodes and inside devices are web servers, and agents crawl the web servers to collect data. We deployed sensor networks at many sites for 8 years using Field Servers to examine that the proposed architecture actually works well for long-term even in extreme environments. The *pure web* architecture could work much better than we expected. The oldest prototype Field Server is still on active service and accessible by the newest browser such as Firefox and Safari. It will be kept on active services for long years as hardware; however network services such as the agent service must be sustainable as network business.

References

1. Chalmers, D.J.: The Conscious Mind: In Search of a Fundamental Theory (Philosophy of Mind Series). Oxford Univ. Press, Oxford (1997)
2. Hirafuji, M.: Arbitrary Function Generating Circuit Using a Simple Operational Element and an Encryption Method Thereof. US Patent: US6,898,616 B2 (2005)
3. Hirafuji, M., Ninomiya, S., Kiura, T., Fukatsu, T., Hu, H., Yuoichi, H., Tanaka, K., Sugahara, K., Watanabe, T., Kameoka, T., Hashimoto, A., Ito, R., Ehsani, R., Shimamura, H.: Field Server Projects. In: Proceedings of Practical Applications of Sensor Networking (SAINT 2007), pp. 15–19 (2007)
4. Ju, H.T., Choi, M.J., Hong, J.W.: An Efficient and Lightweight Embedded Web Server for Web-based Network Element Management. International Journal of Network Management 10(5), 261–275 (2000)
5. Kahn, J.M., Katz, R.H., Pister, K.S.J.: Next century challenges: Mobile networking for "smart dust". In: Proceedings of the 5th Annual ACM/IEEE International Conference on Mobile Computing and Networking, pp. 271–278. ACM, New York (1999)

A Multi-Agent View of the Sensor Web

Quan Bai, Siddeswara Mayura Guru, Daniel Smith,
Qing Liu, and Andrew Terhorst

Abstract. The rapid growth in sensor and ubiquitous device deployments has resulted in an explosion in the availability of data. The concept of the Sensor Web has provided a web-based information sharing platform to allow different organisations to share their sensor offerings. We compare the Open Geospatial Consortium - Sensor Web Enablement (OGC-SWE) with Multi-Agent System (MAS), and identify the similarities between the concepts. These similarities motivate the adoption of MAS based techniques to address related problems in OGC-SWE. Brokerage facilitators commonly used in MAS, the Yellow Pages Agent and Blackboard Agent, are considered to address service interaction issues identified within OGC-SWE. Furthermore, the use of MAS based reputation mechanisms are explored to address potential trust issues between service providers and consumers in OGC-SWE.

1 Introduction

A large amount of times-series spatio-temporal data is collected from the physical environment due to increase in the deployment of sensors. The deployment ranges from satellite-based sensors [24] that monitor large scale environmental and ecological systems, to smart meters [5] that monitor energy usages.

Meanwhile, there is a need for an open platform to allow users from different geographical location to share data for macro monitoring. The Open Geospatial Consortium (OGC), an international standards organisation, has a Sensor Web working group developing Sensor Web Enablement (SWE). The goal of OGC-SWE is to develop interoperable data models and web service

Quan Bai · Siddeswara Mayura Guru · Daniel Smith · Qing Liu · Andrew Terhorst
Tasmanian ICT Centre, CSIRO, GPO Box 1538, Hobart, TAS, 7001, Australia
e-mail: {quan.bai, siddeswara.guru, daniel.v.smith}@csiro.au,
{q.liu, andrew.terhorst}@csiro.au

Q. Bai and N. Fukuta (Eds.): Advances in Practical Multi-Agent Systems, SCI 325, pp. 435–444.
springerlink.com

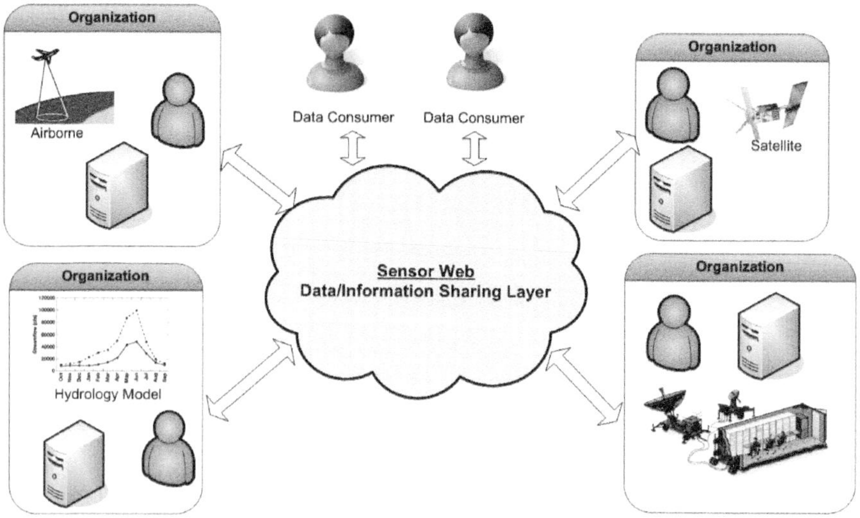

Fig. 1 Data and Information Sharing on the Sensor Web

interface standards to discover, exchange and process sensor observations [1]. Fig. 1 shows the high-level vision of the Sensor Web.

The sensor web enables integration of web-enabled sensors and sensor systems for a global scale collaborative sensing. Following are some of the challenges to be addressed for the realisation of the Sensor Web [11] [10]:

- develop an open standard distributed computing infrastructure which can accommodate all the sensor system in the world,
- automatic discovery of heterogeneous resources needed to fulfil the end-users queries,
- managing data from different sensors with spatio-temporal variability,
- and uniform representation of data from similar sensors.

Sensor Web is not the only field facing the above challenges. Similar challenges are important and have been investigated in Multi-Agent Systems (MASs). In MAS, agents with heterogeneous resources and capabilities could be gathered together toward achieving some tasks [21]. To make such agents work together smoothly, MASs need to have coordination mechanisms to handle communication problems, organisational problems and various conflicts among agents. In this paper, we look at the problems in Sensor Web, specifically OGC-SWE from a Multi-Agent perspective, and explore the use of some of the solutions available in the multi-agent community to solve problems in Sensor Web.

The rest of the paper is organised as follows: Section 2 introduces the concepts of Sensor Web and an overview of OGC-SWE framework data model and services. It also highlight some of the drawbacks of the OGC-SWE with the useability example from hydrology domain. In Section 3, we compare sensor web with MASs, and list some similarities between them. Then, in

Section 4, we propose the use of some multi-agent coordination techniques to address some of the problems in sensor web. Finally, the paper is concluded in Section 5.

2 The Sensor Web

The concept of the Sensor Web has evolved as Internet technologies have matured. The term, Sensor Web, was first coined in NASA's Jet Propulsion Laboratory and defined as "a system of intra-communicating, spatially distributed sensor pods that are communicated with each other and can be deployed to monitor and explore new environments" [4]. It can be argued that this definition draws many parallels with the concept of a web-enabled sensor network. However, the dramatic increase in sensor deployments have led to the Sensor Web concept becoming far more broader and scalable than NASA's original definition. Today, the Sensor Web is perceived as a World Wide Web for sensors.

The Sensor Web consists of sensors and their networks, but just as importantly, Internet technologies. The dominance of the Internet to enable coordination of different resources (data, computation and intelligence) was envisioned in the last decade and presented as one of the top 21 ideas for the 21st century [6]. The basic capability of the Sensor Web is its ability to store and query heterogenous sensor offerings based upon their locations. The output of the queries are then transformed to suit a client's output device. These concepts have gained wider adoption due to the popularity of map tools like Google Maps[1] and NASA World Wind[2].

OGC-SWE provides an open standard framework to exploit web-enabled sensors and sensor systems [15]. The framework is based upon the Service Oriented Architecture (SOA) and enables the discovery, exchange and processing of sensor observations. The SWE standards are built primarily on the following specifications:

1. **Observation and Measurement (O&M)** [13]: provides a standard model based upon an XML schema to represent and exchange observation results.
2. **Sensor Model Language(SensorML)** [16]: provides information models and encodings that enable discovery and tasking of Web-enabled sensors and the exploitation of sensor observations. SensorML defines the models and XML Schema for describing any process, including measurement by a sensor system and post-measurement processing.
3. **Sensor Observation Service (SOS)** [19]: provides a standard web service interface to request, filter and retrieve sensor observations. SOS can also provide metadata information about the associated sensors, platforms

[1] `http://maps.google.com`
[2] `http://worldwind.arc.nasa.gov`

and observations. It is an important component of SWE and acts as a intermediary between a user and a sensor system's observation repository.

4. **Sensor Planning Service (SPS)** [20]: provides a standard web service interface to assist in collection of feasibility plans and to process collection requests for a sensor.
5. **Sensor Alert Service (SAS)** [17]: provides a standard web service interface to publish and subscribe to alerts from sensors. This service may be superseded by Sensor Event Services (SES) in the future releases.
6. **Web Notification Services (WNS)** [14]: provides a standard web service interface for asynchronous message delivery. This service can be used in conjunction with other services like SAS and SPS.
7. **Catalog Service for the Web (CSW)** [2]: supports the ability to publish and search collection of descriptive information (metadata) for data, services, and related information objects.

The advantage of OGC-SWE is its ability to store, query and publish its offerings based upon the locations, boundaries and relationships amongst geographic features and phenomenon. It provides a standard framework to describe phenomenon, measurement types and data types. Conceptually, this makes automatic data discovery and sharing relatively easy.

There are several drawbacks of the current OGC-SWE services including:

- The user (Services or Client) needs to be knowledgeable about OGC-SWE and the operation of its services.
- There are no guidelines for the interaction of different services; it is generally application-specific.
- There has been limited work to address security and trust issues in the interaction between the services. This will be an important issue to build the reputation of a service provider in the Sensor Web.
- Interaction between services is passive, for example, a user can only interact with the SOS, and not vice versa.
- The management of a catalog of services can be cumbersome and doesn't adequately scale as the number of services and/or sensor offerings increases.

These problems can also exist in various Multi-Agent applications and have been addressed in [21]. Consequently, we believe adopting these ideas from MAS could enrich the experience of using OGC-SWE.

Let us consider the useability of OGC-SWE by considering an example of a typical query that a user may want to execute in a hydrology application given in [8]. The query could be "Give the rainfall data from time A to time B in the geographic area X". Anybody familiar with databases would expect the query to be executed with relative ease and the results to be available to the user. However, in the SWE framework, a user needs to select the SOS instances by invoking the GetRecord of the web-catalog service (CSW). This can only be possible if the SOSs are registered in the CSW. The GetCapabilities operation is invoked to identify the capabilities of the

SOSs and to get the offeringid (user has to know the offeringid to run any GetObservation operations), then the GetObservation operation is invoked to get the observation from the sensors.

These steps can be intimidating and complex to follow for a naive user. It may be possible that some SOSs that fulfil the query could be missed. Although there has been some work to develop cascading SOS to aggregate or fuse the SOSs [9], the SOSs still need to be identified by the CSW. There is a need for automation in the query execution. Intelligent coordinators (agents) can interact with the services to pass messages, process requests, submit requests and coordinate activities between the services. This infuses intelligence into the framework and enables the introduction of application-specific services. In the next section, we will outline some of the similarities between MAS and the Sensor Web.

3 Sensor Webs and Multi-Agent Systems

The emergence of MASs was derived from the demands of solving complex problems that require diverse expertise and multiple resources. In a MAS, agents with different goals, heterogeneous resources and different perspectives of the environment are required to work together to complete a complex problem (see Fig. 2). To achieve collaboration, agents need to use communication languages and protocols to exchange information. Traditional agent communication languages are message-based, however, given the increased adoption

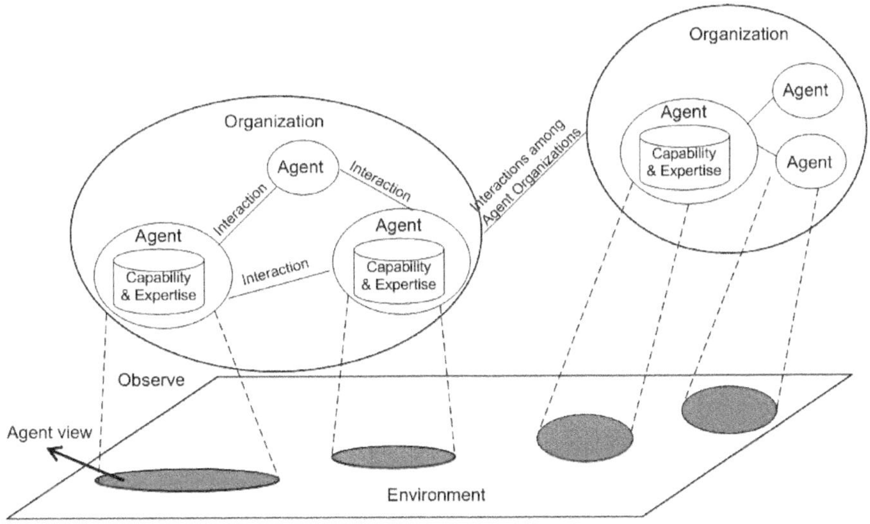

Fig. 2 Multi-Agent System

Table 1 Comparison of Sensor Webs and Multi-Agent Systems

	Sensor Web Systems	**Multi-Agent Systems**
System architecture	Distributed	Distributed
Interaction	XML-based encoding	Agent Communication Language (Message-based or knowledge-based)
Application platform	Web-based	Web-based or other platforms
Components	loosely coupled services	Heterogeneous autonomous entities

of web-based applications, knowledge-based communication languages have also been used by MAS applications.

There are several similarities between MASs and the Sensor Web. Firstly, both are distributed systems that may include a number of heterogeneous entities. The handling of these knowledge representation heterogeneities is a common problem to MASs and the Sensor Web. Secondly, they can both be built with a Service Oriented Architecture (SOA). Thirdly, the Internet is the major application platform for the Sensor Web and MASs. Table 1 summarises the similarities between the Sensor Web and MASs.

4 Using Multi-Agent Coordination Techniques in Sensor Web

As introduced in section 3, there are similarities between the MAS and the Sensor Web. In fact, the Sensor Web may be considered a web-based MAS from the perspective of the agent community. Both the Sensor Web and web-based MAS are comprised of heterogenous services that have a limited ability to understand the environment independently, however, they are capable of working together to fulfil complex user requests.

Compared with the Sensor Web, MAS is a far more mature research area. A number of agent coordination techniques have been developed to handle interaction and heterogeneity problems in MASs. Some of these techniques could potentially be used to address related problems in the Sensor Web.

4.1 Including Multi-Agent Facilitators in Sensor Web

From the hydrology query example given in section 2, query execution in OGC-SWE leaves many onerous tasks to the user, and requires the user to have an ability to search, select and evaluate services on the Sensor Web. Such requirements are difficult to satisfy in applications with a large number of services. In addition, the communication between the CSW, service providers and service consumers are uni-directional. This creates a large communication gap between the service consumers and providers. To overcome these limitations, intelligent brokers are required to facilitate interactions and enable bi-directional communications between the CSW, service consumers and providers. With bi-directional communications, service consumers can not only seek the required services from the CSW, but service providers can have access to the consumer requirements.

There is a set of intelligent facilitators for supporting multi-agent interactions. Two kinds of these multi-agent facilitators, the Yellow Pages Agent and Blackboard Agent, are considered for use in the Sensor Web (see Fig. 3).

Yellow Pages Agent (YPA) is a multi-agent facilitator that processes the advertisements of service providers and assists service consumers to find suitable providers based upon their advertised capabilities [22].

In the Sensor Web, service providers could publish and update their metadata descriptions via the YPA to the CSW. The YPA classifies service providers based upon their capabilities, identifying the offeringid before updating the CSW. For a user executing the example hydrology query, the CSW could provide an getObservation URI of SOSs directly, saving a couple of operations (the getRecord of the CSW and getCapabilities of the SOS). This could help to improve the automation of the query execution process and reduce user involvement.

In some situations, the capabilities of a service provider cannot be fully published to all the users due to security reasons, conflict goals, etc. In this situation, we need to have a brokering service between the consumers and providers to restrict the access of the consumers.

Blackboard Agent (BBA) is a multi-agent facilitator that can collect, hold and publish requests from service consumers [3]. Service providers can seek and select tasks that they can handle from the BBA, and interact with the corresponding service consumers actively. The inclusion of the BBA enables bi-directional interactions between service consumers and providers. This allows service providers to protect their information.

4.2 Building Trust Based on Reputations

On the Sensor Web, different organisations can publish their data as discoverable services. On the other side, consumers can specify their requirements in queries, and find suitable services with the assistance of agent facilitators (as introduced in Subsection 4.1). However, in many situations, a consumer may face a dilemma in making a choice from a bunch of providers which can all offer satisfactory services. In such situations, a consumer needs to know not only what a provider can do, but also how trustable a provider is. This brings the issue of building trust between service consumers and providers. In this paper, we suggest to include a reputation-based mechanism to facilitate service consumers to select trustable providers on the Sensor Web.

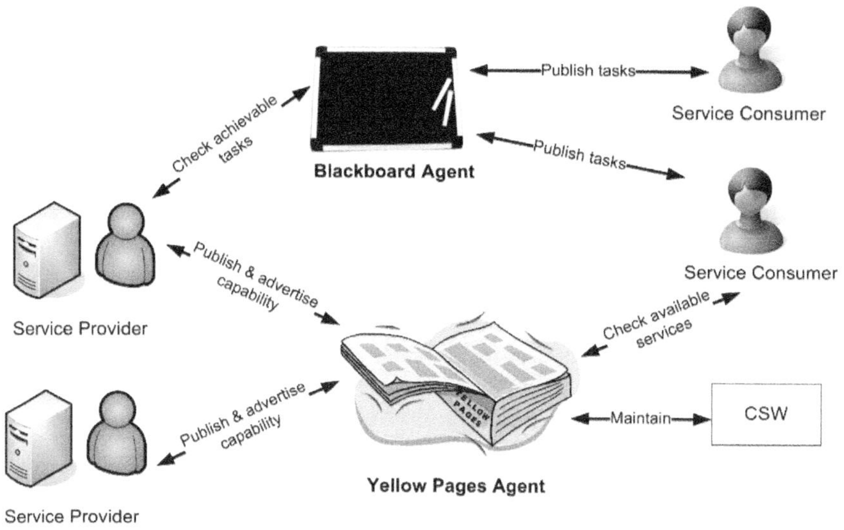

Fig. 3 Including a Yellow-Pages Agent and a Black-Board Agent in the Sensor Web

Reputation refers to the perception that an entity knows another's intentions and norms [12]. Reputation-based mechanisms have been widely used in multi-agent applications, such as agent-based auction systems [18]. Most existing reputation mechanisms are based on simple numerical evaluation mechanisms in which service consumers provide (numerical) "scores" for service providers. These scores will be a reference for other consumers to select suitable services in the future.

In the Sensor Web, reputation information of service providers could be stored in the YPA. After a consumer finds a suitable provider (via YPA) and gets the required sensor offerings, an evaluation of the provider could be generated with respect to the stability, reliability and quality of these offerings. The consumer could provide evaluation feedback to the YPA, and the YPA could update the reputation value of the provider. In environmental applications, it may also be useful for the service provider to provide an evaluation of their own sensor offerings to the YPA.

Reputation provides a reference for service consumers to evaluate and select suitable service providers. It could potentially reduce the impact of poor quality service offerings upon the Sensor Web.

5 Conclusion and Future Work

The Sensor Web promises an information sharing infrastructure that allows end users to access sensor offerings on a global scale. In this paper, we considered OGC-SWE and identified some limitations in realising the vision of the Sensor Web on a macro scale. The major challenge is the interaction between the services to fulfil the user requirement. Based on our observation there are similarities between the Sensor Web and MAS. We propose the use of multi-agent facilitators like YPA to make the communication between the services bi-directional and BBA to make the services more interactive. This will make the OGC-SWE services less complex to use without making any changes to the service interfaces. There is not much work in the OGC-SWE community about developing trust between the service providers and consumers. We have proposed the the use of reputation-based methods to build and manage trust within the Sensor Web. This will improve the quality and reputation of services in the Sensor Web.

We would like to implement some of the agent-based methods discussed in this paper in the hydrological Sensor Web [23]. Our intention is to use agents as separate services in the OGC-SWE framework.

Acknowledgements. This project is jointly funded by the CSIRO Water for a Healthy Country Flagship and the Tasmanian Government. The CSIRO Tasmanian ICT Centre is jointly funded by the Australian Government through the Intelligent Island Program and Australia's Commonwealth Scientific and Industrial Research Organisation (CSIRO). The Intelligent Island Program is administered by the Tasmanian Department of Economic Development, Tourism and Arts.

References

1. Botts, M., Percivall, G., Reed, C., Davidson, J.: OGC Sensor Web Enablement: Overview and High Level Architecture. In: Nittel, S., Labrinidis, A., Stefanidis, A. (eds.) GSN 2006. LNCS, vol. 4540, pp. 175–190. Springer, Heidelberg (2008)
2. Catalog Services for the Web, `http://www.opengeospatial.org/standards/cat` (last visited at September 2009)
3. Decker, K., Sycara, K., Williamson, M.: Middle-Agents for the Internet. In: Proceedings of the Fifteenth International Joint Conference on Artificial Intelligence, Nagoya, Japan, pp. 578–584 (1997)
4. Delin, K.A.: The Sensor Web: A Macro-Instrument for Coordinating Sensing. Sensors 2, 270–285 (2002)
5. Derbel, F.: Trends in Smart Metering. In: Proceedings of the 6th International Multi-Conference on Systems, Signals and Devices, pp. 1–4 (2009)
6. Gross, N.: 21 Ideas for the 21st Century: 14 Internet: The Earth will don an Electronic Skin, `http://www.businessweek.com/1999/99_35/b3644024.htm` (last visited at September 2009)
7. Guru, S.M., Peters, C., Smith, D., Pratt, A., Bai, Q., Terhorst, A.: Sensor web and scientific workflow technology to manage and observe hydrological phenomenon. In: Abstrtact accepted for 9th International Conference on HydroInformatics (2010)

8. Guru, S.M., Taylor, P., Neuhaus, H., Shu, Y., Smith, D., Terhorst, A.: Hydrological sensor web for the south esk catchment in the tasmanian state of australia. In: Fourth IEEE International Conference on eScience [poster], pp. 432–433 (2008)
9. Havlik, D., Bleier, T., Schimak, G.: Sharing sensor data with sensorsa and cascading sensor observation service. Sensors 9(7), 5493–5502 (2009)
10. Jirka, S., Broring, A., Stasch, C.: Discovery mechanisms for the sensor web. Sensor, 2661–2681 (2009)
11. Moodley, D., Simonis, I.: New Architecture for the Sensor Web: the SWAP Framework. In: Cruz, I., Decker, S., Allemang, D., Preist, C., Schwabe, D., Mika, P., Uschold, M., Aroyo, L.M. (eds.) ISWC 2006. LNCS, vol. 4273, Springer, Heidelberg (2006), http://www.ict.csiro.au/ssn06/data/swap.pdf
12. Mui, L., Mohtashemi, M., Halberstadt, A.: Notions of Reputation in Multi-Agents Systems: a Review. In: Proceedings of the First International Joint Conference on Autonomous Agents and Multiagent Systems, pp. 280–287 (2002)
13. Open Geospatial Consortium: Observation Measurement Specification, document 07-022r1, http://www.opengeospatial.org/standards/om (last visited at September 2009)
14. Open Geospatial Consortium: Web Notification Service, OpenGIS Project Document: OGC 03-008r2 Version 0.1.0., http://portal.opengeospatial.org/files/?artifact_id=1367 (2003), (last visited at September 2009)
15. Open Geospatial Consortium: OGC Sensor Web Enablement: Overview and High Level Architecture. Tech. Rep. OGC 07-165 (2007)
16. OpenGIS Sensor Model Language (SensorML), http://www.opengeospatial.org/standards/sensorml (last visited at September 2009)
17. Request for Comment on Sensor Alert Service, http://www.opengeospatial.org/standards/requests/44 (last visited at September 2009)
18. Sabater, J., Sierra, C.: Reputation and Social Network Analysis in Multi-Agent Systems. In: Proceedings of the First International Joint Conference on Autonomous Agents and Multiagent Systems, pp. 475–482 (2002)
19. Sensor Observation Service Specification, http://www.opengeospatial.org/standards/sos (last visited at September 2009)
20. Sensor Planning Service Specification, http://www.opengeospatial.org/standards/sps (last visited at September 2009)
21. Subrahmanian, V., Bonatti, P., Dix, J., Eiter, T., Kraus, S., Ozcan, F., Ross, R.: Heterogeneous Agent Systems. The MIT Press, Cambridge (2000)
22. Sycara, K., Klusch, M., Widoff, S., Lu, J.: Dynamic Service Matchmaking Among Agents in Open Information Environments. SIGMOD Record 28(1), 47–53 (1999)
23. SouthEsk Hydrological Sensor Web, http://152.83.132.197/au.csiro.OgcThinClient/site/OgcThinClient.html (last visited at September 2009)
24. Ye, W., Silva, F., DeSchon, A., Bhatt, S.: Architecture of a Satellite-based Sensor Network for Environment Observation. In: Proceedings of NASA Earth Science Technology Conference (ESTC 2008) (2008) http://www.esto.nasa.gov/conferences/estc2008/papers/Ye_Wei_A8P2.pdf

Author Index

GPSR Compliance

The European Union's (EU) General Product Safety Regulation (GPSR) is a set of rules that requires consumer products to be safe and our obligations to ensure this.

If you have any concerns about our products, you can contact us on ProductSafety@springernature.com

In case Publisher is established outside the EU, the EU authorized representative is:

Springer Nature Customer Service Center GmbH
Europaplatz 3
69115 Heidelberg, Germany

Batch number: 10755461

Printed by Printforce, the Netherlands –